Cytoskeletal Regulation
of Membrane Function

Society of General Physiologists Series • Volume 52

Cytoskeletal Regulation of Membrane Function

Society of General Physiologists • 50th Annual Symposium

Edited by
Stanley C. Froehner
University of North Carolina at Chapel Hill,
Chapel Hill, North Carolina
and
Vann Bennett
Duke University, Durham, North Carolina

Marine Biological Laboratory
Woods Hole, Massachusetts

5–7 September 1996

© The Rockefeller University Press
New York

Copyright © 1997 by The Rockefeller University Press
All rights reserved
Library of Congress Catalog Card Number 97-66364
ISBN 0-87470-059-0
Printed in the United States of America

Contents

Preface

This volume is a collection of invited contributions from speakers at the 50th Annual Symposium of the Society of General Physiologists on Cytoskeletal Regulation of Membrane Function. The meeting, which was held September 6–8, 1996 at the Marine Biological Laboratory in Woods Hole, Massachusetts, marked a half-century for the society which was founded on this site. A theme that connected the diverse speakers was a fascination with the molecular interactions at the interface between the plasma membrane and cytoplasm of metazoan cells. This research area was once primarily of interest to biochemists and basic cell biologists, but, as exemplified at this meeting, is of importance to physiologists, neuroscientists, and clinicians as well.

Progress in understanding organization, regulation, and assembly pathways for specialized membrane domains was presented. These domains include the sarcolemma of striated and cardiac muscle, the neuromuscular junction, which is the best understood model for synapses, the node of Ranvier of myelinated axons, desmosomes, focal adhesions, tight junctions, basolateral domains of epithelial cells, and septate junctions of *Drosophila*. As expected, each of these membrane assemblies exhibit a complex molecular architecture with much remaining to be discovered. However, unifying concepts are emerging that promise to provide a logical framework underlying the apparent diversity among membrane domains. Dystrophin and associated proteins, for example, have related genes which encode key proteins at the neuromuscular junction. Tight junctions, synapses, and septate junctions share proteins of the MAGUK family with PDZ domains that associate with recently defined regions of a variety of ion channels. These developments encourage the optimistic prediction that the signaling pathways involved in assembly of focal adhesions also will have broad relevance.

A dynamic interplay between clinical and basic science areas was another theme of this meeting. Dystrophin was initially characterized based on its role in Duchenne and Becker's muscular dystrophy and has led to the subsequent discovery of the sarcoglycans, which perform a key function in maintaining normal muscle. Elucidation of the basic biology of intermediate filament-linking proteins has resulted in a connection between blistering diseases and neuron dysfunction. The combination of human genetics, newly deciphered proteins, and gene-knockout techniques in mice promises to provide continued exciting developments.

The symposium was supported by grants from the National Institute of General Medical Sciences, the National Institute of Arthritis and Musculoskeletal and Skin Disease, the National Institute of Neurological Disorders and Stroke, Axon Instruments, Inc., Merck Research Laboratories, SmithKline Beecham Pharmaceuticals, and Cytoskeleton Inc. Special thanks are due to Sue Judd and Jane MacNeil for help in organizing the meeting, to John Burris and the Marine Biological Laboratories for their support and hospitality, and to Joseph White at The Rockefeller University Press for editorial assistance.

Stanley C. Froehner
Vann Bennett

Chapter 1

Actin–Membrane Interactions

Actin-binding Membrane Proteins Identified by F-actin Blot Overlays

Elizabeth J. Luna,* Kersi N. Pestonjamasp,* Richard E. Cheney,‡
Christopher P. Strassel,* Tze Hong Lu,* Catherine P. Chia,* Anne L. Hitt,*
Marcus Fechheimer,§ Heinz Furthmayr,‖ and Mark S. Mooseker‡

*Worcester Foundation for Biomedical Research, Shrewsbury, Massachusetts 01545;
‡Department of Biology, Yale University, New Haven, Connecticut 06520;
§Department of Cellular Biology, University of Georgia, Athens, Georgia 30602; and
‖Department of Pathology, Stanford University, Stanford, California 94305

Actin and associated proteins at the cytoskeleton-plasma membrane interface stabilize the membrane bilayer, control cell shape, and delimit specialized membrane domains. To identify membrane proteins that bind directly to F-actin, we have developed a blot overlay assay with ^{125}I-labeled F-actin. In the soil amoebae, *Dictyostelium discoideum*, the major proteins reactive in this assay are p30a, a 34-kD peripheral membrane protein that is concentrated in filopodia and at sites of cell–cell adhesion, and ponticulin, a 17-kD transmembrane glycoprotein required for efficient chemotaxis and for control of pseudopod dynamics. Proteins with apparent molecular masses of ∼34- and ∼17-kD also are observed on F-actin blot overlays of many mammalian cell lines. However, in mammalian cells, the most prominent F-actin binding proteins in this assay exhibit apparent molecular masses of 78-, 80-, 81-, ∼120-, and 205-kD. Bovine neutrophils contain the 78-, 81-, and 205-kD proteins, all of which co-isolate with a plasma membrane–enriched fraction. We have previously identified the 78-, 80-, and 81-kD proteins as moesin, radixin, and ezrin, respectively. These proteins, which are members of the protein 4.1 superfamily, co-localize with actin in cell surface extensions and have been implicated in the protrusion of microvilli, filopodia, and membrane ruffles. The 205-kD protein (p205) appears to be absent from current databases, and its characteristics are still under investigation. We here report that the 120-kD protein is drebrin, a submembranous actin-binding protein originally identified as a developmentally regulated brain protein. Thus, it appears that F-actin blot overlays provide an efficient assay for simultaneous monitoring of a subset of F-actin binding proteins, including p30a, ponticulin, moesin, radixin, ezrin, p205, and drebrin.

Richard E. Cherney's current address is Physiology Department, University of North Carolina at Chapel Hill, Chapel Hill, North Carolina 27599; Tze Hong Lu's current address is Department of Medicine, University of Massachusetts Medical Center, Worcester, Massachusetts 01655; Catherine P. Chia's current address is School of Biological Sciences, University of Nebraska-Lincoln, Lincoln, Nebraska 68588.

Cytoskeletal Regulation of Membrane Function © 1997 by The Rockefeller University Press

Introduction

Interactions between the plasma membrane and the actin-based cytoskeleton are important for the control of cell shape, maintenance of cellular integrity, and the organization of membrane proteins into functional domains (Campbell, 1995; Fox, 1993; Hitt and Luna, 1994; Luna and Hitt, 1992; Lux and Palek, 1995; Mohandas and Evans, 1994). Such functional domains include synapses (Burns and Augustine, 1995), focal adhesions (Hynes, 1992; Jockusch and Rüdiger, 1996), adherens junctions (Mays et al., 1994; Tsukita et al., 1993), and other sites of plasma membrane attachment to a basement membrane or to another cell (Fox, 1993; Tidball, 1991). The structures of dynamic membrane domains, which include microvilli, filopodia, pseudopods, and other cell surface extensions, are less well understood (Bretscher, 1993; Furthmayr et al., 1992; Luna and Hitt, 1992; Mooseker, 1985; Ridley, 1994). Interactions between the actin cortex and the plasma membrane in these transient domains play important roles during chemotaxis, phagocytosis, and the early stages of cell–cell adhesion (Condeelis, 1993; Hynes and Lander, 1992; Stossel, 1993). For instance, adhesion molecules on microvilli are thought to be required for cell surface attachment during the initial stages of cell adhesion (Berlin et al., 1995; Choi and Siu, 1987; Law, 1994; Moore et al., 1995; Picker et al., 1991).

Most of the evidence for the role of the membrane skeleton in dynamic processes is based on the phenotypes of cells expressing abnormal levels or types of various membrane-associated actin-binding proteins. For example, antisense oligonucleotides that reduce intracellular levels of the structurally related MER proteins (moesin, ezrin, and radixin) are reported to reduce both cell–cell adhesion and the presence of microvilli, filopodia, and ruffles (Takeuchi et al., 1994). Conversely, high expression levels of partial cDNAs encoding carboxy-terminal sequences of either *Drosophila* moesin (Edwards et al., 1994) or murine radixin (Henry et al., 1995) influence the formation of actin-rich plasma membrane protuberances. Similarly, overexpression of drebrin in cells that normally lack this protein induces the formation of highly branched cell extensions (Shirao, 1995) and stabilizes cell-substratum attachments in the presence of cytoskeleton-disrupting drugs (Ikeda et al., 1995).

Another well-characterized actin-binding membrane protein in a dynamic cell is ponticulin, a transmembrane glycoprotein that accounts for most of the high-affinity actin-binding and nucleating activities associated with plasma membranes from the soil amoeba, *Dictyostelium discoideum* (Chia et al., 1993; Hitt et al., 1994*a*, *b*; Shariff and Luna, 1990; Wuestehube and Luna, 1987). Mutant amoeba lacking ponticulin exhibit major aberrations during translocation, chemotaxis, and multicellular development (Hitt et al., 1994*a*; Shutt et al., 1995). The inefficiencies exhibited by these cells during directed cell movement are apparently due to a loss of cellular control over pseudopod dynamics (Shutt et al., 1995). Surprisingly, under conditions that require the cells to migrate in order to undergo multicellular development, ponticulin-minus *Dictyostelium* amoebae aggregate into mounds faster than the parental cells (Hitt et al., 1994*a*), rather than slower as would be expected if chemotaxis were the only defective process. Thus, it is possible that ponticulin serves as a negative regulator of another cellular function, perhaps adhesion, that also is required for early development.

Components of the *Dictyostelium* cell–cell adhesion sites include ponticulin and a number of other membrane-associated polypeptides, including a 34-kD actin filament crosslinking protein (p30a) that is also present in filopodia, phagocytic cups, and cleavage furrows (Fechheimer, 1987; Fechheimer et al., 1994; Furukawa and Fechheimer, 1994, 1996). Cells lacking the 34-kD protein show an increased persistence of motility during chemotaxis, lose bits of cytoplasm during locomotion, and exhibit strain-dependent aberrations in the numbers and lengths of filopodia (Rivero et al., 1996).

To determine which proteins eluting from F-actin affinity columns (Luna et al., 1982; Miller and Alberts, 1989) interact directly with actin and to provide a sensitive assay for these proteins, we have developed a procedure in which proteins fractionated on sodium dodecylsulfate (SDS)-gels are electrotransferred to nitrocellulose, fixed by gentle heating, and overlaid with ^{125}I-labeled F-actin (Chia et al., 1991; Pestonjamasp et al., 1995). We have now identified ponticulin, p30a, moesin, ezrin, radixin, and drebrin as six of the major polypeptides that bind to F-actin in this simple assay. Interestingly, these membrane-associated proteins, all of which have been implicated in either the formation or the stabilization of cell surface extensions, contain actin-binding sequences that are relatively easily renatured after detergent solubilization.

Materials and Methods
Cells

Dictyostelium discoideum amoebae (strain AX3, a gift from Dr. R. Kessin, Columbia University, NY) were grown in rotating suspensions at 20°C in HL-5 medium (Cocucci and Sussman, 1970). Cells in exponential growth phase were harvested by centrifugation and fractionated as described in Chia et al. (1991).

MCF-7 breast carcinoma cells were obtained from Dr. David Kupfer, SHSY5Y neuroblastoma cells were provided by Dr. Alonzo H. Ross, and NIH 3T3 and NRK fibroblasts were gifts of Dr. Yu-Li Wang, all of the Worcester Foundation. These cells were harvested by centrifugation for 3 min at 1,000 g, resuspended in phosphate-buffered saline, pH 7.4, and immediately solubilized in Laemmli Sample Buffer (Laemmli, 1970) for SDS-polyacrylamide gel electrophoresis (SDS-PAGE).

Suspension cultures of HeLa-S3 were grown in Joklik minimal essential medium (Irvine Scientific, Santa Ana, CA) supplemented with 5% fetal calf serum and 5% newborn calf serum and generously provided by Dr. Thoru Pederson of the Worcester Foundation for Biomedical Research. These cells were grown at 37°C in spinner flasks until harvesting as described (Pederson, 1972). HeLa plasma membrane ghosts and vesicles were purified by rate zonal centrifugation on a 30–45% step sucrose gradient as described by Atkinson (1973).

Neutrophils were purified from bovine blood, obtained fresh from a local abattoir, lysed by nitrogen cavitation, and fractionated on Percoll gradients as described previously (Pestonjamasp et al., 1995). Cell surface proteins were labeled with 5 mM sulfo-NHS-biotin (Pierce Chemical, Rockford, IL) at 4°C for 10 min and visualized with ^{125}I-labeled streptavidin (Goodloe-Holland and Luna, 1987; Ingalls et al., 1986; Pestonjamasp et al., 1995).

Purification of Myosin-V and p120

Myosin-V and p120 were purified from 120 brains of freshly hatched chicks by the method of Cheney et al. (1993) with the following modifications: Batches of 10 brains were homogenized in 40 ml ice-cold Homogenization Buffer (40 mM HEPES, 10 mM K-EDTA, 5 mM disodium ATP, 2 mM DTT, 1 mM Pefabloc-SC, 1 mM benzamidine, and 2 μg/ml aprotinin, pH 7.7) by 10 up and down strokes of a 55-ml Potter-Elvejhem homogenizer attached to a 1/4 inch electric drill operated at high speed. The pooled homogenates were centrifuged for 40 min at 35,000 g in two Sorvall SS-34 rotors. The supernatants were collected, and 4 M NaCl was added to a final concentration of 600 mM. The solution was incubated on ice for 1 h without stirring and was then centrifuged as before. The clear and gelatinous pellets of ~0.5 ml each were resuspended in a total of 45 ml of Wash Buffer (25 mM HEPES, 2 mM K-EDTA, 2 mM K-EGTA, 2 mM DTT, pH 7.2) using a 40-ml Dounce homogenizer. Then, 5 ml 10% Triton X-100 (SurfactAmps; Pierce Chemical Co.) was added to yield a final concentration of 1%. The resuspended pellets were incubated 2–3 min in a 37°C water bath with occasional mixing to solubilize membranes. The pellets, which contain primarily myosin-V and actin, were collected by centrifugation for 20 min at 35,000 g. Residual detergent was removed by resuspending the pellet in 25 ml Wash Buffer and centrifuging for 20 min at 35,000 g. The rinsed pellet was resuspended in 8 ml of S-500 Buffer (25 mM HEPES, 600 mM NaCl, 5 mM MgCl$_2$, 2 mM Na-EGTA, 5 mM disodium ATP, 2 mM DTT, pH 8.0) to which additional ATP had been added to a final concentration of 10 mM. The suspension was incubated for ~20 min on ice to solubilize the myosin-V and then centrifuged for 30 min at 150,000 g. The supernatant from this step, which contained primarily myosin-V and actin, as well as smaller amounts of the 120-kD protein, was loaded onto a 1.5 × 100 cm Sephacryl S-500HR column equilibrated in S-500 buffer. The column was run at a flow rate of approximately 15 ml/h, and 2.5-ml fractions were collected and analyzed by SDS-PAGE. The fractions containing myosin-V and the 120-kD protein were pooled and diluted into 2 volumes of TMAE Buffer (20 mM triethanolamine, 1 mM Na-EGTA, 2 mM DTT, pH 7.5), and the pH of this solution was adjusted to 7.5 at room temperature using a pH 6.7 stock of 10% triethanolamine-HCl. The protein solution was then loaded onto a 5 ml column of Fractogel EMD TMAE 650(S) (EM Separations, Gibbstown, NJ) pre-equilibrated in TMAE Buffer with 230 mM NaCl. The column was then washed with ~20 ml of the same buffer and was eluted with a 60 ml linear gradient of 230 to 500 mM NaCl in TMAE buffer. Fractions of 1 ml were collected at a flow rate of 10 ml/h. An aliquot of each fraction was analyzed by SDS-PAGE. The 120-kD protein eluted slightly ahead (at ~265 mM NaCl) of the main myosin-V peak (~275 mM NaCl). By repeating the TMAE column step with a 230–430 mM NaCl gradient, myosin-V could be obtained which was free from p120 detectable by Coomassie staining. The fractions enriched for the 120-kD protein were stored at 0°C before TCA precipitation, denaturation in Laemmli Sample Buffer, and analysis by SDS-polyacrylamide gel electrophoresis (SDS-PAGE), electrotransfer, and F-actin blot overlay assays.

F-Actin Blot Overlays

Cells and cell fractions were denatured at 70°C for 10 min in Laemmli Sample Buffer and electrophoresed into discontinuous SDS-polyacrylamide gels as described (Laemmli, 1970). *Dictyostelium* cell and membrane fractions were solubilized in Laemmli Sample Buffer lacking a disulfide-reducing agent. Proteins were electrotransferred to nitrocellulose at 6 V/cm for 16–20 h at 4°C (Towbin et al., 1979).

When necessary for the visualization of low molecular mass proteins ($\leq$34 kD), the nitrocellulose blots were heat-fixed by incubation at 50°C in 150 mM NaCl, 10 mM sodium phosphate, pH 7.4 (Chia et al., 1991). Nitrocellulose blots were then blocked in 5% milk, 90 mM NaCl, 0.5% (v/v) Tween-20, 10 mM Tris-HCl, pH 7.5 (TBST), and probed with 50 μg/ml gelsolin-capped, phalloidin-stabilized, ^{125}I-labeled F-actin in TBST for 2 h at $\sim$21°C (Chia et al., 1991; Pestonjamasp et al., 1995). The ^{125}I-labeled F-actin solution was recovered for re-use and was stored at 0–4°C between uses until depletion or decay of the radioactivity (usually 3 to 4 months). The blot was washed 4 to 5 times (2 min/wash) with TBST, air dried, and exposed either at 21°C to a phosphorimager screen or at −80°C to film in the presence of an intensifying screen.

Amino Acid Sequencing

Fractions enriched in the 120-kD protein were separated on a 5% SDS-gel, transferred to nitrocellulose (0.45-μm pore size), and stained briefly with Ponceau S. The area containing the 120-kD polypeptide was excised and washed 10 times with sterile deionized distilled water in a Spin-X™ centrifuge filter unit (0.22 μm porosity; Costar®; Costar Corp., Cambridge, MA). Endo-Lys-C digestion, microbore HPLC separation of proteolytic fragments, and subsequent microsequencing of the major product were carried out by Dr. John Leszyk in the William M. Keck Protein Chemistry Facility at the Worcester Foundation for Biomedical Research. The resulting sequence was compared with those in the nonredundant GenBank protein database, Release 97.0, using the BLASTP search algorithm (Altschul et al., 1990).

Results

Actin-binding Membrane Proteins in *Dictyostelium discoideum*

A number of other investigators have overlaid SDS-polyacrylamide gels or nitrocellulose blots with labeled G-actin to identify proteins that bind actin monomers (Bärmann et al., 1986; Mitchell et al., 1986; Schleicher et al., 1984; Snabes et al., 1983; Stratford and Brown, 1985; Tanaka et al., 1994; Walker et al., 1984). Here, we describe a blot overlay procedure that visualizes a number of proteins that bind to actin filaments (F-actin). In experiments with cell fractions from *Dictyostelium discoideum* (Fig. 1), we adjusted the conditions for detergent solubilization and other important experimental steps to minimize F-actin binding to the majority of the cytoplasmic F-actin binding proteins (Fig. 1, lane *2*), while retaining binding to prominent membrane-associated proteins (Fig. 1, lanes *3* and *4*). As shown below, this procedure appears to preferentially recognize a subset of F-actin binding membrane proteins in a variety of cell types.

In blot overlay assays with gelsolin-capped, phalloidin-stabilized ^{125}I-labeled F-actin, only two major polypeptides were detected in *Dictyostelium* whole cell extracts (Fig. 1, lane *1*), Triton-resistant cytoskeletons (Fig. 1, lane *2*), and crude plasma membranes (Fig. 1, lane *3*). The first was a 30- to 34-kD cytoskeletal protein that co-sedimented with actin filaments in cytoskeletons (Fig. 1, lane *2*) and crude membranes (Fig. 1, lane *3*), but did not co-purify with plasma membranes from log-phase cells after the actin had been removed by dialysis against a low ionic strength buffer (Fig. 1, lane *4*). Based on immunological cross-reactivity and binding of

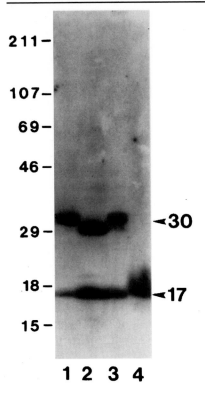

Figure 1. *Dictyostelium* proteins visualized by ^{125}I-labeled F-actin after SDS-polyacrylamide gel electrophoresis and electrotransfer to nitrocellulose. Autoradiogram of 40-μg aliquots of whole cell extract (lane *1*), actin-rich cytoskeletons resistant to extraction by 1% Triton X-100 (lane *2*), crude membranes (lane *3*), and highly purified plasma membranes (lane *4*) from log-phase cells. Developmentally induced cell aggregation increases by 8- to 26-fold the amount of the 30-kD protein that co-isolates with plasma membranes, especially with membranes from regions of cell-cell contact (Fechheimer, et al., 1994). Migration positions of molecular mass standards, in kD, are shown on the left. Reproduced from Chia et al., *Cell Motility and the Cytoskeleton*, (1991, Vol. 18:164–179). Copyright © 1991, John Wiley & Sons.

[125]I-labeled actin to purified protein (Chia et al., 1991; Fechheimer and Furukawa, 1993), this polypeptide is p30a, a calcium-sensitive actin bundling protein (Fechheimer and Taylor, 1984) that is concentrated in filopodia (Fechheimer, 1987), cell–cell contact regions (Fechheimer et al., 1994), and at phagocytotic cups and cleavage furrows (Furukawa et al., 1992; Furukawa and Fechheimer, 1994). In vitro, p30a stabilizes actin filaments to disassembly (Zigmond et al., 1992) and stabilizes the association of F-actin with isolated cell–cell contact regions (Fechheimer et al., 1994). Tight binding of p30a to isolated membranes requires F-actin, and the presence of membranes stabilizes the association between p30a and F-actin (Fechheimer et al., 1994). Thus, p30a and F-actin co-assemble onto the plasma membrane with positive cooperativity, where they form selectively stabilized cytoskeletal domains that, in turn, may help organize membrane proteins involved in cell–cell adhesion, cytokinesis, and/or phagocytosis.

The second prominent F-actin-binding *Dictyostelium* protein detected on blot overlays was a 17-kD polypeptide that was enriched in both cytoskeletons (Fig. 1, lane *2*) and plasma membranes (Fig. 1, lanes *3* and *4*). This polypeptide has been identified as ponticulin (Chia et al., 1991), a 17-kD transmembrane glycoprotein that binds directly and specifically to F-actin on affinity columns (Chia et al., 1993; Wuestehube and Luna, 1987) and nucleates actin filament assembly (Chia et al., 1993; Shariff and Luna, 1990). Disruption of the single-copy ponticulin gene by homologous recombination leads to cells that contain an order of magnitude less actin associated with their plasma membranes (Hitt et al., 1994*a*), as well as correlated defects in pseudopod dynamics, chemotaxis, and cell aggregation (Hitt et al., 1994*a*; Shutt et al., 1995).

Actin-binding Membrane Proteins in Mammalian Cell Lines

F-actin blot overlay assays also visualized proteins with sizes similar to those of p30a and ponticulin in whole cell extracts from various mammalian cell lines (Fig. 2). As in log-phase *Dictyostelium* amoebae, the ~34-kD protein (p34) did not co-isolate with purified membranes, whereas the ~16-kD protein was enriched in plasma membranes from neutrophils (Wuestehube et al., 1989) and HeLa S3 cells (not shown). Although not yet demonstrated, it is possible that p34 and p16 in Fig. 2 represent the immunocrossreactive mammalian homologues of *Dictyostelium* p30a (Johns et al., 1988) and ponticulin (Wuestehube et al., 1989), respectively.

Mammalian cells also contained even more prominent F-actin binding poly-peptides with molecular masses between ~60- and ~205-kD (Fig. 2). No cell line contained all of these proteins, and no one protein was found in high abundance in all the cells. The overall appearance was of seven to nine major F-actin binding pro-teins that were each found in a number of different types of cells.

We have recently identified most of the major actin-binding proteins in these mammalian cell lines. Using bovine neutrophil plasma membranes, which were en-riched in the 78-, 81-, and 205-kD proteins (Fig. 3), we found that these polypeptides

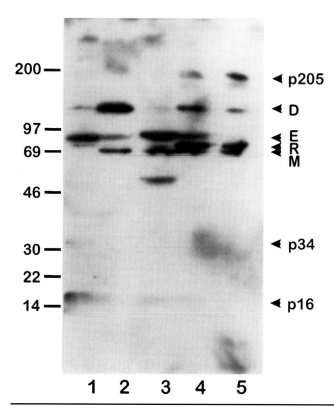

Figure 2. Proteins in mammalian cell lines that are visualized by ^{125}I-labeled F-actin after SDS-polyacrylamide gel electrophoresis and electrotransfer to nitrocellulose. Autoradiogram of whole cell extracts (10[6] cells/lane) from MCF-7 breast carcinoma (lane *1*), SHSY5Y neuroblastoma (lane *2*), NIH 3T3 fibroblasts (lane *3*), NRK fibroblasts (lane *4*), and HeLa S3 cervical carcinoma (lane *5*) cells. Migration positions of molecular mass standards, in kD, are denoted on the left. Arrows on the right denote the positions of a 16-kD protein (*p16*), a 34-kD protein (*p34*), moesin (*M*), radixin (*R*), ezrin (*E*), drebrin (*D*), and p205.

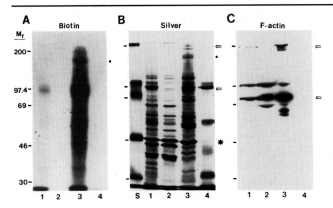

Figure 3. Proteins in bovine neutrophils that are labeled extracellularly by a charged biotinylating agent and visualized by ^{125}I-labeled streptavidin (*A*) or which bind directly to ^{125}I-labeled F-actin (*C*) after SDS-polyacrylamide gel electrophoresis and electrotransfer to nitrocellulose. Total proteins were visualized on SDS-gels stained with silver (*B*). Each lane was loaded with 10 μg (*A* and *B*) or 100 μg (*C*) of whole cell extract (lanes *1*), cytosol (lanes *2*), a plasma membrane-enriched fraction (lanes *3*), and pooled granules (lanes *4*). The lower edges of the molecular mass standards (lane *S*) are denoted on the left of each panel. Approximate migration positions of a prominent cell surface biotinylated protein (●), actin (*), and major membrane-associated actin-binding proteins of ∼205- and ∼80-kD (*arrows*) are shown to the right of the relevant panels. Reproduced from *Molecular Biology of the Cell* (1995, Vol. 6:247–259), with permission of the American Society for Cell Biology.

were, respectively, moesin, ezrin, and a novel protein (p205) that was not in the sequence databanks (Pestonjamasp et al., 1995). Although all three proteins co-purified with plasma membrane markers during cell fractionation (Fig. 3, *A* and *C*, lanes *3*), significant amounts of moesin and ezrin also were found in the cytosol (Fig. 3 *C*, lane *2*). By contrast, nearly all the p205 was found tightly associated with the plasma membrane (Fig. 3 *C*, lane *3*).

The F-actin binding visualized in Fig. 3 *C* appeared to be specific by a number of criteria (Pestonjamasp et al., 1995). First, binding of ^{125}I-labeled actin was dependent upon the assembly state of actin, i.e., F-actin bound, but G-actin did not. Second, other negatively charged biopolymers, including microtubules, heparin, and polyaspartic acid, did not bind to the F-actin binding proteins. Finally, the interaction was specific for the sides of the actin filaments because myosin heads competed for binding in an ATP-dependent fashion, whereas proteins that bind to the barbed ends of actin filaments did not interfere with the binding of the already capped ^{125}I-labeled actin filaments (Pestonjamasp et al., 1995).

Based on immunocrossreactivity, co-immunoprecipitation with specific antisera, and ^{125}I-labeled F-actin binding to bacterially-expressed protein, the ∼80-kD F-actin binding polypeptide, seen most prominently in Fig. 2, lanes *4* and *5*, is radixin (Pestonjamasp et al., 1995). Along with moesin and ezrin, radixin is the third member of the so-called MER family of membrane-associated proteins. We (Pestonjamasp et al., 1995) and others (Turunen et al., 1994) have localized the actin-binding site in these proteins to a short carboxy-terminal sequence that is very highly conserved (≥85% identity) among these three proteins. Actin binding is isoform-dependent since beta-actin binds with even higher avidity than does alpha-actin (Shuster and Herman, 1995; Yao et al., 1996). Because the carboxy-terminal F-actin binding site is apparently blocked in MER proteins in the cytoplasm (Bretscher,

1983; Fazioli et al., 1993; Krieg and Hunter, 1992), but is uncovered by truncation (Algrain et al., 1993; Edwards et al., 1994) or denaturation (Gary and Bretscher, 1995; Pestonjamasp et al., 1995) of highly conserved sequences in the amino-terminus, it has been proposed that the accessibility of the carboxy-terminus, and thus its ability to bind F-actin, is controlled in the cell (Algrain et al., 1993; Berryman et al., 1995; Bretscher et al., 1995; Furthmayr et al., 1992; Martin et al., 1995; Pestonjamasp et al., 1995).

The ~205-kD F-actin binding protein (p205) observed in a number of cell types, including NRK fibroblasts (Fig. 2, lane *4*), HeLa cervical carcinoma cells (Fig. 2, lane *5*), and bovine neutrophils (Fig. 3 *C*, lane *3*) appears to be a previously undescribed protein (Pestonjamasp et al., unpublished observations). First, p205 was not recognized by antibodies against myosin II, fodrin, talin, or tensin (Pestonjamasp et al., 1995). Second, p205 is probably not an unconventional myosin (Hasson and Mooseker, 1996; Mooseker and Cheney, 1995) because ^{125}I-labeled F-actin bound neither to myosin II (Fig. 1, lane *2*), nor to any of several *Dictyostelium* or chicken myosin I proteins (not shown), nor to purified chick brain myosin-V (Fig. 4, lane *1*) under the conditions of our assay. Finally, none of six amino acid microsequences obtained from tryptic and Endo-Lys-C digests of p205 was represented in the GenBank (release 97.0) or dbEST (release 102696) databases (E.J. Luna, unpublished observation). Work is in progress in the Luna laboratory to characterize the primary structure, native F-actin binding activities, intracellular localization, and tissue distribution of this protein.

Identification of Drebrin as a Major 120-kD F-actin Binding Protein

Although chick brain myosin-V did not bind to F-actin in the overlay assay, a minor contaminant of highly purified myosin-V that has an apparent molecular weight of 120-kD was recognized (Fig. 4, lane *1*). This 120-kD polypeptide co-migrated with a Coomassie blue–stained protein that co-purified with myosin-V during precipita-

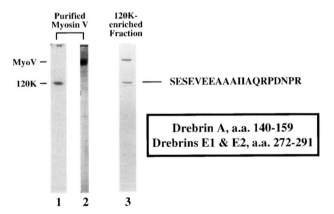

Figure 4. Identification of the ~120-kD F-actin binding protein as drebrin. A 120-kD protein, first discovered as a major F-actin-binding protein in an F-actin blot overlay assay (lane *1*) of highly purified myosin-V from chick brains is virtually undetectable by staining for total protein with Coomassie blue (lane *2*). This 120-kD protein was recovered as a major Coomassie blue–stained component from adjacent fractions of the TMAE anion exchange column (lane *3*). The amino acid sequence of a proteolytic fragment from this polypeptide, shown to the right, corresponds exactly to sequences from the differentially spliced isoforms A, E1, and E2 of drebrin.

tion with F-actin, extraction with Triton X-100, and gel filtration (Fig. 4, lane *3*). The 120-kD protein and actin were usually the only major contaminants of myosin-V visible by Coomassie blue staining at intermediate stages of myosin-V purification. This 120-kD protein was separated from the bulk of the myosin-V on a TMAE anion exchange column during the final step of myosin-V purification.

The 120-kD protein was identified by sequencing a proteolytic fragment obtained after preparative SDS-PAGE. In searches of the non-redundant protein databases, the derived sequence, SESEVEEAAAIIAQRPDNPRR (Fig. 4), was an exact match for residues 140–159 of chicken drebrin A and for residues 272–291 of rat drebrin A, human drebrin E, and chicken drebrin E. This identity was highly significant ($P = 2.7 \times 10^{-6}$), suggesting that most, if not all, of the differentially spliced isoforms of the drebrin genes (Kojima et al., 1993; Shirao et al., 1992), which encode highly homologous proteins, bind F-actin on blot overlays. Although drebrin is often said to be specific for neurons and neuroblastomas (Ikeda et al., 1995; Shirao et al., 1992), we observed that the 120-kD protein was also present, albeit at lower levels, in a number of cell lines (Fig. 2). To confirm the presence of drebrin in non-neuronal tissues, a nitrocellulose blot containing whole cell extracts was first stained with ^{125}I-labeled F-actin and then with a monoclonal antibody against drebrin plus an alkaline phosphatase–conjugated secondary antibody (Fig. 5). The 120-kD polypeptide that bound ^{125}I-labeled F-actin (Fig. 5 *A*) in NRK fibroblasts (lane *3*), as well as the drebrin isoforms in brain (lanes *1* and *2*) and SHSY5Y neuroblastomas (lane *4*), were all recognized by the anti-drebrin antibody (Fig. 5 *B*). Thus, it is likely that drebrin is present at some level in many cell lines.

Discussion

F-actin blot overlays are a sensitive, efficient method by which to identify and track the subset of F-actin binding proteins that are recognized in this assay. Only two to nine major polypeptides in any particular cell type react strongly with ^{125}I-labeled F-actin under the conditions described here (initial 10-min denaturation at 70°C in

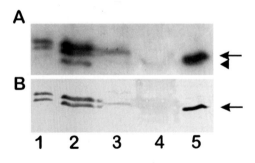

Figure 5. Recognition of the 120-kD F-actin binding protein with an antibody against drebrin. F-actin blot overlay (*A*) and anti-drebrin antibody-stained (*B*) proteins from the same lanes of an SDS-gel loaded with 100 μg of bovine brain extract (lane *1*), 300 μg of bovine brain extract (lane *2*), 4×10^6 NRK fibroblasts (lane *3*), 4×10^6 HL60 cells (lane *4*), and 4×10^6 SHSY5Y neuroblastoma cells (lane *5*). The nitrocellulose blot was stained first with ^{125}I-labeled F-actin, exposed to film, and then stained with monoclonal antibody M2F6 against drebrin (MBL International Corporation, Watertown, MA) and visualized using an alkaline phosphatase–conjugated rabbit anti–mouse antibody (Sigma Chemical Co., St. Louis, MO). Arrows designate the position of drebrin isoforms; the arrowhead marks the position of a higher mobility protein not recognized by the M2F6 monoclonal antibody.

Laemmli Sample Buffer; minimal renaturation after electrotransfer). Their characteristic molecular masses permit each polypeptide to be identified and monitored in a single, simple assay. While the sensitivity of detection may vary for each polypeptide, ^{125}I-labeled F-actin blot overlays are at least as sensitive as immunoassays with primary antibody and ^{125}I-labeled protein A for the detection of ponticulin (Chia et al., 1991). Furthermore, the ^{125}I-labeled F-actin overlay solution can be re-used for several months, albeit with loss of sensitivity due to depletion and/or radioactive decay. Since F-actin is relatively inexpensive to prepare, F-actin blot overlays are more efficient, cheaper, and faster than a series of immunoblot assays with antibodies against each of the reactive proteins.

We have now identified six of these polypeptides as ponticulin, p30a, moesin, radixin, ezrin, and drebrin. Although ponticulin is a transmembrane protein whereas the other five proteins are peripheral proteins, all six proteins associate with the plasma membrane, where they appear to play roles in the formation and/or control of cell surface extensions. While the function and subcellular distribution of p205 are still under investigation, this protein co-purifies with plasma membranes from both bovine neutrophils and human cervical carcinoma cells, suggesting that p205 also plays a role at the membrane-cytoskeleton boundary.

This group of polypeptides does not contain those proteins originally identified by virtue of their binding to labeled G-actin in gel or blot overlay assays. These G-actin binding proteins include gelsolin and villin (Snabes, et al., 1983), hisactophilin (Schleicher et al., 1984), comitin (Stratford and Brown, 1985; Weiner et al., 1993), annexin VI (Tanaka et al., 1994), a 50-kD liver membrane protein (Bärmann et al., 1986), and a 43-kD protein associated with the acetylcholine receptor (Walker et al., 1984). As mentioned above, our assay also does not visualize myosin II (Mitchell et al., 1986), probably due to the absence of the extensive renaturation procedure that is required to restore native myosin structure after SDS treatment (Muhlrad and Morales, 1984).

The apparent correlation between membrane association and reactivity with ^{125}I-labeled F-actin on blot overlays raises the question of whether or not the F-actin binding site is conserved among these proteins. As a first step towards answering this question, we identified the F-actin binding site recognized by ^{125}I-labeled F-actin in moesin, ezrin, and radixin as a highly conserved sequence at the carboxy terminus of these proteins (Pestonjamasp et al., 1995). In our study, truncation of the carboxy-terminal 22 amino acids in bacterially-expressed moesin completely eliminated binding, but binding was retained in constructs expressing only the last 48 residues (Pestonjamasp et al., 1995). These results were in gratifying agreement with the studies of Turunen et al. (Turunen et al., 1994), who reported binding of endogenous actin to GST fusion proteins containing the carboxy-terminal 34 residues of ezrin.

Although the F-actin binding sequence in moesin, ezrin, and radixin is ≥85% identical among all three family members (Lankes et al., 1993), it does not appear to be present in the other F-actin binding proteins visualized by ^{125}I-labeled F-actin blot overlays. First, preliminary experiments with an affinity-purified antibody directed against the conserved MER family carboxy terminus do not recognize any of the other proteins observed on F-actin blot overlays (Pestonjamasp, Strassel, and Luna, unpublished observations). Second, sequence comparisons with drebrin (Shirao et al., 1992) and p30a (Fechheimer et al., 1991) do not reveal any significant homologies with the MER F-actin binding sequence. Thus, these F-actin binding sites apparently share only the capability to be easily renatured after SDS treatment, implying that the requisite amino acid sequences are short and contiguous. An apparent exception to this rule may be the F-actin binding site in ponticulin, which is not readily renatured after reduction with thiols (Chia et al., 1991). Thus, this F-actin binding site may contain a more extensive secondary or tertiary structure than those in the other identified F-actin binding membrane proteins. Definitive comparisons of these diverse F-actin binding sites await the identification of the responsible sequences in ponticulin, p30a, drebrin, and p205.

Acknowledgments

This research was supported by National Institutes of Health grants GM33048 and CA54885 to E.J. Luna, by NSF grant MCB 94-05738 to M. Fechheimer, and by a basic research grant from the Muscular Dystrophy Association to M.S. Mooseker. This research also benefited from grants to the Worcester Foundation for Biomedical Research from the J. Aron Charitable Foundation and the Stork Foundation.

References

Algrain, M., O. Turunen, A. Vaheri, D. Louvard, and M. Arpin. 1993. Ezrin contains cytoskeleton and membrane binding domains accounting for its proposed role as a membrane-cytoskeletal linker. *J. Cell Biol.* 120:129–139.

Altschul, S.F., W. Gish, W. Miller, E.W. Myers, and D.J. Lipman. 1990. Basic local alignment search tool. *J. Mol. Biol.* 215:403–410.

Atkinson, P.H. 1973. HeLa cell plasma membranes. *Methods Cell Biol.* 7:157–188.

Bärmann, M., J. Wadsack, and M. Frimmer. 1986. A 50 kDa, actin-binding protein in plasma membranes of rat hepatocytes and of rat liver tumors. *Biochim. Biophys. Acta.* 859:110–116.

Berlin, C., R.F. Bargatze, J.J. Campbell, U.H. von Andrian, M.C. Szabo, S.R. Hasslen, R.D. Nelson, E.L. Berg, S.L. Erlandsen, and E.C. Butcher. 1995. α4 integrins mediate lymphocyte attachment and rolling under physiologic flow. *Cell.* 80:413–422.

Berryman, M., R. Gary, and A. Bretscher. 1995. Ezrin oligomers are major cytoskeletal components of placental microvilli: a proposal for their involvement in cortical morphogenesis. *J. Cell Biol.* 131:1231–1242.

Bretscher, A. 1983. Purification of an 80,000-dalton protein that is a component of the isolated microvillus cytoskeleton, and its localization in nonmuscle cells. *J. Cell Biol.* 97:425–432.

Bretscher, A. 1993. Microfilaments and membranes. *Curr. Opin. Cell Biol.* 5:653–660.

Bretscher, A., R. Gary, and M. Berryman. 1995. Soluble ezrin purified from placenta exists as stable monomers and elongated dimers with masked C-terminal ezrin-radixin-moesin association domains. *Biochemistry.* 34:16830–16837.

Burns, M.E., and G.J. Augustine. 1995. Synaptic structure and function: dynamic organization yields architectural precision. *Cell.* 83:187–194.

Campbell, K.P. 1995. Three muscular dystrophies: loss of cytoskeleton-extracellular matrix linkage. *Cell.* 80:675–679.

Cheney, R.E., M.K. O'Shea, J.E. Heuser, M.V. Coelho, J.S. Wolenski, E.M. Espreafico, P. Forscher, R.E. Larson, and M.S. Mooseker. 1993. Brain myosin-V is a two-headed unconventional myosin with motor activity. *Cell.* 75:13–23.

Chia, C.P., A.L. Hitt, and E.J. Luna. 1991. Direct binding of F-actin to ponticulin, an integral plasma membrane glycoprotein. *Cell Motil. Cytoskeleton.* 18:164–179.

Chia, C.P., A. Shariff, S.A. Savage, and E.J. Luna. 1993. The integral membrane protein, ponticulin, acts as a monomer in nucleating actin assembly. *J. Cell Biol.* 120:909–922.

Choi, A.H.C., and C.-H. Siu. 1987. Filopodia are enriched in a cell cohesion molecule of M_r 80,000 and participate in cell-cell contact formation in *Dictyostelium discoideum. J. Cell Biol.* 104:1375–1387.

Cocucci, S.M., and M. Sussman. 1970. RNA in cytoplasmic and nuclear fractions of cellular slime mold amoebas. *J. Cell Biol.* 45:399–407.

Condeelis, J. 1993. Life at the leading edge: the formation of cell protrusions. *Annu. Rev. Cell Biol.* 9:411–444.

Edwards, K.A., R.A. Montague, S. Shepard, B.A. Edgar, R.L. Erikson, and D.P. Kiehart. 1994. Identification of *Drosophila* cytoskeletal proteins by induction of abnormal cell shape in fission yeast. *Proc. Natl. Acad. Sci. USA.* 91:4589–4593.

Fazioli, F., W.T. Wong, S.J. Ullrich, K. Sakaguchi, E. Appella, and P.P. Di Fiore. 1993. The ezrin-like family of tyrosine kinase substrates: receptor-specific pattern of tyrosine phosphorylation and relationship to malignant transformation. *Oncogene.* 8:1335–1345.

Fechheimer, M. 1987. The *Dictyostelium discoideum* 30,000-dalton protein is an actin filament-bundling protein that is selectively present in filopodia. *J. Cell Biol.* 104:1539–1551.

Fechheimer, M., H.M. Ingalls, R. Furukawa, and E.J. Luna. 1994. Association of the *Dictyostelium* 30,000 M_r actin bundling protein with contact regions. *J. Cell Sci.* 107:2393–2401.

Fechheimer, M., D. Murdock, M. Carney, and C.V.C. Glover. 1991. Isolation and sequencing of cDNA clones encoding the *Dictyostelium discoideum* 30,000-dalton actin-bundling protein. *J. Biol. Chem.* 266:2883–2889.

Fechheimer, M., and R. Furukawa. 1993. A 27,000 dalton core of the *Dictyostelium* 34,000 dalton protein retains Ca^{+2}-regulated actin cross-linking but lacks bundling activity. *J. Cell Biol.* 120:1169–1176.

Fechheimer, M., and D.L. Taylor. 1984. Isolation and characterization of a 30,000-dalton calcium-sensitive actin cross-linking protein from *Dictyostelium discoideum. J. Biol. Chem.* 259:4514–4520.

Fox, J.E.B. 1993. Regulation of platelet function by the cytoskeleton. *In* Mechanisms of Platelet Activation and Control. K.S. Authi, et al., editor. Plenum Press, New York. 175–185.

Furthmayr, H., W. Lankes, and M. Amieva. 1992. Moesin, a new cytoskeletal protein and constituent of filopodia: its role in cellular functions. *Kidney Int.* 41:665–670.

Furukawa, R., S. Butz, E. Fleischmann, and M. Fechheimer. 1992. The *Dictyostelium discoideum* 30,000 dalton protein contributes to phagocytosis. *Protoplasma.* 169:18–27.

Furukawa, R., and M. Fechheimer. 1994. Differential localization of α-actinin and the 30 kD actin-bundling protein in the cleavage furrow, phagocytic cup, and contractile vacuole of *Dictyostelium discoideum. Cell Motil. Cytoskeleton.* 29:46–56.

Furukawa, R., and M. Fechheimer. 1996. Role of the *Dictyostelium* 30 kDa protein in actin bundle formation. *Biochemistry.* 35:7224–7232.

Gary, R., and A. Bretscher. 1995. Ezrin self-association involves binding of an N-terminal domain to a normally masked C-terminal domain that includes the F-actin binding site. *Mol. Biol. Cell.* 6:1061–1075.

Goodloe-Holland, C.M., and E.J. Luna. 1987. Purification and characterization of *Dictyostelium discoideum* plasma membranes. *Methods Cell Biol.* 28:103–128.

Hasson, T., and M.S. Mooseker. 1996. Vertebrate unconventional myosins. *J. Biol. Chem.* 271:16431–16434.

Henry, M.D., C.G. Agosti, and F. Solomon. 1995. Molecular dissection of radixin: distinct and interdependent functions of the amino- and carboxy-terminal domains. *J. Cell Biol.* 129:1007–1022.

Hitt, A.L., J.H. Hartwig, and E.J. Luna. 1994a. Ponticulin is the major high-affinity link between the plasma membrane and the cortical actin network in *Dictyostelium*. *J. Cell Biol.* 126:1433–1444.

Hitt, A.L., T.H. Lu, and E.J. Luna. 1994b. Ponticulin is an atypical membrane protein. *J. Cell Biol.* 126:1421–1431.

Hitt, A.L., and E.J. Luna. 1994. Membrane interactions with the actin cytoskeleton. *Curr. Opin. Cell Biol.* 6:120–130.

Hynes, R.O. 1992. Integrins: versatility, modulation, and signaling in cell adhesion. *Cell.* 69:11–25.

Hynes, R.O., and A.D. Lander. 1992. Contact and adhesive specificities in the associations, migrations, and targeting of cells and axons. *Cell.* 68:303–322.

Ikeda, K., T. Shirao, M. Toda, H. Asada, S. Toya, and K. Uyemura. 1995. Effect of a neuron-specific actin-binding protein, drebrin A, on cell-substratum adhesion. *Neurosci. Lett.* 194:197–200.

Ingalls, H.M., C.M. Goodloe-Holland, and E.J. Luna. 1986. Junctional plasma membrane domains isolated from aggregating *Dictyostelium discoideum* amebae. *Proc. Natl. Acad. Sci USA.* 83:4779–4783.

Jockusch, B.M., and M. Rüdiger. 1996. Crosstalk between cell adhesion molecules: vinculin as a paradigm for regulation by conformation. *Trends Cell Biol.* 6:311–315.

Johns, J.A., A.M. Brock, and J.D. Pardee. 1988. Colocalization of F-actin and 34-kilodalton actin bundling protein in *Dictyostelium* amoebae and cultured fibroblasts. *Cell Motil. Cytoskeleton.* 9:205–218.

Kojima, N., T. Shirao, and K. Obata. 1993. Molecular cloning of a developmentally regulated brain protein, chicken drebrin A and its expression by alternative splicing of the drebrin gene. *Mol. Brain Res.* 19:101–114.

Krieg, J., and T. Hunter. 1992. Identification of the two major epidermal growth factor-induced tyrosine phosphorylation sites in the microvillar core protein ezrin. *J. Biol. Chem.* 267:19258–19265.

Laemmli, U.K. 1970. Cleavage of structural proteins during the assembly of the head of bacteriophage T4. *Nature (Lond.).* 227:680–685.

Lankes, W.T., R. Schwartz-Albiez, and H. Furthmayr. 1993. Cloning and sequencing of porcine moesin and radixin cDNA and identification of highly conserved domains. *Biochim. Biophys. Acta.* 1216:479–482.

Law, D. 1994. Adhesion and its role in the virulence of enteropathogenic *Escherichia coli. Clin. Microbiol. Rev.* 7:152–173.

Luna, E.J., and A.L. Hitt. 1992. Cytoskeleton-plasma membrane interactions. *Science.* 258:955–964.

Luna, E.J., Y.-L. Wang, E.W. Voss, Jr., D. Branton, and D.L. Taylor. 1982. A stable, high capacity, F-actin affinity column. *J. Biol. Chem.* 257:13095–13100.

Lux, S.E., and J. Palek. 1995. Disorders of the red cell membrane. *In* Blood: Principles and Practice of Hematology. R.I. Handin, S.E. Lux, and T.P. Stossel, editors. J.B. Lippincott Company, Philadelphia. 1701–1818.

Martin, M., C. Andréoli, A. Sahuquet, P. Montcourrier, M. Algrain, and P. Mangeat. 1995. Ezrin NH_2-terminal domain inhibits the cell extension activity of the COOH-terminal domain. *J. Cell Biol.* 128:1081–1093.

Mays, R.W., K.A. Beck, and W.J. Nelson. 1994. Organization and function of the cytoskeleton in polarized epithelial cells: a component of the protein sorting machinery. *Curr. Opin. Cell Biol.* 6:16–24.

Miller, K.G., and B.M. Alberts. 1989. F-actin affinity chromatography: technique for isolating previously unidentified actin-binding proteins. *Proc. Natl Acad. Sci. USA.* 86:4808–4812.

Mitchell, E.J., R. Jakes, and J. Kendrick-Jones. 1986. Localisation of light chain and actin binding sites on myosin. *Eur. J. Biochem.* 161:25–35.

Mohandas, N., and E. Evans. 1994. Mechanical properties of the red cell membrane in relation to molecular structure and genetic defects. *Annu. Rev. Biophys. Biomol. Struct.* 23:787–818.

Moore, K.L., K.D. Patel, R.E. Bruehl, L. Fugang, D.A. Johnson, H.S. Lichenstein, R.D. Cummings, D.F. Bainton, and R.P. McEver. 1995. P-selectin glycoprotein ligand-1 mediates rolling of human neutrophils on P-selectin. *J. Cell Biol.* 128:661–671.

Mooseker, M.S. 1985. Organization, chemistry, and assembly of the cytoskeletal apparatus of the intestinal brush border. *Annu. Rev. Cell Biol.* 1:209–241.

Mooseker, M.S., and R.E. Cheney. 1995. Unconventional myosins. *Annu. Rev. Cell Develop. Biol.* 11:633–675.

Muhlrad, A., and M.F. Morales. 1984. Isolation and partial renaturation of proteolytic fragments of the myosin head. *Proc. Natl. Acad. Sci. USA.* 81:1003–1007.

Pederson, T. 1972. Chromatin structure and the cell cycle. *Proc. Natl. Acad. Sci. USA.* 69:2224–2228.

Pestonjamasp, K., M.R. Amieva, C.P. Strassel, W.M. Nauseef, H. Furthmayr, and E.J. Luna. 1995. Moesin, ezrin, and p205 are actin-binding proteins associated with neutrophil plasma membranes. *Mol. Biol. Cell.* 6:247–259.

Picker, L.J., R.A. Warnock, A.R. Burns, C.M. Doerschuk, E.L. Berg, and E.C. Butcher. 1991. The neutrophil selectin LECAM-1 presents carbohydrate ligands to the vascular selectins ELAM-1 and GMP-140. *Cell.* 66:921–933.

Ridley, A.J. 1994. Membrane ruffling and signal transduction. *BioEssays.* 16:321–327.

Rivero, F., R. Furukawa, A.A. Noegel, and M. Fechheimer. 1996. *Dictyostelium discoideum* cells lacking the 34,000 dalton actin binding protein can grow, locomote, and develop, but exhibit defects in regulation of cell structure and movement: a case of partial redundancy. *J. Cell Biol.* 135:965–980.

Schleicher, M., G. Gerisch, and G. Isenberg. 1984. New actin-binding proteins from *Dictyostelium discoideum. EMBO J.* 3:2095–2100.

Shariff, A., and E.J. Luna. 1990. *Dictyostelium discoideum* plasma membranes contain an actin-nucleating activity that requires ponticulin, an integral membrane glycoprotein. *J. Cell Biol.* 110:681–692.

Shirao, T. 1995. The roles of microfilament-associated proteins, drebrins, in brain morphogenesis: a review. *J. Biochem.* 117:231–236.

Shirao, T., N. Kojima, and K. Obata. 1992. Cloning of drebrin A and induction of neurite-like processes in drebrin-transfected cells. *NeuroReport.* 3:109–112.

Shuster, C.B., and I.M. Herman. 1995. Indirect association of ezrin with F-actin: isoform specificity and calcium sensitivity. *J. Cell Biol.* 128:837–848.

Shutt, D.C., D. Wessels, K. Wagenknecht, A. Chandrasekhar, A.L. Hitt, E.J. Luna, and D.R. Soll. 1995. Ponticulin plays a role in the positional stabilization of pseudopods. *J. Cell Biol.* 131:1495–1506.

Snabes, M.C., A.E. Boyd, III, and J. Bryan. 1983. Identification of G actin-binding proteins in rat tissues using a gel overlay technique. *Exp. Cell Res.* 146:63–70.

Stossel, T.P. 1993. On the crawling of animal cells. *Science.* 260:1086–1094.

Stratford, C.A., and S. Brown. 1985. Isolation of an actin-binding protein from membranes of *Dictyostelium discoideum. J. Cell Biol.* 100:727–735.

Takeuchi, K., N. Sato, H. Kasahara, N. Funayama, A. Nagafuchi, S. Yonemura, S. Tsukita, and S. Tsukita. 1994. Perturbation of cell adhesion and microvilli formation by antisense oligonucleotides to ERM family members. *J. Cell Biol.* 125:1371–1384.

Tanaka, K., T. Tashiro, S. Sekimoto, and Y. Komiya. 1994. Axonal transport of actin and actin-binding proteins in the rat sciatic nerve. *Neurosci. Res.* 19:295–302.

Tidball, J.G. 1991. Force transmission across muscle cell membranes. *J. Biomechanics.* 24:43–52.

Towbin, H., T. Stahelin, and J. Gordon. 1979. Electrophoretic transfer of proteins from polyacrylamide gels to nitrocellulose sheets: procedure and some applications. *Proc. Natl. Acad. Sci. USA.* 76:4350–4354.

Tsukita, S., M. Itoh, A. Nagafuchi, S. Yonemura, and S. Tsukita. 1993. Submembranous junctional plaque proteins include potential tumor suppressor molecules. *J. Cell Biol.* 123:1049–1053.

Turunen, O., T. Wahlström, and A. Vaheri. 1994. Ezrin has a COOH-terminal actin-binding site that is conserved in the ezrin protein family. *J. Cell Biol.* 126:1445–1453.

Walker, J.H., C.M. Boustead, and V. Witzemann. 1984. The 43-K protein, v_1, associated with acetylcholine receptor containing membrane fragments is an actin-binding protein. *EMBO J.* 3:2287–2290.

Weiner, O.H., J. Murphy, G. Griffiths, M. Schleicher, and A.A. Noegel. 1993. The actin-binding protein comitin (p24) is a component of the Golgi apparatus. *J. Cell Biol.* 123:23–34.

Wuestehube, L.J., C.P. Chia, and E.J. Luna. 1989. Immunofluorescence localization of ponticulin in motile cells. *Cell Motil. Cytoskeleton.* 13:245–263.

Wuestehube, L.J., and E.J. Luna. 1987. F-actin binds to the cytoplasmic surface of ponticulin, a 17-kD integral glycoprotein from *Dictyostelium discoideum. J. Cell Biol.* 105:1741–1751.

Yao, X., L. Cheng, and J.G. Forte. 1996. Biochemical characterization of ezrin-actin interaction. *J. Biol. Chem.* 271:7224–7229.

Zigmond, S.H., R. Furukawa, and M. Fechheimer. 1992. Inhibition of actin filament depolymerization by the *Dictyostelium* 30,000-D actin-bundling protein. *J. Cell Biol.* 119:559–567.

Interactions between Dystrophin and the Sarcolemma Membrane

Jeffrey S. Chamberlain, Kathleen Corrado, Jill A. Rafael, Gregory A. Cox, Michael Hauser, and Carey Lumeng

Department of Human Genetics, The University of Michigan Medical School, Ann Arbor, Michigan 48109-0618

Dystrophin serves as a link between the subsarcolemmal cytoskeleton and the extracellular matrix. The NH_2 terminus attaches to the cytoskeleton, while the COOH terminus attaches to the dystrophin associated protein (DAP) complex, which can be separated into the dystroglycan, sarcoglycan, and syntrophin subcomplexes. While the function of each DAP is not known, the dystroglycan complex binds laminin in the extracellular matrix, and binds the dystrophin COOH terminus in vitro. The syntrophins also bind the dystrophin COOH terminus in vitro, but no evidence has been reported for an interaction between dystrophin and the sarcoglycans. Human mutations have been found in dystrophin, the sarcoglycans and laminin, all of which lead to various types of muscular dystrophy. We have been studying the dystrophin domains necessary for formation of a functional complex by generating transgenic *mdx* (dystrophin minus) mice expressing internally truncated dystrophins. These mice provide in vivo models to study the localization of truncated dystrophin isoforms, the association of the truncated proteins with the DAP complex, and the functional capacity of the assembled DAP complexes. Expression of a dystrophin deleted for most of the NH_2-terminal domain in *mdx* mice leads to only a mild dystrophy, indicating that dystrophin can attach to the cytoskeleton by multiple mechanisms. Truncation of the central rod domain leads to normal DAP complex formation and almost fully prevents development of dystrophy. Deletion analysis of the COOH-terminal regions indicates that a broad cysteine-rich domain is indispensable for dystrophin function. This region coincides with the in vitro identified β-dystroglycan binding domain. Mice lacking this latter domain express very low levels of the sarcoglycans, indicating that the sarcoglycan complex binds dystrophin via dystroglycan. All deletion constructs tested lead to normal expression of the syntrophins, indicating that syntrophin associates with the DAP complex via multiple binding partners.

Introduction

Duchenne muscular dystrophy (DMD) is an X-linked lethal genetic disorder arising from mutations in the dystrophin gene (Monaco et al., 1986; Emery, 1993). Studies of patient mutations have revealed many features of the dystrophin protein and have implicated certain regions of the molecule as being required for functional activity either because of a required association with members of the dystrophin as-

Cytoskeletal Regulation of Membrane Function © 1997 by The Rockefeller University Press

sociated protein complex or for stability of the protein (Beggs et al., 1991; Ohlendieck et al., 1993; Kramarcy et al., 1994; Bies et al., 1992). Analysis of patient deletions has indicated that DMD generally arises from mutations that lead to production of COOH-terminally truncated proteins that are generally unstable and do not accumulate within muscle fibers. In contrast, many in-frame deletions lead to production of internally truncated molecules that display more stability in muscle and lead to a milder Becker muscular dystrophy (BMD) phenotype (Koenig et al., 1989). While these studies indicate that COOH-terminal sequences are important for dystrophin function, the large degree of mRNA and protein instability associated with COOH-terminal deletions has prevented a clear description of the precise functional roles of different parts of the molecule (McCabe et al., 1989). For example, some regions are likely important for association with the DAP members, while others are not required or may only be needed to confer stability on the protein. We have addressed these questions by generating a series of transgenic mice on the *mdx* mouse background, which produces no dystrophin in muscle. These mice express levels of dystrophin sufficient to prevent dystrophy if the generated protein is functional. The results implicate certain regions of the COOH terminus as absolutely required for dystrophin function, while the amino terminus appears to be important for protein stability and to a lesser extent association with the cytoskeleton. In contrast, the central rod domain can be deleted with minimal impact on the stability or function of dystrophin.

Methods

Transgenic mice were generated using modifications of the full-length murine dystrophin cDNA (Chamberlain et al., 1991; Lee et al., 1991). Each construct utilized promoter/enhancer sequences from the murine muscle creatine kinase (MCK) gene (Johnson et al., 1989; Cox et al., 1993), the SV40 polyadenylation signal, and various introns derived either from adenovirus, SV40, or MCK (Rafael et al., 1996). Transgenic mice were generated and analyzed as previously described (Hogan et al., 1986). Antisera against the NH_2 and COOH termini were prepared as described (Cox et al., 1994; Rafael et al., 1996). Histological 4-μm sections were prepared from muscle tissue fixed in 2% formaldehyde and 2% glutaraldehyde, embedded in glycol methacrylate, and stained with haematoxylin and eosin. Immunostaining of unfixed 7-μm muscle cryosections was performed as previously described (Phelps et al., 1995). For central nuclei counts, histological sections from 3 to 4 month old mice were photographed, and the percentage of centrally nucleated myofibers was determined by dividing the number of myofibers containing one or more centrally located nuclei by the total number of nucleated fibers. Myofibers with no nuclei in the plane of section were not counted. An average of 1,400 myofibers were counted for each muscle group. Serum was isolated from blood obtained from the retroorbital sinus of 4 week old mice, stored at $-70°C$, and assayed for the muscle isozyme of pyruvate kinase as described (Phelps et al., 1995). Force generation of diaphragm muscle strips was analyzed using small bundles of intact fibers removed from the diaphragm muscles of 3–4 month old mice. Forces were determined during maximum isometric tetanic contraction in vitro at $25°C$, then normalized to total cross-sectional area (specific force) (McCully and Faulkner, 1984; Cox et al., 1993).

Results

To study the function of various dystrophin domains a series of dystrophin expression vectors was prepared to generate moderate to high levels of modified dystrophin proteins in striated muscle (Fig. 1). Each of these constructs was injected into mouse embryos to generate transgenic animals. F_0 mice were assayed for the presence of transgene sequences by PCR, and positive mice were bred onto the *mdx* background to produce lines of animals expressing dystrophin only from the transgene and not from the endogenous dystrophin gene. The *mdx* mouse contains a point mutation in the dystrophin gene and produces no dystrophin in muscle tissues (Sicinski et al., 1989; Im et al., 1996). Mice that inherited both the transgene and the *mdx* mutation (Amalfitano and Chamberlain, 1996) were analyzed for levels of dystrophin expression, localization of the expressed dystrophin, localization of the

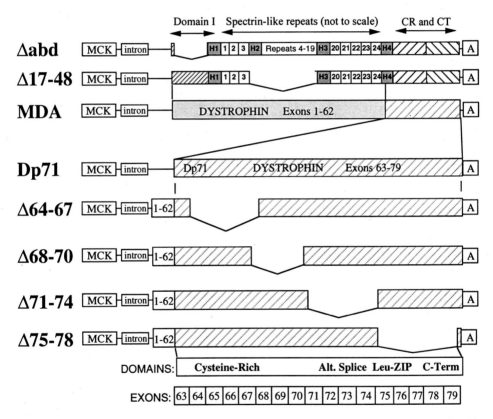

Figure 1. Expression vectors used to generate transgenic mice. *MCK*: muscle creatine kinase enhancer plus promoter; *A*: SV40 polyadenylation sequence. *Domain I*: NH₂-terminal actin binding domain; *CR*: cysteine-rich region; *CT*: COOH-terminal domain. Exon numbers are indicated for the COOH-terminal regions (*bottom*). At the top four hinges are noted (*H1–H4*), as are 24 spectrin-like repeats (*1, 2*, etc.). *Δabd*: actin-binding domain deletion, see Table II; *Δ17-48*: rod domain deletion, see Table I; *MDA*: full-length dystrophin cDNA, see Fig. 2; *Dp71*: COOH-terminal regions, see Fig. 3; The bottom four constructs are deletions of the COOH-terminal regions, data from *Δ64-67* is presented in Fig. 4.

known DAP complex members, morphology of the transgenic muscle including the presence of dystrophic pathology, and contractile properties of the diaphragm muscles.

The initial construct we tested expressed a full-length mouse dystrophin cDNA. These studies allowed us to ask whether expression of the major muscle isoform of dystrophin in the transgenic system was able to prevent dystrophy and restore normal expression of the DAPs (in *mdx* muscle the DAPs are unstable and accumulate at low levels [Ohlendieck and Campbell, 1991]). Mice expressing fifty-fold higher than normal levels of the full-length dystrophin displayed a normal muscle morphology and a lack of dystrophy (Fig. 2). These animals also displayed normal localization and expression of dystrophin and the DAPs, and normal contractile properties (Cox et al., 1993). Similar results were obtained with animals expressing levels of dystrophin between 20% and 5× of wild-type levels (Phelps et al., 1995).

Studies of DMD patients with deletions in the dystrophin gene have suggested that expression of the COOH terminus is critical for normal dystrophin function (Hoffman et al., 1991). In addition, several patient studies have suggested that the COOH-terminal domain may be the region that binds to the DAP complex (Suzuki et al., 1994). To address these issues, we generated transgenic mice expressing only the COOH-terminal domains of dystrophin (Fig. 1). These mice generate a protein equivalent to the COOH-terminal 71-kD isoform of dystrophin known as Dp71 (Cox et al., 1994). Expression of Dp71 in skeletal muscle restored expression of all the DAP proteins to control levels, and the Dp71 protein was localized properly to the sarcolemma (Fig. 3). However, examination of the muscle morphology revealed the animals to have a dystrophic phenotype, which at early ages was more severe than in the dystrophin negative mice (Fig. 3; Cox et al., 1994). These data indicate that although the COOH terminus binds to the DAP complex in muscle, restoration of expression of the COOH terminus and the DAP complex is not sufficient to prevent dystrophy. Instead, additional sequences at the NH$_2$ terminus of dystrophin must also be required for normal function.

To examine additional portions of the dystrophin sequence we next asked whether deletion of the dystrophin central rod domain encoded by exons 17–48 would affect the localization or function of dystrophin. This deletion is based on a genomic deletion identified in a mildly affected patient with BMD (England et al.,

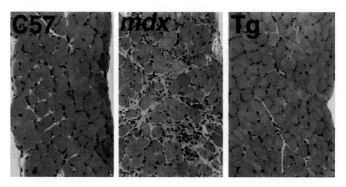

Figure 2. Histological analysis of quadriceps sections of transgenic mouse MDA, which expresses 50-fold higher than normal levels of dystrophin in striated muscle. The transgenic muscle does not display a dystrophic morphology. *C57*: control C57BL/10 muscle; *mdx*: mutant dystrophin minus muscle; *Tg*: transgenic muscle on the *mdx* background.

1990). Sixteen of the 24 spectrin-like repeats in dystrophin are removed from this protein. Our results indicated that expression of moderate to high levels of this construct almost completely prevent dystrophic symptoms in *mdx* mice. The expressed dystrophin properly localizes to the sarcolemma (Phelps et al., 1995), and similar results from Wells et al. indicate that this protein restores normal expression of the DAPs (Wells et al., 1995). Transgenic mice expressing the exon 17–48 deletion display contractile properties not significantly different from control animals, suggesting that the central part of the dystrophin molecule plays a minor role in the function of this protein (Table I).

Dystrophin is thought to bind to the actin cytoskeleton immediately below the sarcolemma via the amino terminal domain, a region with a high degree of sequence conservation with α-actinin and β-spectrin (Koenig et al., 1988). Although portions of the NH_2-terminal domain bind actin in vitro (Way et al., 1992; Jarrett and Foster, 1995), no region has been shown to be indispensable for binding to actin (Corrado et al., 1994). To determine the effect of an NH_2-terminal deletion of

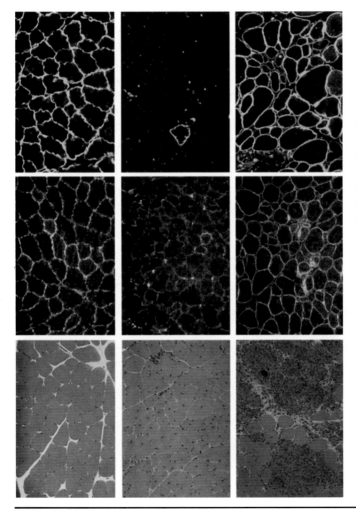

Figure 3. Immunofluorescence and morphological analysis of transgenic mouse line pDp71, which expresses the Dp71 isoform of dystrophin in muscle. *Left*: C57BL/10 muscle; *middle*: *mdx* muscle; *right*: transgenic line Dp71. *Top*: immunostaining with COOH-terminal anti-dystrophin antisera; *middle*: immunostaining with anti–α-dystroglycan antisera; *bottom*: haematoxylin and eosin stained sections. All panels show quadriceps muscle.

TABLE I

Levels of Expression and the Corresponding Phenotypes of Transgenic *mdx* Mice Expressing a Rod Domain Deleted Dystrophin

| | QUADRICEPS | | | DIAPHRAGM | | |
Mouse strain	Level of expression*	%Central nuclei	M-pyruvate‡ kinase (U/L)	Level of expression	%Central nuclei	Specific force‡ (kN/M²)
C57Bl/10*mdx*	—	88.7%	11,427 ± 1,626 (6)	—	54.8%	126 ± 19 (5)
C57Bl/10	1×	0.7%	503 ± 152 (6)	1.0×	0.19%	237 ± 23 (5)
11922	>10×	9.1%	731 ± 146 (9)	0.2×	8.3%	192 ± 11 (4)
11956	>10×	7.1%	N/D	0.9×	0.14%	N/D

*The overall levels of dystrophin expression were estimated from at least three independent immunoblots and are expressed relative to control C57BL/10 levels. ‡Mean ± standard deviation is shown, followed by the number of animals tested. N/D, not determined.

dystrophin in vivo, we generated transgenic mice expressing a dystrophin molecule missing amino acids 45–273 (Fig. 1; from near the beginning of exon 3 to the middle of exon 8). Few animals were obtained that expressed high levels of this construct, suggesting that the truncated protein was highly unstable in muscle (Corrado et al., 1996). However, those animals with moderate levels of expression displayed a very mild phenotype, with little fibrosis, a small number of centrally nucleated fibers (a measure of the amount of degeneration and regeneration occurring in the muscle), and a moderate decrease in force and power generating capacity (Table II). These data suggest the NH_2 terminus is not critical for dystrophin function; that dystrophin can still attach to the subsarcolemmal cytoskeleton in the absence of these deleted amino acids, and that this region of the molecule is likely critical for maintaining the stability of the protein (Beggs et al., 1991; Corrado et al., 1996).

The final series of constructs tested in transgenic mice contained a staggered series of deletions through the COOH-terminal domain (Fig. 1). Deletion of regions between exons 70 and 78 had little effect on the muscle morphology, function or expression of DAPs (Rafael et al., 1996). In contrast, several deletions that removed regions between exons 64 and 70 resulted in dramatic effects on the muscle. Deletion of exons 68–70 resulted in a phenotype indistinguishable from the *mdx*

TABLE II

Phenotype of Transgenic *mdx* Mice Expressing an NH_2-terminally Deleted Dystrophin

| | | | Percentage of centrally nucleated fibers | |
Line	Specific force* (kN/M²)	Normalized power* (W/kg)	Diaphragm	Quadriceps
C57BL	250 ± 29	60 ± 18	0.2	0.7
Δabd1F‡	182 ± 15	42 ± 6	11	5.3
mdx	108 ± 22	20 ± 6	55	89

*Data is expressed as the mean followed by the value for one standard error of the mean. ‡Δabd1F is the actin binding domain deletion transgenic mouse.

mouse, indicating that this region of dystrophin was critical for its function. A more severe effect was observed in the mice missing exons 64–70. These animals expressed control levels of dystrophin that localized properly to the sarcolemma, however the levels of dystroglycan and the sarcoglycans were considerably less than in *mdx* animals (Fig. 4). Similarly to the Dp71 transgenic animals, the 64–67 deletion mice displayed a phenotype slightly more severe than in *mdx* animals. These data point to the critical importance of this part of the molecule, which is a highly cysteine-rich region. Binding studies in vitro have implicated this region as being required for binding to β-dystroglycan (Suzuki et al., 1994; Jung et al., 1995), and removal of this portion of the protein appears to render the molecule nonfunctional, although it still localizes to the sarcolemma. Interestingly, all of the transgenic animals we have examined to date, including each of those displayed in Fig. 1, express normal levels of the DAP syntrophin (e.g., Fig. 4). Although syntrophin binds to dystrophin in vitro in a region encoded by exon 74 (Yang et al., 1995; Suzuki et al., 1995; Ahn et al., 1996), this region is not required for proper localization of syntrophin in vivo (Rafael et al., 1994). Syntrophin might be held in the complex by association with dystrobrevin, or by other as yet unidentified molecules (Fig. 5).

Discussion

While patient studies have provided considerable insight into structure/function correlates in the dystrophin molecule, many key questions have remained unanswered. By generating transgenic animals that express mutant dystrophin molecules

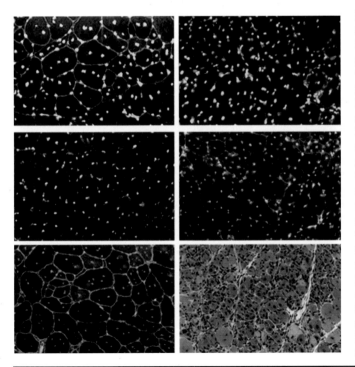

Figure 4. Immunofluorescence and morphological analysis of transgenic mouse line pΔ64-67, which expresses a dystrophin molecule missing the NH$_2$-terminal half of the cysteine-rich domain (see Fig. 1). *Top left*: immunostaining with NH$_2$-terminal anti-dystrophin antisera; *top right*: immunostaining with anti–α-dystroglycan antisera; *middle left*: immunostaining with anti–α-sarcoglycan antisera; *middle right*: immunostaining with anti–β-sarcoglycan antisera; *bottom left*: immunostaining with anti–syntrophin antisera; *bottom right*: haematoxylin and eosin stained quadriceps muscle.

from a strong, muscle-specific promoter, it is possible to produce sufficient levels of the mutant proteins to determine their functional capacity. Applying this approach to the various domains of dystrophin has confirmed a number of inferences from the patient data, but has also identified several unexpected results.

While the NH_2-terminal domain of dystrophin appears to bind directly to actin filaments, it has become clear that this region of the molecule is not absolutely required for binding to the cytoskeleton. Whereas deletion of the COOH terminus severs the link to the extracellular matrix and results in a severe dystrophy (Fig. 4), deletion of the NH_2 terminus results in a mild dystrophy (Table I). These data indicate that dystrophin must have multiple sites for binding to the cytoskeleton, only one of which is the NH_2-terminal regions between exons 3 and 8 (Corrado et al.,

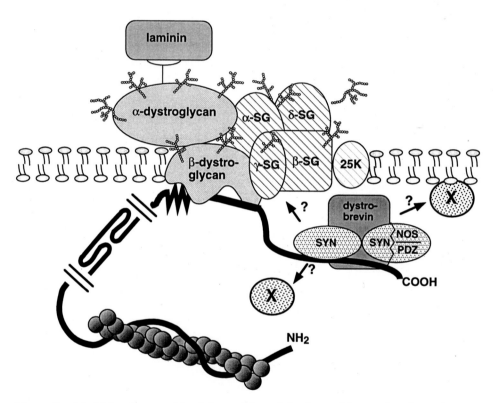

Figure 5. Model for the assembly of dystrophin and the dystrophin associated protein complex. A broad region at the NH_2 terminus of dystrophin interacts with the subsarcolemmal cytoskeleton. The NH_2 terminus is linked to the COOH terminus via a large rod domain composed of 24 spectrin-like repeats (only two of which are shown). The COOH-terminal domain is subdivided into a cysteine rich domain and a COOH-terminal region. The cysteine-rich domain binds to β-dystroglycan, which links to laminin via α-dystroglycan. The sarcoglycans (*SG*) stabilize the interaction between β-DG and laminin by an unknown mechanism. The syntrophins (*SYN*) bind to the COOH-terminal region, and also may bind additional PDZ domain proteins such as nitric oxide synthase (*NOS*). Dystrobrevin also binds to dystrophin and/or SYN. Unknown interactions are indicated by *?*, and possible unidentified protein are indicated by *X*. For simplicity dystrophin is represented as a monomer, although it may exist in the cell as a homodimer.

1996). In agreement with many patient studies, the transgenic model system demonstrated the relatively minor contribution of the central rod domain of dystrophin to the function of this protein. However, portions of the COOH-terminal domain were observed to be indispensable for function. The region between exons 64–70 are critical for dystrophin function. Deletion of this region not only makes the protein relatively unstable (Rafael et al., 1996), but also severs the link to β-dystroglycan. Without a link to dystroglycan, the sarcoglycan complex is destabilized, and fails to accumulate normally (Fig. 4). In contrast, the regions between exons 70–78, including the syntrophin binding site, appear less important for normal function of dystrophin. Deletions of these regions have no obvious effect on DAP expression, dystrophin localization, or the morphology of the muscle. These results, combined with numerous studies by many labs of the various DAPs, can be used to modify models of the DAP complex assembly as outlined in Fig. 5.

Together these studies have clarified a number of aspects of dystrophin function, however, numerous questions remain to be answered. In particular, it is not clear how syntrophin is able to associate with dystrophin in vivo, particularly in the absence of sequences encoded on exon 74. Nor is it clear what role syntrophin plays in the DAP complex. Does syntrophin play an important role in maintaining the integrity of the sarcolemmal membrane, or could it play a regulatory role in modulating assembly of the DAP complex? The minor effect observed by deletions of the dystrophin syntrophin contact site suggests that the function of syntrophin may be less structural than regulatory, but more direct data will be needed to clarify this point. Finally, it is still unclear what regulates assembly of the full complex, and what role phosphorylation may play in this process. The COOH terminus of dystrophin is phosphorylated in a tissue specific manner (Cox et al., 1994), and several DAPs contain consensus phosphorylation sites. Identification of the sites of post-translational modification and mutation of these sites in vivo may be required to clarify these issues.

Acknowledgments

We thank Stephanie Phelps, Patti Mills, Erica Adkins, and Tressia Hutchinson for excellent technical assistance, Sally Camper for generation of transgenic animals, and the U.M. morphology core for histochemical staining.

Supported by National Institutes of Health grant AR40864 and by The Muscular Dystrophy Association (USA).

References

Ahn, A.H., C.A. Freener, E. Gussoni, M. Yoshida, E. Ozawa, and L.M. Kunkel. 1996. The three human syntrophin genes are expressed in diverse tissues, have distinct chromosomal locations, and each bind to dystrophin and its relatives. *J. Biol. Chem.* 271:2724–2730.

Amalfitano, A., and J.S. Chamberlain. 1996. The mdx-ARMS assay: a rapid PCR method for genotyping mdx mice. *Musc. Nerve.* 19:1549–1553.

Beggs, A.H., E.P. Hoffman, J.R. Snyder, K. Arahata, L. Specht, F. Shapiro, C. Angelini, H. Sugita, and L.M. Kunkel. 1991. Exploring the molecular basis for variability among patients with Becker muscular dystrophy: dystrophin gene and protein studies. *Am. J. Hum. Genet.* 49:54–67.

Bies, R.D., C.T. Caskey, and R. Fenwick. 1992. An intact cysteine-rich domain is required for dystrophin function. *J. Clin. Invest.* 90:666–672.

Chamberlain, J.S., J.A. Pearlman, D.M. Muzny, A. Civetello, N.J. Farwell, R. Malek, P. Powaser, A.A. Reeves, C. Lee, and C.T. Caskey. 1991. Mouse dystrophin cDNA sequence. *Genbank accession number* M68859.

Corrado, K., P.L. Mills, and J.S. Chamberlain. 1994. Deletion analysis of the dystrophin-actin binding domain. *FEBS Lett.* 344:255–260.

Corrado, K., J.A. Rafael, P.L. Mills, N.M. Cole, J.A. Faulkner, K. Wang, and J.S. Chamberlain. 1996. Transgenic *mdx* mice expressing dystrophin with a deletion in the actin-binding domain display a "mild Becker" phenotype. *J. Cell Biol.* 134:873–884.

Cox, G.A., N.M. Cole, K. Matsumura, S.F. Phelps, S.D. Hauschka, K.P. Campbell, J.A. Faulkner, and J.S. Chamberlain. 1993. Overexpression of dystrophin in transgenic *mdx* mice eliminates dystrophic symptoms without toxicity. *Nature.* 364:725–729.

Cox, G.A., Y. Sunada, K.P. Campbell, and J.S. Chamberlain. 1994. Dp71 can restore the dystrophin-associated glycoprotein complex in muscle but fails to prevent dystrophy. *Nature. Genet.* 8:333–339.

Emery, A.E.H. 1993. Duchenne Muscular Dystrophy. Oxford Monographs on Medical Genetics No. 24. Oxford Medical Publications, Oxford. pp. 392.

England, S.B., L.V. Nicholson, M.A. Johnson, S.M. Forrest, D.R. Love, E.E. Zubrzycka-Gaarn, D.E. Bulman, J.B. Harris, and K.E. Davies. 1990. Very mild muscular dystrophy associated with the deletion of 46% of dystrophin. *Nature.* 343:180–182.

Hoffman, E.P., C.A. Garcia, J.S. Chamberlain, C. Angelini, J.R. Lupski, and R. Fenwick. 1991. Is the carboxyl-terminus of dystrophin required for membrane association? A novel, severe case of Duchenne muscular dystrophy. *Ann. Neurol.* 30:605–610.

Hogan, B., F. Constantini, and E. Lacey. 1986. Manipulating the Mouse Embryo: A Laboratory Manual. Cold Spring Harbor Laboratory, Cold Spring Harbor. pp. 332.

Im, W.B., S.F. Phelps, E.H. Copen, E.G. Adams, J.L. Slightom, and J.S. Chamberlain. 1996. Differential expression of dystrophin isoforms in multiple strains of *mdx* mice with different mutations. *Hum. Mol. Genet.* 5:1149–1153.

Jarrett, H.W., and J.L. Foster. 1995. Alternate binding of actin and calmodulin to multiple sites on dystrophin. *J. Biol. Chem.* 270:5578–5586.

Johnson, J.E., B.J. Wold, and S.D. Hauschka. 1989. Muscle creatine kinase sequence elements regulating skeletal and cardiac muscle expression in transgenic mice. *Mol. Cell Biol.* 9:3393–3399.

Jung, D., B. Yang, J. Meyer, J.S. Chamberlain, and K.P. Campbell. 1995. Identification and characterization of the dystrophin anchoring site on β-dystroglycan. *J. Biol. Chem.* 270:27305–27310.

Koenig, M., A.H. Beggs, M. Moyer, S. Scherpf, K. Heindrich, T. Bettecken, G. Meng, C.R. Muller, M. Lindlof, H. Kaariainen, et al. 1989. The molecular basis for Duchenne versus Becker muscular dystrophy: correlation of severity with type of deletion. *Am. J. Hum. Genet.* 45:498–506.

Koenig, M., A.P. Monaco, and L.M. Kunkel. 1988. The complete sequence of dystrophin predicts a rod-shaped cytoskeletal protein. *Cell.* 53:219–226.

Kramarcy, N.R., A. Vidal, S.C. Froehner, and R. Sealock. 1994. Association of utrophin and multiple dystrophin short forms with the mammalian M_r 58,000 dystrophin-associated protein (syntrophin). *J. Biol. Chem.* 269:2870–2876.

Lee, C.C., J.A. Pearlman, J.S. Chamberlain, and C.T. Caskey. 1991. Expression of recombinant dystrophin and its localization to the cell membrane. *Nature.* 349:334–336.

McCabe, E.R., J. Towbin, J. Chamberlain, L. Baumbach, J. Witkowski, G.J. van Ommen, M. Koenig, L.M. Kunkel, and W.K. Seltzer. 1989. Complementary DNA probes for the Duchenne muscular dystrophy locus demonstrate a previously undetectable deletion in a patient with dystrophic myopathy, glycerol kinase deficiency, and congenital adrenal hypoplasia. *J. Clin. Invest.* 83:95–99.

McCully, K.K., and J.A. Faulkner. 1984. Length-tension relationship of mammalian diaphragm muscles. *J. Appl. Physiol.* 54:1681–1686.

Monaco, A.P., R.L. Neve, C. Coletti-Feener, C.J. Bertelson, D.M. Kurnit, and L.M. Kunkel. 1986. Isolation of candidate cDNA clones for portions of the Duchenne muscular dystrophy gene. *Nature.* 323:646–650.

Ohlendieck, K., and K.P. Campbell. 1991. Dystrophin-associated proteins are greatly reduced in skeletal muscle from mdx mice. *J. Cell Biol.* 115:1685–1694.

Ohlendieck, K., K. Matsumura, V.V. Ionasescu, J.A. Towbin, E.P. Bosch, S.L. Weinstein, S.W. Sernett, and K.P. Campbell. 1993. Duchenne muscular dystrophy: deficiency of dystrophin-associated proteins in the sarcolemma. *Neurology.* 43:795–800.

Phelps, S.F., M.A. Hauser, N.M. Cole, J.A. Rafael, R.T. Hinkle, J.A. Faulkner, and J.S. Chamberlain. 1995. Expression of full-length and truncated dystrophin mini-genes in transgenic *mdx* mice. *Hum. Mol. Genet.* 4:1251–1258.

Rafael, J.A., G.A. Cox, K. Corrado, D. Jung, K.P. Campbell, and J.S. Chamberlain. 1996. Forced expression of dystrophin deletion constructs reveals structure-function correlations. *J. Cell Biol.* 134:93–102.

Rafael, J.A., Y. Sunada, N.M. Cole, K.P. Campbell, J.A. Faulkner, and J.S. Chamberlain. 1994. Prevention of dystrophic pathology in *mdx* mice by a truncated dystrophin isoform. *Hum. Mol. Genet.* 3:1725–1733.

Sicinski, P., Y. Geng, A.S. Ryder-Cook, E.A. Barnard, M.G. Darlison, and P.J. Barnard. 1989. The molecular basis of muscular dystrophy in the mdx mouse: a point mutation. *Science.* 244:1578–1580.

Suzuki, A., M. Yoshida, K. Hayashi, Y. Mizuno, Y. Hagiwara, and E. Ozawa. 1994. Molecular organization at the glycoprotein-complex-binding site of dystrophin: three dystrophin-associated proteins bind directly to the carboxy-terminal portion of dystrophin. *Eur. J. Biochem.* 220:283–292.

Suzuki, A., M. Yoshida, and E. Ozawa. 1995. Mammalian α1- and β1-syntrophin bind to the alternative splice-prone region of the dystrophin COOH terminus. *J. Cell Biol.* 128:373–381.

Way, M., B. Pope, R.A. Cross, J. Kendrick-Jones, and A.G. Weeds. 1992. Expression of the N-terminal domain of dystrophin in *E. coli* and demonstration of binding to F-actin. *FEBS Lett.* 301:243–245.

Wells, D.J., K.E. Wells, E.A. Asante, G. Turner, Y. Sunada, K.P. Campbell, F.S. Walsh, and G. Dickson. 1995. Expression of human full-length and minidystrophin in transgenic *mdx* mice: implications for gene therapy of Duchenne muscular dystrophy. *Hum. Mol. Genet.* 4:1245–1250.

Yang, B., D. Jung, J.A. Rafael, J.S. Chamberlain, and K.P. Campbell. 1995. Identification of α-syntrophin binding to syntrophin triplet, dystrophin, and utrophin. *J. Biol. Chem.* 270:4975–4978.

A Multiple Site, Side Binding Model for the Interaction of Dystrophin with F-Actin

James M. Ervasti,*‡ Inna N. Rybakova,* and Kurt J. Amann‡

*Department of Physiology and ‡Graduate Program in Cellular and Molecular Biology, University of Wisconsin Medical School, Madison, Wisconsin 53706

Introduction

The dystrophin-glycoprotein complex of striated muscle has emerged as a physiologically important nexus between the actin-based cortical cytoskeleton and the extracellular matrix (Ervasti and Campbell, 1993b; Campbell, 1995; Worton, 1995). The transmembrane linkage formed by the dystrophin-glycoprotein complex may play a structural role in maintaining sarcolemmal membrane integrity during contraction (Menke and Jockusch, 1991; Stedman et al., 1991; Petrof et al., 1993; Pasternak et al., 1995). As an alternative, or perhaps additional function, the dystrophin-glycoprotein complex may further serve to organize or modulate other proteins involved in signal transduction (Sealock and Froehner, 1994; Yang et al., 1995; Brenman et al., 1996). In working to better understand the cellular role(s) of dystrophin, recent work has been directed towards the identification of its functionally important domains as well as the role of these domains in binding dystrophin-associated proteins.

Striated muscle dystrophin is predominantly expressed as a 427-kD, four domain protein with the first three domains exhibiting significant sequence homology with the cytoskeletal proteins α-actinin and spectrin (Koenig et al., 1988; Matsudaira, 1991): an amino-terminal, putative actin binding domain; a rod-like domain comprised of 24 triple helical coiled coil repeats; a cysteine-rich domain, and a carboxy-terminal domain. The functional importance of the cysteine-rich and carboxy-terminal domains of dystrophin is evident from several studies that correlate specific mutations or deletions in these domains with presentation of the most severe forms of muscular dystrophy (Hoffman et al., 1991; Helliwell et al., 1992; Matsumura et al., 1993). Localization of the binding sites for dystrophin-associated proteins to the cysteine-rich and carboxy-terminal domains (Suzuki et al., 1992, 1994; Kramarcy et al., 1994) explained why these two domains are essential for normal dystrophin function in skeletal muscle. On the other hand, a patient expressing a dystrophin lacking the amino-terminal half of the protein presented with a dystrophic phenotype of intermediate severity (Takeshima et al., 1994) while transgenic expression of a nonmuscle isoform of dystrophin (DP71) lacking the amino-terminal and rod domains in dystrophin-deficient muscle fails to correct the dystrophic phenotype (Cox et al., 1994; Greenberg et al., 1994). These data indicate that the cysteine-rich and carboxy-terminal domains of dystrophin alone are insufficient to confer normal muscle function and leave unresolved a functional role for the amino-terminal and rod-like dystrophin domains.

Certainly, the strong sequence homology of the amino-terminal 246 amino acids of dystrophin with the actin binding domains of several well characterized F-actin crosslinking proteins (Koenig et al., 1988; Matsudaira, 1991) and accumulated data (Hemmings et al., 1992; Way et al., 1992; Fabbrizio et al., 1993; Corrado et al., 1994; Jarrett and Foster, 1995) support its hypothesized role in binding F-actin. In addition, several studies have noted a correlation between mutations or deletions in the amino-terminal domain with the expression of intermediate to severe forms of muscular dystrophy (Beggs et al., 1991; Prior et al., 1993; Winnard et al., 1993; Comi et al., 1994; Muntoni et al., 1994). However, it was not clear from these studies whether the more severe phenotypes presented by patients with defects or deletions in the amino-terminal domain were due to the absence of an important functional domain or simply to the low abundance of an unstable truncated protein. In fact, transgenic *mdx* mice expressing normal levels of a dystrophin construct lacking amino acids 45–273 present with a benign phenotype, indicating that the amino-terminal domain is not essential for normal dystrophin function (Corrado et al., 1996). The functional importance of the dystrophin rod domain has also been in question due to the very mild phenotype of patients expressing dystrophin with as much as 66% of the rod domain deleted (England et al., 1990; Passos-Bueno et al., 1994). In this chapter, we review recent evidence (Rybakova et al., 1996) suggesting that both the amino-terminal and rod domains of dystrophin may be mutually involved in binding to actin filaments through the concerted effect of two or more redundant actin binding sites.

Dystrophin-Glycoprotein Complex Binding to F-Actin

We examined the actin-binding properties of purified dystrophin-glycoprotein complex (Ervasti et al., 1990; Ervasti et al., 1991; Ervasti and Campbell, 1991; Ervasti and Campbell, 1993a) because it was the best characterized and highest yielding preparation of intact dystrophin currently available and because of its documented importance as a functional unit in skeletal muscle (Campbell, 1995; Worton, 1995). Incubation of submicromolar concentrations of purified dystrophin-glycoprotein complex with F-actin followed by high speed sedimentation (100,000 g for 20 min) resulted in the cosedimentation of a significant fraction of dystrophin-glycoprotein complex with the F-actin pellet (Fig. 1 *A*). Even at the highest concentrations examined (>2 μM), virtually no dystrophin-glycoprotein complex sedimented in the absence of F-actin (Fig. 1 *A*). These data indicated that the sedimentation of purified dystrophin-glycoprotein complex resulted from a specific interaction of dystrophin-glycoprotein complex with F-actin. Incubation of increasing amounts of dystrophin-glycoprotein complex with a fixed amount of F-actin followed by high speed sedimentation demonstrated that dystrophin-glycoprotein complex bound F-actin in a saturable manner (Fig. 1 *B*). Nonlinear regression analysis of the binding data from three independent experiments with different dystrophin-glycoprotein complex and actin preparations (Fig. 1 *B*) yielded an average dissociation constant (K_d) of 0.5 μM and saturation (B_{max}) at 0.042 $\pm$ 0.005 mol/mol, which corresponds to one dystrophin per 24 actin monomers. The shape of the curve as well as the calculated Hill coefficient (1.40 $\pm$ 0.33) suggested that dystrophin in the glycoprotein complex bound F-actin with little or no cooperativity.

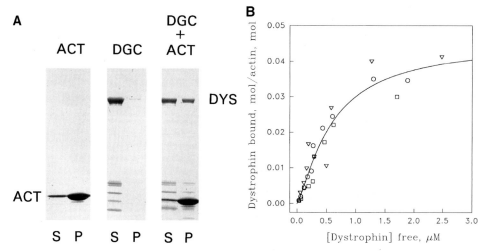

Figure 1. Cosedimentation of the dystrophin-glycoprotein complex with F-actin. Shown in *A* are Coomassie blue–stained SDS-polyacrylamide gels of 100,000 *g* supernatants (*S*) and pellets (*P*) of dystrophin-glycoprotein complex incubated in the presence (*DGC* + *ACT*) or absence (*DGC*) of F-actin, or F-actin alone (*ACT*). The concentration of dystrophin-glyco-protein complex was 0.21 mg/ml (0.25 μM dystrophin) in DGC + ACT and 1.83 mg/ml (2.14 μM dystrophin) in DGC; F-actin was present at 6 μM in all panels. The molecular weight standards ($\times 10^{-3}$) are indicated on the left. (*B*) Increasing amounts of dystrophin-glycopro-tein complex were incubated with 6.5 μM F-actin with subsequent centrifugation at 100, 000 *g*. The amount of free and bound dystrophin was determined densitometrically from Coomassie blue–stained gels of 100,000 *g* supernatant and pellet fractions as shown in *A*. The binding data were fitted using nonlinear regression analysis. Different symbols denote the data of three independent experiments performed with different dystrophin-glycoprotein complex and F-actin preparations. Adapted from *The Journal of Cell Biology* (Rybakova et al., 1996) by copyright permission of The Rockefeller University Press.

Other studies using purified *Torpedo* dystrophin (Lebart et al., 1995), maltose binding protein/dystrophin chimeras comprising the first 90 to 385 amino acids of dystrophin (Corrado et al., 1994; Jarrett and Foster, 1995), or nonfusion recombi-nants encoding only the first 246 amino acids of dystrophin (Way et al., 1992) have reported K_d values ranging from 0.1–44 μM. However, the studies reporting the highest affinities for dystrophin estimated binding constants using a direct ELISA assay which, for a variety of reasons (Goldberg and Djavadi-Ohaniance, 1993), is generally considered inappropriate for measuring ligand binding affinity. Further-more, it is unclear what conformational effect the >40-kD maltose binding protein portion of the chimeras may have exerted on the normally unconstrained amino terminus of native dystrophin. Therefore, we prepared a dystrophin amino-termi-nal fragment initiated at the start methionine. The recombinant protein DYS246 comprised the amino-terminal 246 amino acids of dystrophin with a small (~5 kD), carboxy-terminal poly-histidine tag which facilitated rapid purification to >95% purity. DYS246 clearly bound F-actin in a relevant manner as dystrophin-glycopro-tein complex cosedimentation with F-actin was inhibited by the presence of 50- or

100-fold molar excesses of DYS246 (Fig. 2 *A*). However, DYS246 appeared to bind F-actin in the high-speed sedimentation assay with at least an order of magnitude lower affinity and substantially higher capacity (Fig. 2 *B*) than was observed for purified dystrophin-glycoprotein complex (K_d = 0.5 μM, B_{max} = 1:24, Fig. 1). Our

A

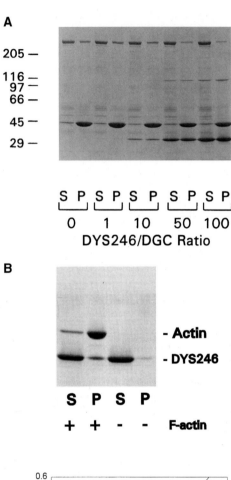

B

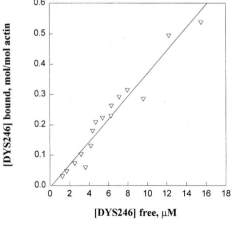

Figure 2. Cosedimentation of DYS246 with F-actin and effect on dystrophin-glycoprotein complex binding to F-actin. Shown in *A* is a Coomassie blue–stained SDS-polyacrylamide gel of F-actin cosedimentation assays performed with 6.5 μM F-actin and 0.171 mg/ml dystrophin-glycoprotein complex (0.2 μM dystrophin) in the absence or presence of 0.2, 2, 10, or 20 μM DYS246 to give the indicated molar ratios of DYS246 relative to intact dystrophin. Shown in the upper panel of *B* is a Coomassie blue–stained SDS-polyacrylamide gel of supernatants (*S*) and pellets (*P*) of recombinant DYS246 sedimented at 100,000 *g* in the presence or absence of F-actin. In the lower panel, increasing amounts of DYS246 was incubated with 6.5 μM F-actin with subsequent centrifugation at 100,000 *g*. The amount of free and bound protein was determined densitometrically from Coomassie blue–stained gels of 100,000 *g* loaded with equal volumes of supernatant and pellet fractions. Adapted from *The Journal of Cell Biology* (Rybakova et al., 1996) by copyright permission of The Rockefeller University Press.

results are most consistent with those reported for the binding of a nonfusion recombinant corresponding to dystrophin amino acids 1-246 to F-actin with 1:1 stoichiometry and an apparent K_d of 44 µM as determined in the high-speed sedimentation assay (Way et al., 1992). Taken together, these data suggest that the amino-terminal, putative actin binding domain of dystrophin is alone insufficient to explain the F-actin binding properties of full-length dystrophin.

A Novel F-Actin Binding Site in the Dystrophin Rod Domain

While several studies have documented the actin binding properties of the amino-terminal domain of dystrophin, the possible existence of additional F-actin binding sites located in other domains of dystrophin has not been explored. As a precedent for the necessity of such studies, tensin has been shown to exhibit homology with the consensus site found in the F-actin crosslinking superfamily of proteins (Matsudaira, 1991), yet nonhomologous sequences appeared to be responsible for its actin binding activity (Lo et al., 1994). It was previously shown that limited calpain digestion of purified dystrophin-glycoprotein complex yielded a stable series of dystrophin fragments, including a 31-kD fragment corresponding to the amino-terminal, putative actin-binding domain of dystrophin (Suzuki et al., 1992). Therefore, we sought to identify dystrophin fragments that cosedimented with F-actin after limited calpain digestion of dystrophin-glycoprotein complex (Fig. 3 *A*). Surprisingly, the 31-kD dystrophin fragment recognized by antibodies specific for the amino-terminal 15 amino acids of dystrophin (Fig. 3 *B*) did not cosediment with F-actin when 0.4 µM digested dystrophin-glycoprotein complex was analyzed in the high-speed sedimentation assay (Fig. 3 *A*). However, dystrophin monoclonal antibody XIXC2 (Ervasti et al., 1990), which maps to an epitope contained in dystrophin amino acids 1416–1494 (Fig. 3 *B*) did identify a 50-kD fragment of the dystrophin rod domain that retained F-actin binding activity (Fig. 3 *A*). The 50-kD, actin binding fragment of dystrophin was further localized within the rod domain by the absence of reactivity with monoclonal antibody DYS1 (Fig. 3 *A*), which recognizes an epitope located within dystrophin amino acids 1030–1388 (Fig. 3 *B*). Nitrocellulose transfers stained with monoclonal antibodies specific for adhalin/α-sarcoglycan and β-dystroglycan indicated that neither of these integral membrane components of the dystrophin-glycoprotein complex was digested by m-calpain or cosedimented with F-actin after calpain digestion of the dystrophin-glycoprotein complex (not shown). These results suggest that a portion of the dystrophin rod domain may also play a role in the actin binding activity of dystrophin. In support of this result, the first α-helical repeat of β-spectrin was recently shown to participate in the binding of its amino-terminal domain to F-actin (Li and Bennett, 1996). In addition, the domain of c-Abl tyrosine kinase responsible for its actin binding activity shows weak homology with the triple helical repeats common to dystrophin, α-actinin, and spectrin (Van Etten et al., 1994). Interestingly, the portion of the dystrophin rod domain that exhibits actin binding activity (amino acids 1416–1880), contains repeats 10 and 14, which conform poorly with the stereotypical repeat pattern (Koenig and Kunkel, 1990; Winder et al., 1995).

To confirm the actin binding activity of the 50-kD fragment of the dystrophin rod domain (Fig. 3 *A*), a recombinant protein corresponding to dystrophin amino

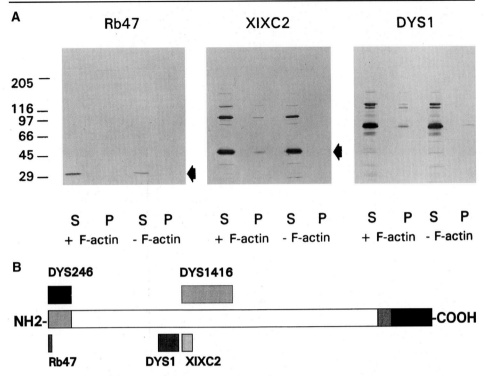

Figure 3. Cosedimentation of dystrophin fragments with F-actin after limited digestion of dystrophin-glycoprotein complex with calpain. Shown in *A* are identical nitrocellulose transfers stained with rabbit polyclonal antibodies raised against a synthetic peptide corresponding to the first 15 amino acids of dystrophin (*Rb 47*), or monoclonal antibodies XIXC2 and DYS1 that are specific for adjacent (but not overlapping) epitopes located in the rod domain of dystrophin (refer to *B*). The nitrocellulose transfers contained electrophoretically separated supernatants (*S*) and pellets (*P*) of calpain-digested dystrophin-glycoprotein complex incubated in the presence (*+ F-actin*) or absence (*− F-actin*) of F-actin and sedimented at 100,000 *g*. Arrows identify the 31-kD and 50-kD dystrophin fragments detected by rabbit 47 and XIXC2 antibodies, respectively. The concentration of calpain-digested dystrophin-glycoprotein complex was 0.34 mg/ml (0.4 µM dystrophin) and F-actin was present at 6.5 µM. The molecular weight standards ($\times 10^{-3}$) are indicated on the left. Shown in *B* is a diagram illustrating the location of epitopes for Rb 47 (amino acids 1–15), XIXC2 (amino acids 1416–1494), and DYS1 (amino acids 1181–1388) in the dystrophin primary sequence as well as the relative location of the dystrophin sequences encoded by recombinant proteins DYS246 (amino acids 1–246) and DYS1416 (amino acids 1416–1880). *f*Adapted from *The Journal of Cell Biology* (Rybakova et al., 1996) by copyright permission of The Rockefeller University Press.

acids 1416–1880 (DYS1416) was evaluated for interaction with F-actin using the high-speed sedimentation assay. While DYS1416 specifically cosedimented with F-actin (Fig. 4), it was apparent from the binding data obtained over the measurable concentration range that DYS1416 bound F-actin with substantially lower affinity and higher capacity (Fig. 4) than was observed with purified dystrophin-gly-

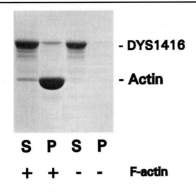

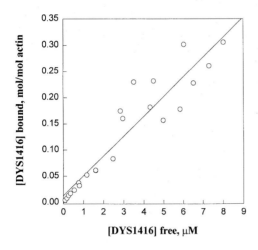

Figure 4. Cosedimentation of DYS1416 with F-actin. Shown in the upper panel is a Coomassie blue–stained SDS-polyacrylamide gel of supernatants (*S*) and pellets (*P*) of recombinant DYS1416 sedimented at 100,000 *g* in the presence or absence of F-actin. In the lower panel, increasing amounts of DYS1416 was incubated with 6.5 μM F-actin with subsequent centrifugation at 100,000 *g*. The amount of free and bound protein was determined densitometrically from Coomassie blue–stained gels of 100,000 *g* supernatant and pellet fractions as shown in top panel. Adapted from *The Journal of Cell Biology* (Rybakova et al., 1996) by copyright permission of The Rockefeller University Press.

coprotein complex (Fig. 1). These results could be explained by a model whereby intact dystrophin binds with 0.5 μM affinity to F-actin through the combined effect of multiple lower affinity contact sites.

Dystrophin in the Glycoprotein Complex Does Not Cross-link F-Actin

Dystrophin is commonly modeled as an antiparallel dimer based primarily on its sequence homology with spectrin (Koenig et al., 1988). If dystrophin exists as a dimer in the glycoprotein complex it would be expected to cross-link actin filaments into supermolecular networks or bundles. Alternatively, the presence of additional actin binding sites within dystrophin (Figs. 3 and 4) might enable a dystrophin monomer to crosslink actin filaments in a manner similar to fimbrin (Matsudaira, 1991). To test these possibilities, the effect of dystrophin-glycoprotein complex on gelation of actin-filament solutions was examined by falling-ball viscometry. α-Actinin increased the viscosity of F-actin more than 2.5-fold when present at a 1:20 molar ratio with F-actin and formed a gel when α-actinin and actin were mixed at a 1:5 molar ratio (Fig. 5). In contrast with α-actinin, dystrophin-glycoprotein complex had no effect on the gelation of F-actin as the viscosity of F-actin did not change in the presence of dystrophin-glycoprotein complex (Fig. 5) at molar ratios as high as 1:5. We also failed to observe dystrophin-glycoprotein complex crosslinking of F-actin by two additional methods, electron microscopy and low-speed sedimentation, which were also performed in parallel with α-actinin as a positive control (Rybakova et al., 1996). Taken together, these results indicate that purified dystrophin-glycoprotein complex is unable to cross-link different actin filaments into supermolecular networks or bundles.

The Dystrophin-Glycoprotein Complex Protects F-Actin from Depolymerization

The stoichiometry of dystrophin-glycoprotein complex binding to F-actin (1 dystrophin per 24 actin monomers, Fig. 1), the evidence for an additional actin binding site located in the rod domain (Figs. 3 and 4), and the lack of cross-linking activity (Fig. 5) raise the possibility that full-length dystrophin may bind alongside an actin filament through interaction with multiple actin monomers. It is noteworthy that the actin side-binding protein tropomyosin binds to F-actin with a stoichiometry of 1 tropomyosin per 7 actin monomers and also protects F-actin from depolymerization (Broschat, 1990). Based on the stoichiometry of dystrophin-glycoprotein complex binding to F-actin (Fig. 1), a dystrophin molecule has the capacity to interact with 24 actin monomers which would span a length of 130 nm if arranged in a single strand of a filament or one-half of this length when comprising both strands of the filament. While both possibilities would be sufficient to allow linear contact with amino acids 1–246 and 1416–1880 of an extended dystrophin molecule with a length of 120–140 nm (Cullen et al., 1990; Pons et al., 1990), they also lead to the prediction that dystrophin should prevent or retard the dissociation of actin monomers

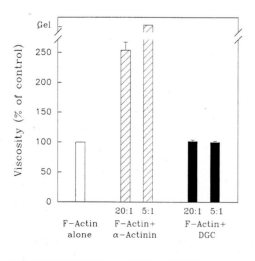

Figure 5. Viscometric analysis of α-actinin and dystrophin-glycoprotein complex on the gelation of F-actin. Average viscosity measurements (±SEM, $n \geq 3$) obtained by falling-ball viscometry at 25°C are expressed as the percent of that obtained for 6 μM F-actin alone (60.37 ± 2.01 cP). Gel indicates a viscosity at which the ball does not fall. The concentrations of α-actinin were 0.06 mg/ml (0.3 μM) and 0.24 mg/ml (1.2 μM). Dystrophin-glycoprotein complex was present at concentrations 0.26 mg/ml and 1 mg/ml (0.3 and 1.2 μM dystrophin, respectively). F-actin was present at 6 μM in all assays. Adapted from *The Journal of Cell Biology* (Rybakova et al., 1996) by copyright permission of The Rockefeller University Press.

from the ends of filaments to which it is bound (Broschat, 1990; Schafer and Cooper, 1995). To test this hypothesis, F-actin alone, or F-actin preequilibrated with dystrophin-glycoprotein complex was rapidly diluted into buffer conditions that favored actin depolymerization. At various times post-dilution, the amount of F-actin remaining was determined by high-speed sedimentation analysis (Fig. 6). As hypothesized, preequilibration of dystrophin-glycoprotein complex with F-actin significantly slowed the depolymerization of F-actin when compared to F-actin alone (Fig. 6). For example, 4 h after dilution, 4.6 ± 1.9% of F-actin alone pelleted while 14.9 ± 0.2% pelleted when dystrophin-glycoprotein complex was present at a 1:5 molar ratio with F-actin. We further determined the molar ratio of dystrophin that cosedimented with actin using the densitometrically determined fraction of each protein that pelleted and the total protein concentrations. Over the time range of 50–80 minutes post-dilution, the average molar ratio of dystrophin that cosedimented with the actin pellet was 0.13 ± 0.02. These results suggest that an intact dystrophin molecule in the glycoprotein complex has the capacity to interact directly with multiple actin monomers within a filament. Interestingly, while dystrophin is highly susceptible to proteolysis in vitro (Koenig and Kunkel, 1990; Suzuki et al., 1992), the amino-terminal half of the molecule appears to be protected from proteolysis in situ (Hori et al., 1995), perhaps through a lateral association with F-actin.

Functional Implications of an Actin Side Binding Model for Dystrophin

Our recent experimental findings (Rybakova et al., 1996) suggest that dystrophin interacts with F-actin in a manner that dramatically distinguishes it from the other proteins that comprise the actin cross-linking superfamily (Matsudaira, 1991). In

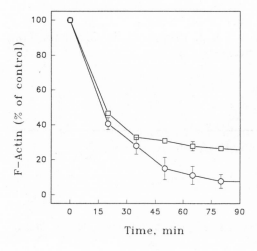

Figure 6. Dystrophin-glycoprotein complex protects F-actin from depolymerization. 7.5 μM F-actin in G-buffer containing 10 mM NaCl and 0.2 mM $MgCl_2$ was incubated 30 min at room temperature alone (*circles*) or in the presence of dystrophin-glycoprotein complex (*squares*) present at a 1:5 molar ratio with respect to actin. The samples were rapidly diluted threefold into G-buffer and centrifuged at 100,000 g for 20 minutes after various incubation times post-dilution. The fraction of actin (% *F-actin*) remaining in the pellet was determined densitometrically from Coomassie blue–stained gels loaded with equal volumes of supernatants and pellets. Time points include centrifugation time and represent the average (±SEM) of three or more independent determinations. Adapted from *The Journal of Cell Biology* (Rybakova et al., 1996) by copyright permission of The Rockefeller University Press.

light of these results, we propose a model in which the amino-terminal and rod do-mains both participate in dystrophin binding along an actin filament through the concerted effect of two or more low affinity binding sites. Of course, many aspects of this model await experimental confirmation. However, an actin side-binding function for dystrophin could prove particularly important in stabilizing lateral as-sociations between thin filaments and the sarcolemma at myotendinous junctions (Tidball and Law, 1991). In addition, dystrophin may play a structural role at cos-tameres (Porter et al., 1992) through a side binding interaction with peripheral actin filaments emanating from the Z- and M-lines of myofibers (Bard and Franzini-Armstrong, 1991). Both of these specialized sites of force transmission are dis-rupted in dystrophin-deficient *mdx* mouse muscle (Tidball and Law, 1991; Ridge et al., 1994). While the concerted effect of multiple, low affinity binding sites can ex-plain the relatively high affinity of intact dystrophin for F-actin at equilibrium in vitro, the association/dissociation rates at any particular site are likely to be rapid. Therefore, it is possible that the ensemble mechanical properties of the dystrophin/F-actin interaction in vivo could be either rigid (elastic) or deformable (fluid) de-pending on the rate at which a stress is applied to the sarcolemma membrane. This concept is derived from the model of multiple, rapidly rearranging cross-links between α-actinin and actin filaments proposed to explain the deformation rate-dependent differences in mechanical behavior of the cortical cytoskeleton (Sato et al., 1987).

The redundancy provided by a multiple actin binding site model may help to explain why restoration of a truncated dystrophin comprising only the cysteine-rich and carboxy-terminal domains is insufficient to correct the pathologies observed in dystrophic muscle (Cox et al., 1994; Greenberg et al., 1994), yet no specific se-quences within the amino-terminal (Corrado et al., 1996) and rod domains (En-gland et al., 1990; Passos-Bueno et al., 1994) also appear to be essential for normal dystrophin function. Perhaps some mild forms of muscular dystrophy are due to gene deletions that allow the expression of sufficient portions of the amino-termi-nal and/or rod domains (in any combination with cysteine-rich and carboxy-termi-nal domains) to conserve at least partial actin binding activity. The presence of ac-tin binding sites within the rod domain of dystrophin may further provide a basis by which dystrophin isoforms lacking the amino-terminal domain (Blake et al., 1992; Byers et al., 1993; Lidov et al., 1995; D'Souza et al., 1995) would be able to function in a manner analogous to full-length dystrophin.

Acknowledgments

This work was supported by grants from the National Institutes of Health (AR42423 and AR01985) and the Muscular Dystrophy Association.

References

Bard, F., and C. Franzini-Armstrong. 1991. Extra actin filaments at the periphery of skeletal muscle myofibrils. *Tissue Cell*. 23:191–197.

Beggs, A.H., E.P. Hoffman, J.R. Snyder, K. Arahata, L. Specht, F. Shapiro, C. Angelini, H. Sugita, and L.M. Kunkel. 1991. Exploring the molecular basis for variability among patients with Becker muscular dystrophy: dystrophin gene and protein studies. *Am. J. Hum. Genet.* 49:54–67.

Blake, D.J., D.R. Love, J. Tinsley, G.E. Morris, H. Turley, K. Gatter, G. Dickson, Y.H. Edwards, and K.E. Davies. 1992. Characterization of a 4.8kb transcript from the Duchenne muscular dystrophy locus expressed in Schwannoma cells. *Hum. Mol. Genet.* 1:103–109.

Brenman, J.E., D.S. Chao, S.H. Gee, A.W. McGee, S.E. Craven, D.R. Santillano, Z.Q. Wu, F. Huang, H.H. Xia, M.F. Peters, S.C. Froehner, and D.S. Bredt. 1996. Interaction of nitric oxide synthase with the postsynaptic density protein PSD-95 and α1-syntrophin mediated by PDZ domains. *Cell.* 84:757–767.

Broschat, K.O. 1990. Tropomyosin prevents depolymerization of actin filaments from the pointed end. *J. Biol. Chem.* 265:21323–21329.

Byers, T.J., H.G.W. Lidov, and L.M. Kunkel. 1993. An alternative dystrophin transcript specific to peripheral nerve. *Nature Genet.* 4:77–81.

Campbell, K.P. 1995. Three muscular dystrophies: loss of cytoskeleton-extracellular matrix linkage. *Cell.* 80:675–679.

Comi, G.P., A. Prelle, N. Bresolin, M. Moggio, A. Bardoni, A. Gallanti, G. Vita, A. Toscano, M.T. Ferro, A. Bordoni, et al. 1994. Clinical variability in Becker muscular dystrophy. Genetic, biochemical and immunohistochemical correlates. *Brain.* 117:1–14.

Corrado, K., P.L. Mills, and J.S. Chamberlain. 1994. Deletion analysis of the dystrophin-actin binding domain. *FEBS Lett.* 344:255–260.

Corrado, K., J.A. Rafael, P.L. Mills, N.M. Cole, J.A. Faulkner, K. Wang, and J.S. Chamberlain. 1996. Transgenic mdx mice expressing dystrophin with a deletion in the actin-binding domain display a "mild Becker" phenotype. *J. Cell Biol.* 134:873–884.

Cox, G.A., Y. Sunada, K.P. Campbell, and J.S. Chamberlain. 1994. Dp71 can restore the dystrophin-associated glycoprotein complex in muscle but fails to prevent dystrophy. *Nature Genet.* 8:333–339.

Cullen, M.J., J. Walsh, L.V.B. Nicholson, and J.B. Harris. 1990. Ultrastructural localization of dystrophin in human muscle by using gold immunolabeling. *Proc. R. Soc. Lond.* 240:197–210.

D'Souza, V.N., N.T. Man, G.E. Morris, W. Karges, D.M. Pillers, and P.N. Ray. 1995. A novel dystrophin isoform is required for normal retinal electrophysiology. *Hum. Mol. Genet.* 4:837–842.

England, S.B., L.V.B. Nicholson, M.A. Johnson, S.M. Forrest, D.R. Love, E.E. Zubrzycka-Gaarn, D.E. Bulman, J.B. Harris, and K.E. Davies. 1990. Very mild muscular dystrophy associated with the deletion of 46% of dystrophin. *Nature.* 343:180–182.

Ervasti, J.M., K. Ohlendieck, S.D. Kahl, M.G. Gaver, and K.P. Campbell. 1990. Deficiency of a glycoprotein component of the dystrophin complex in dystrophic muscle. *Nature.* 345:315–319.

Ervasti, J.M., S.D. Kahl, and K.P. Campbell. 1991. Purification of dystrophin from skeletal muscle. *J. Biol. Chem.* 266:9161–9165.

Ervasti, J.M., and K.P. Campbell. 1991. Membrane organization of the dystrophin-glycoprotein complex. *Cell.* 66:1121–1131.

Ervasti, J.M., and K.P. Campbell. 1993a. A role for the dystrophin-glycoprotein complex as a transmembrane linker between laminin and actin. *J. Cell Biol.* 122:809–823.

Ervasti, J.M., and K.P. Campbell. 1993b. Dystrophin and the membrane skeleton. *Curr. Opin. Cell Biol.* 5:82–87.

Fabbrizio, E., A. Bonet-Kerrache, J.J. Leger, and D. Mornet. 1993. Actin-dystrophin inter-face. *Biochemistry.* 32:10457–10463.

Goldberg, M.E., and L. Djavadi-Ohaniance. 1993. Methods for measurement of antibody/antigen affinity based on ELISA and RIA. *Curr. Opin. Immunol.* 5:278–281.

Greenberg, D.S., Y. Sunada, K.P. Campbell, D. Yaffe, and U. Nudel. 1994. Exogenous Dp71 restores the levels of dystrophin associated proteins but does not alleviate muscle damage in mdx mice. *Nature Genet.* 8:340–344.

Helliwell, T.R., J.M. Ellis, R.C. Mountford, R.E. Appleton, and G.E. Morris. 1992. A trun-cated dystrophin lacking the C-terminal domain is localized at the muscle membrane. *Am. J. Hum. Genet.* 58:508–514.

Hemmings, L., P.A. Kuhlman, and D.R. Critchley. 1992. Analysis of the actin-binding do-main of α-actinin by mutagenesis and demonstration that dystrophin contains a functionally homologous domain. *J. Cell Biol.* 116:1369–1380.

Hoffman, E.P., C.A. Garcia, J.S. Chamberlain, C. Angelini, J.R. Lupski, and R. Fenwick. 1991. Is the carboxyl-terminus of dystrophin required for membrane association? A novel, severe case of Duchenne muscular dystrophy. *Ann. Neurol.* 30:605–610.

Hori, S., S. Ohtani, N. Man, and G.E. Morris. 1995. The N-terminal half of dystrophin is pro-tected from proteolysis *in situ. Biochem. Biophys. Res. Commun.* 209:1062–1067.

Jarrett, H.W., and J.L. Foster. 1995. Alternate binding of actin and calmodulin to multiple sites on dystrophin. *J. Biol. Chem.* 270:5578–5586.

Koenig, M., and L.M. Kunkel. 1990. Detailed analysis of the repeat domain of dystrophin re-veals four potential hinge segments that may confer flexibility. *J. Biol. Chem.* 265:4560–4566.

Koenig, M., A.P. Monaco, and L.M. Kunkel. 1988. The complete sequence of dystrophin pre-dicts a rod-shaped cytoskeletal protein. *Cell.* 53:219–228.

Kramarcy, N.R., A. Vidal, S.C. Froehner, and R. Sealock. 1994. Association of utrophin and multiple dystrophin short forms with the mammalian M_r 58,000 dystrophin-associated pro-tein (syntrophin). *J. Biol. Chem.* 269:2870–2876.

Lebart, M.C., D. Casanova, and Y. Benyamin. 1995. Actin interaction with purified dystro-phin from electric organ of *Torpedo marmorata*: possible resemblance with filamin actin in-terface. *J. Musc. Res. Cell Motil.* 16:543–552.

Li, X., and V. Bennett. 1996. Identification of the spectrin subunit and domains required for formation of spectrin/adducin/actin complexes. *J. Biol. Chem.* 271:15695–15702.

Lidov, H.G.W., S. Selig, and L.M. Kunkel. 1995. Dp140: a novel 140 kDa CNS transcript from the dystrophin locus. *Hum. Mol. Genet.* 4:329–335.

Lo, S.H., P.A. Janmey, J.H. Hartwig, and L.B. Chen. 1994. Interactions of tensin with actin and identification of its three distinct actin-binding domains. *J. Cell Biol.* 125:1067–1075.

Matsudaira, P. 1991. Modular organization of actin crosslinking proteins. *Trends Biochem. Sci.* 16:87–92.

Matsumura, K., F.M.S. Tome, V. Ionasescu, J.M. Ervasti, R.D. Anderson, N.B. Romero, D. Simon, D. Recan, J.-C. Kaplan, M. Fardeau, and K.P. Campbell. 1993. Deficiency of dystro-phin-associated proteins in Duchenne muscular dystrophy patients lacking COOH-terminal domains of dystrophin. *J. Clin. Invest.* 92:866–871.

Menke, A., and H. Jockusch. 1991. Decreased osmotic stability of dystrophin-less muscle cells from the mdx mouse. *Nature.* 349:69–71.

Muntoni, F., P. Gobbi, C. Sewry, T. Sherratt, J. Taylor, S.K. Sandhu, S. Abbs, R. Roberts, S.V. Hodgson, M. Bobrow, and V. Dubowitz. 1994. Deletions in the 5' region of dystrophin and resulting phenotypes. *J. Med. Genet.* 31:843–847.

Passos-Bueno, M.R., M. Vainzof, S.K. Marie, and M. Zatz. 1994. Half the dystrophin gene is apparently enough for a mild clinical course: confirmation of its potential use for gene therapy. *Hum. Mol. Genet.* 3:919–922.

Pasternak, C., S. Wong, and E.L. Elson. 1995. Mechanical function of dystrophin in muscle cells. *J. Cell Biol.* 128:355–361.

Petrof, B.J., J.B. Shrager, H.H. Stedman, A.M. Kelly, and H.L. Sweeney. 1993. Dystrophin protects the sarcolemma from stresses developed during muscle contraction. *Proc. Natl. Acad. Sci. USA.* 90:3710–3714.

Pons, F., N. Augier, R. Heilig, J. Leger, D. Mornet, and J.J. Leger. 1990. Isolated dystrophin molecules as seen by electron microscopy. *Proc. Natl. Acad. Sci. USA.* 87:7851–7855.

Porter, G.A., G.M. Dmytrenko, J.C. Winkelmann, and R.J. Bloch. 1992. Dystrophin colocalizes with β-spectrin in distinct subsarcolemmal domains in mammalian skeletal muscle. *J. Cell Biol.* 117:997–1005.

Prior, T.W., A.C. Papp, P.J. Snyder, A.H.M. Burghes, C. Bartolo, M.S. Sedra, L.M. Western, and J.R. Mendell. 1993. A missense mutation in the dystrophin gene in a Duchenne muscular dystrophy patient. *Nature Genet.* 4:357–360.

Ridge, J.C., J.G. Tidball, K. Ahl, D.J. Law, and W.L. Rickoll. 1994. Modifications in myotendinous junction surface morphology in dystrophin-deficient mouse muscle. *Exp. Mol. Pathol.* 61:58–68.

Rybakova, I.N., K.J. Amann, and J.M. Ervasti. 1996. A new model for the interaction of dystrophin with F-actin. *J. Cell Biol.* 135:661–672.

Sato, M., W.H. Schwarz, and T.D. Pollard. 1987. Dependence of the mechanical properties of actin/α-actinin gels on deformation rate. *Nature.* 325:828–830.

Schafer, D.A., and J.A. Cooper. 1995. Control of actin assembly at filament ends. *Annu. Rev. Cell Biol.* 11:497–518.

Sealock, R., and S.C. Froehner. 1994. Dystrophin-associated proteins and synapse formation: is α-dystroglycan the agrin receptor. *Cell.* 77:617–619.

Stedman, H.H., H.L. Sweeney, J.B. Shrager, H.C. Maguire, R.A. Panettieri, B. Petrof, M. Narusawa, J.M. Leferovich, J.T. Sladky, and A.M. Kelly. 1991. The mdx mouse diaphragm reproduces the degenerative changes of Duchenne muscular dystrophy. *Nature.* 352:536–539.

Suzuki, A., M. Yoshida, K. Hayashi, Y. Mizuno, Y. Hagiwara, and E. Ozawa. 1994. Molecular organization at the glycoprotein-complex-binding site of dystrophin. Three dystrophin-associated proteins bind directly to the carboxy-terminal portion of dystrophin. *Eur. J. Biochem.* 220:283–292.

Suzuki, A., M. Yoshida, H. Yamamoto, and E. Ozawa. 1992. Glycoprotein-binding site of dystrophin is confined to the cysteine-rich domain and the first half of the carboxy-terminal domain. *FEBS Lett.* 308:154–160.

Takeshima, Y., H. Nishio, N. Narita, H. Wada, Y. Ishikawa, R. Minami, H. Nakamura, and M. Matsuo. 1994. Amino-terminal deletion of 53% of dystrophin results in an intermediate Duchenne-Becker muscular dystrophy phenotype. *Neurology.* 44:1648–1651.

Tidball, J.G., and D.J. Law. 1991. Dystrophin is required for normal thin filament-membrane associations at myotendinous junctions. *Am. J. Pathol.* 138:17–21.

Van Etten, R.A., P.K. Jackson, D. Baltimore, M.C. Sanders, P.T. Matsudaira, and P.A. Janmey. 1994. The COOH terminus of the c-Abl tyrosine kinase contains distinct F- and G-actin binding domains with bundling activity. *J. Cell Biol.* 124:325–340.

Way, M., B. Pope, R.A. Cross, J. Kendrick-Jones, and A.G. Weeds. 1992. Expression of the N-terminal domain of dystrophin in *E. coli* and demonstration of binding to F-actin. *FEBS Lett.* 301:243–245.

Winder, S.J., T.J. Gibson, and J. Kendrick-Jones. 1995. Dystrophin and utrophin: the missing links. *FEBS Lett.* 369:27–33.

Winnard, A.V., C.J. Klein, D.D. Coovert, T. Prior, A. Papp, P. Snyder, D.E. Bulman, P.N. Ray, P. McAndrew, W. King, R.T. Moxley, J.R. Mendell, and A.H.M. Burghes. 1993. Characterization of translational frame exception patients in Duchenne/Becker muscular dystrophy. *Hum. Mol. Genet.* 2:737–744.

Worton, R. 1995. Muscular dystrophies: diseases of the dystrophin-glycoprotein complex. *Science.* 270:755–756.

Yang, B., D. Jung, D. Motto, J. Meyer, G. Koretzky, and K.P. Campbell. 1995. SH3 domain-mediated interaction of dystroglycan and Grb2. *J. Biol. Chem.* 270:11711–11714.

Chapter 2

Establishment of Epithelial Cell Polarity

Roles of the Membrane Cytoskeleton in Protein Sorting

W. James Nelson, Kenneth A. Beck, and Peter A. Piepenhagen

Department of Molecular and Cellular Physiology, Beckman Center for Molecular and Genetic Medicine, Stanford University School of Medicine, Stanford, California 94305-5426

Segregation of specific subclasses of membrane proteins and lipids into discrete membrane domains is a fundamental aspect of cell structure and function. Specialized cells, such as transporting epithelia, require functionally distinct membrane domains to regulate vectorial transport of ions and solutes or unidirectional propagation of electrical stimuli (Rodriguez-Boulan and Nelson, 1989; Rodriguez-Boulan and Powell, 1992). In the secretory pathway, membrane proteins are transported through different membrane compartments which are capable of maintaining their own distinctive composition of resident proteins while, at the same time, allowing for a continual flux of nonresident proteins (Machamer, 1991). Thus, the formation of membrane domains is essential not only in cell surface organization of polarized cells but also for the balance of membrane protein retention and transport that facilitates the genesis of distinct organellar membranes.

We have been interested in the roles of the membrane cytoskeleton in regulating the sorting and distribution of proteins in specific membrane domains. Intracellular sites for sorting proteins in polarized epithelial cells appear to be located in the Golgi complex and at the plasma membrane (Rodriguez-Boulan and Nelson, 1989; Rodriguez-Boulan and Powell, 1992). We have studied the interactions of membrane-cytoskeletal complexes with membrane proteins delivered to the apical and basal-lateral membrane domains of polarized epithelial cells. More recently, we have found evidence for a membrane cytoskeleton associated with the Golgi complex. Our results raise the interesting possibility that the membrane cytoskeleton may play a broad role in regulating the protein composition of membrane domains in different stages of the secretory pathway.

Properties of the Spectrin-based Membrane Cytoskeleton

Proteins of the spectrin-based membrane cytoskeleton form a 2-dimensional, oligomeric lattice that binds to specific classes of integral membrane proteins (Bennett, 1990a, b; Bennett and Gilligan, 1993). The major structural component of the membrane cytoskeleton is spectrin, a flexible, rod-shaped molecule composed of homologous, but non-identical, α- and β-subunits. Spectrin heterodimers self-associated end-to-end to form heterotetramers (length 200 nm), which, in erythrocytes, are cross linked by a complex of short actin oligomers, protein 4.1 and adducin.

The spectrin-based membrane cytoskeleton is tightly coupled to the membrane through direct binding to a diverse subset of membrane proteins. Some of

Cytoskeletal Regulation of Membrane Function © 1997 by The Rockefeller University Press

these interactions occur through the membrane-cytoskeleton protein ankyrin (Bennett, 1990*a*; Bennett, 1992). Ankyrin is a bifunctional molecule with separate binding sites for membrane proteins and β-spectrin. Membrane proteins also bind to protein 4.1 and spectrin, the latter having at least two distinct membrane protein binding domains. Thus, the membrane cytoskeleton has the potential for binding a large repertoire of membrane proteins.

Several membrane proteins have been shown to interact with the membrane cytoskeleton. In erythrocytes, ankyrin and protein 4.1 bind to the anion exchanger (band 3) and glycophorin, respectively (Bennett and Gilligan, 1993). In non-erythroid cells, membrane-cytoskeleton proteins bind to Na/K-ATPase (Koob et al., 1987; Nelson and Veshnock, 1987; Morrow et al., 1989; Nelson and Hammerton, 1989; Davis and Bennett, 1990), voltage-gated Na^+ channel (Srinivasan et al., 1988), amiloride-sensitive Na^+ channel (Smith et al., 1991), H/K-ATPase (Smith et al., 1993), CD44 (Lokeshwar and Bourguignon, 1992; Lokeshwar et al., 1994), L1/Ng-CAM (Davis and Bennett, 1994), E-cadherin (Nelson et al., 1990), and IP_3-receptor (Bourguignon and Jin, 1995). Significantly, many of these proteins, like the membrane cytoskeleton, exhibit restricted cell surface distributions in polarized cells (Rodriguez-Boulan and Nelson, 1989). Biophysical studies have shown that proteins bound to the membrane cytoskeleton have reduced lateral diffusion in the plane of the lipid bilayer (Sheetz et al., 1980; Tsuji and Ohnishi, 1986), suggesting that the membrane cytoskeleton can physically restrict the distribution of membrane proteins within a membrane compartment. Furthermore, studies of membrane protein half-lives in different membrane domains in polarized epithelial cells show that interactions with the membrane cytoskeleton may sequester proteins from internalization and degradation, thereby regulating their retention in the plasma membrane (Hammerton et al., 1991; Marrs et al., 1994).

Roles of the Membrane Cytoskeleton in Protein Sorting in the Plasma Membrane

In polarized Madin Darby canine kidney (MDCK) epithelial cell, ankyrin and fodrin become restricted to the sites of cell–cell contact, and their detergent insolubility and metabolic half-life increase indicating assembly into a cytoskeletal complex (Nelson, 1992). Significantly, assembly of ankyrin and fodrin at cell–cell contacts in "nonpolarized" fibroblasts (McNeill et al., 1990) and RPE cells (Marrs et al., 1994) is dependent upon expression of E-cadherin. Deletion of the cytoplasmic domain of E-cadherin in transfected fibroblasts inhibits fodrin accumulation at sites of cell–cell contact, indicating that linkage to the cytoskeleton is important in protein recruitment to the adhesion site (McNeill et al., 1990).

Na/K-ATPase and E-cadherin are directly linked to different membrane-cytoskeleton complexes. Na/K-ATPase binds directly to ankyrin, which in turn binds to fodrin; in vitro studies with purified proteins demonstrated the specificity and high affinity binding of these proteins (Nelson and Veshnock, 1987; Morrow et al., 1989; Davis and Bennett, 1990), and cell fractionation studies demonstrated that these complexes exist in MDCK cells (Nelson and Hammerton, 1989), and other polarized epithelial cells (Marrs et al., 1994). A complex of Na/K-ATPase, ankyrin and fodrin has been detected in association with E-cadherin following fractionation of

MDCK cell extracts (Nelson et al., 1990), but the protein interactions involved in the linkage between the Na/K-ATPase complex and E-cadherin are not known. In MDCK cells, E-cadherin directly associates with two cytoplasmic proteins, termed α- and β-catenin (Kemler and Ozawa, 1989). Alpha-catenin binds to the E-cadherin/β-catenin complex at the cell surface coincident with the titration of the complex into a detergent insoluble fraction (Hinck et al., 1994). These findings indicate that α-catenin is a candidate protein for linking the cadherin/catenin complex to the Na/K-ATPase/ankyrin/fodrin complex, perhaps by binding directly to fodrin (Lombardo et al., 1993) or through actin (Rimm et al., 1995).

Selective incorporation of specific membrane proteins (e.g., Na/K-ATPase) into the membrane cytoskeleton may restrict protein diffusion away from those sites, resulting in their localized retention and accumulation (Nelson, 1992). Loss of membrane-cytoskeleton protein interactions in erythrocytes (Sheetz et al., 1980) and RBL cells (Dahl et al., 1994) increases the diffusion of proteins in the membrane. Also, accumulation of Na/K-ATPase at E-cadherin adhesion sites in MDCK (Nelson and Veshnock, 1986) and RPE cells (Marrs et al., 1994) correlates with >20-fold longer half-lives of cadherin, Na/K-ATPase and membrane-cytoskeletal proteins at sites of cell-cell contact, than at the noncontacting (apical) membrane.

The distribution of the membrane cytoskeleton has been determined in detail in polarized epithelial cells of transporting epithelia in the kidney nephron (Piepenhagen et al., 1995a,b). Using a combination of morphological analysis and double staining with segment-specific antibodies, different nephron segments that expressed Na/K-ATPase could be identified. In proximal tubules, much stronger Na/K-ATPase staining was detected in the convoluted (S1 and S2 segments) than in the straight portion (S3 segment). In descending and ascending thin limbs of the loop of Henle, only background staining for Na/K-ATPase was observed. In ascending thick limbs, Na/K-ATPase staining was more intense than that in proximal convoluted tubules. In distal convoluted tubules, Na/K-ATPase staining similar in intensity to that in ascending thick limbs was found. In connecting tubules and collecting ducts, the intensity of Na/K-ATPase staining was moderate (Piepenhagen et al., 1995a,b).

Using an antibody raised against the α-subunit of fodrin, epithelial cells of the nephron stained in a pattern very similar to that of Na/K-ATPase. The relative intensity of ankyrin staining in different segments of the nephron was also similar to that of Na/K-ATPase and was detected using a polyclonal antiserum specific for isoform 3 of ankyrin. These data indicate that ankyrin and fodrin expressed in the renal tubular epithelial cells differ from those expressed in erythrocytes and endothelial cells. Within collecting ducts, Na/K-ATPase staining was primarily restricted to basal-lateral membranes of principal cells. In contrast, ankyrin staining of basal-lateral membranes was strongest in intercalated cells. Fodrin staining in collecting ducts was restricted to basal-lateral membranes, was present in both principal and intercalated cells, and was relatively more intense than that in other nephron segments compared to ankyrin or Na/K-ATPase.

Taken together, results of studies with kidney epithelial cells in vitro and in vivo demonstrate that the membrane cytoskeleton plays important roles in establishing and maintaining polarized distributions of membrane proteins in functionally specific membrane domains in transporting epithelial cells.

Potential Roles of the Membrane Cytoskeleton in the Golgi Complex

The Golgi apparatus is a highly organized organelle composed of spatially and functionally distinct compartments responsible for the processing and sorting of newly synthesized membrane and secreted proteins (Pfeffer and Rothman, 1987; Rothman and Orci, 1992). Each compartment maintains a unique membrane protein composition under conditions that allow continuous, sequential transit of newly synthesized proteins. This implies the existence of a molecular machinery that can distinguish and segregate resident and transported membrane proteins. Little is known about the maintenance of overall Golgi morphology, the retention of resident proteins within different Golgi compartments, and the sorting of newly synthesized membrane proteins within the trans Golgi network (TGN). Several known protein complexes can be viewed as candidate membrane protein sorting machines, including: clathrin/AP complex (Keen, 1990; Pearce and Robinson, 1990; Robinson, 1994); coatomer (COP) complex (Rothman and Orci, 1992; Kreis and Pepperkok, 1994; Rothman and Warren, 1994); caveolae (Anderson, 1993); p200 (Narula et al., 1992; Narula and Stow, 1995); and spectrin (Bennett and Gilligan, 1993; Beck et al., 1994; Beck and Nelson, 1996).

Specific components of the membrane cytoskeleton localize to the Golgi complex (Beck et al., 1994). Immunofluorescence staining of MDCK cells with an antiserum directed against the β-subunit of canine erythrocyte spectrin showed that homologues of erythroid β-spectrin localizes to cytoplasmic, reticular structures that co-stain with marker proteins of the Golgi complex. The distribution of erythroid β-spectrin is different from that of the non-erythroid isoform, which localizes to the basal-lateral plasma membrane of these cells. Within the Golgi complex, the distribution of β-spectrin partially overlaps that of the medial Golgi marker mannosidase II, indicating that β-spectrin has a restricted distribution within the Golgi complex (Beck et al., 1994). More recent studies have shown that isoforms of ankyrin (Devarajan et al., 1996; Beck et al., 1997) also co-localize with the Golgi complex. Although protein–protein interactions between β-spectrin and ankyrin in the Golgi have not been investigated, the observation that these proteins co-localize indicate that a membrane-cytoskeleton complex associates with a distinct region(s) of the Golgi complex.

The association of β-spectrin with Golgi membranes is dynamic. During mitosis, when the Golgi complex fragments and disperses throughout the cytoplasm (Warren, 1993), Golgi spectrin dissociates from Golgi membranes (Beck et al., 1994). Treatment of cells with Brefeldin A (BFA), which disrupts Golgi structure and function (Orci et al., 1991; Klausner et al., 1992), also leads to dissociation of Golgi spectrin from membranes (Beck et al., 1994). The kinetics of dissociation of Golgi spectrin are rapid (2 min after addition of BFA) and are comparable to those observed for other Golgi coat protein complexes (e.g., coatomer and p200). Since the effect of BFA on the distribution of Golgi coat proteins is thought to result from an inhibition of their cycle of assembly and dissociation (Donaldson et al., 1990; Donaldson et al., 1992), these observations suggest that Golgi spectrin exists in a similar dynamic state.

The identification of Golgi isoforms of membrane cytoskeleton proteins greatly expands the potential contribution of the membrane cytoskeleton to the organization of cellular membranes. Since the membrane cytoskeleton provides structural support for the erythrocyte plasma membrane, it is possible that Golgi spectrin could perform a similar function. The Golgi complex has an ordered structural organization, but little is known about how it is maintained. The membrane cytoskeleton lattice could provide structural support required to maintain Golgi organization. There is also a requirement within the Golgi complex for membrane domain formation and membrane protein sorting at a number of levels including: the maintenance of compartment composition, and the retention of resident proteins and the sorting of classes of proteins at the level of the TGN. The Golgi membrane cytoskeleton could play a role in these processes as well (Beck and Nelson, 1996).

Acknowledgments

Work from this laboratory is supported by grants to W.J. Nelson from the NIH and March-of-Dimes. K.A. Beck was also supported by a postdoctoral fellowship from the NIH, and P.A. Piepenhagen was also supported by predoctoral fellowships from the NSF and the Lieberman Foundation.

References

Anderson, R. 1993. Caveolae: where incoming and outgoing messengers meet. *Proc. Natl. Acad. Sci. USA.* 90:10909–10913.

Beck, K.A., V. Malhotra, and W.J. Nelson. 1994. Golgi spectrin: identification of an erythroid beta-spectrin homologue associated with the Golgi complex. *J. Cell Biol.* 127:707–723.

Beck, K.A., and W.J. Nelson. 1996. The spectrin-based membrane skeleton as a membrane protein sorting machine. *Am. J. Physiol.* 270:C1263–C1270.

Beck, K.A., J. Buchanan, and W.J. Nelson. 1997. Golgi membrane skeleton: identification, localization and oligomerization of a 195 KDa ankyrin isoform associated with the Golgi complex. *J. Cell Sci.* In press.

Bennett, V. 1990a. Spectrin-based membrane skeleton: a multipotential adaptor between plasma membrane and cytoplasm. *Physiol. Rev.* 70:1029–1065.

Bennett, V. 1990b. Spectrin: a structural mediator between diverse plasma membrane proteins and the cytoplasm. *Curr. Opin. Cell Biol.* 2:51–56.

Bennett, V. 1992. Ankyrins: adaptors between diverse plasma membrane proteins and the cytoplasm. *J. Biol. Chem.* 267:8703–8706.

Bennett, V., and D. Gilligan. 1993. The spectrin-based membrane skeleton and micron-scale organization of the plasma membrane. *Annu. Rev. Cell Biol.* 9:27–66.

Bourguignon, L., and H. Jin. 1995. Identification of the ankyrin-binding domain of the mouse T-lymphoma cell inositol 1,4,5-triphosphate (IP3) receptor and its role in the regulation of IP3-mediated internal Ca^{2+} release. *J. Biol. Chem.* 270:7257–7260.

Dahl, S.C., R.W. Greib, M.T. Fox, M. Edidin, and D. Branton. 1994. Rapid capping in alpha-spectrin deficient MEL cells from mice afflicted with hereditary hemolytic anemia. *J. Cell Biol.* 125:1057–1066.

Davis, J., and V. Bennett. 1990. The anion exchanger and Na^+,K^+-ATPase interact with distinct sites on ankyrin in in vitro assays. *J. Biol. Chem.* 265:17252–17256.

Davis, J., and V. Bennett. 1994. Ankyrin binding activity shared by the neurofascin/L1/NrCAM family of nervous system cell adhesion molecules. *J. Biol. Chem.* 269:27163–27166.

Devarajan, P., P. Stabach, A. Mann, T. Ardito, M. Kashgarian, and J. Morrow. 1996. Identification of a small cytoplasmic ankyrin (AnkG119) in the kidney and muscle that binds BIE*spectrin and associates with the Golgi complex. *J. Cell Biol.* 133:819–830.

Donaldson, J.G., D. Cassel, R. Kahn, and R. Klausner. 1992. ADP-ribosylation factor, a small GTP-binding protein, is required for binding of the coatomer protein beta-COP to Golgi membranes. *Proc. Natl. Acad. Sci. USA.* 89:6408–6412.

Donaldson, J.G., J. Lippincott-Schwartz, G.S. Gloom, T.E. Kreis, and R.D. Klausner. 1990. Dissociation of a 110-kD peripheral membrane protein from the Golgi apparatus is an early event in Brefeldin A action. *J. Cell Biol.* 111:2295–2306.

Hammerton, R.W., K.A. Krzeminski, R.W. Mays, D.A. Wollner, and W.J. Nelson. 1991. Mechanism for regulating cell surface distribution of Na/K-ATPase in polarized epithelial cells. *Science.* 254:847–850.

Hinck, L., I.S. Nathke, J. Papkoff, and W.J. Nelson. 1994. Dynamics of cadherin/catenin complex formation: novel protein interactions and pathways of complex formation. *J. Cell Biol.* 125:1327–1340.

Keen, J.H. 1990. Clathrin and associated assembly and disassembly proteins. *Annu. Rev. Biochem.* 59:415–438.

Kemler, R., and M. Ozawa. 1989. Uvomorulin-catenin complex: cytoplasmic anchorage of a Ca^{2+}-dependent cell adhesion molecule. *Bioessays.* 11:88–91.

Klausner, R.D., J.G. Donaldson, and J. Lippincott-Schwartz. 1992. Brefeldin A: insights into the control of membrane traffic and organelle structure. *J. Cell Biol.* 116:1071–1080.

Koob, R., M. Zimmerman, W. Schoner, and D. Drenkhahn. 1987. Co-localization and co-precipitation of ankyrin and Na^+,K^+-ATPase in kidney epithelial cells. *Eur. J. Cell Biol.* 45:230–237.

Kreis, T., and R. Pepperkok. 1994. Coat proteins in intracellular membrane transport. *Curr. Opin. Cell Biol.* 6:533–537.

Lokeshwar, V., N. Fregien and L. Bourguignon. 1994. Ankyrin-binding domain of CD44 (GP85) is required for the expression of hyaluronic acid-mediated adhesion function. *J. Cell Biol.* 126:1099–1109.

Lokeshwar, V.B., and L.Y. Bourguignon. 1992. The lymphoma transmembrane glycoprotein GP85 (CD44) is a novel guanine nucleotide-binding protein which regulates GP85 (CD44)-ankyrin interaction. *J. Biol. Chem.* 267:17252–17256.

Lombardo, C.R., D.L. Rimm, S.P. Kennedy, B.G. Forget, and J.S. Morrow. 1993. Ankyrin independent membrane sites for non-erythroid spectrin. *Mol. Biol. Cell.* 4:57a.

Machamer, C. 1991. Golgi retention signals: do membranes hold the key? *Trends Cell Biol.* 1:141–144.

Marrs, J.A., C. Andersson-Fisone, M.C. Jeong, I.R. Nabi, C. Zurzolo, E. Rodriguez-Boulan, and W.J. Nelson. 1994. Plasticity in epithelial cell phenotype: modulation by differential expression of cadherin cell adhesion molecules. *J. Cell Biol.* 129:509–517.

McNeill, H., M. Ozawa, R. Kemler, and W.J. Nelson. 1990. Novel function of the cell adhesion molecule uvomorulin as an inducer of cell surface polarity. *Cell.* 62:309–316.

Morrow, J.S., C.D. Cianci, T. Ardito, A.S. Mann, and M. Kashgarian. 1989. Ankyrin links fodrin to the alpha subunit of Na^+,K^+-ATPase in Madin-Darby canine kidney cells and in intact renal tubule cells. *J. Cell Biol.* 108:455–465.

Narula, N., I. McMorrow, G. Plopper, J. Doherty, K. Matlin, B. Burke, and J. Stow. 1992. Identification of a 200-kD, brefeldin-sensitive protein on Golgi membranes. *J. Cell Biol.* 117:27–38.

Narula, N., and J. Stow. 1995. Distinct coated vesicles labeled for p200 bud from trans-Golgi network membranes. *Proc. Natl. Acad. Sci. USA.* 92:2874–2878.

Nelson, W., E. Shore, A. Wang, and R.W. Hammerton. 1990. Identification of a membrane-cytoskeletal complex containing the cell adhesion molecule uvomorulin (E-cadherin), ankyrin, and fodrin in Madin-Darby canine kidney epithelial cells. *J. Cell Biol.* 110:349–357.

Nelson, W., and P.J. Veshnock. 1986. Dynamics of membrane-skeleton (fodrin) organization during development of polarity in Madin-Darby canine kidney epithelial cells. *J. Cell Biol.* 103:1751–1765.

Nelson, W.J. 1992. Regulation of cell surface polarity from bacteria to mammals. *Science.* 258:948–955.

Nelson, W.J., and R.W. Hammerton. 1989. A membrane-cytoskeletal complex containing Na+,K+-ATPase, ankyrin, and fodrin in Madin-Darby canine kidney (MDCK) cells: implications for the biogenesis of epithelial cell polarity. *J. Cell Biol.* 108:893–902.

Nelson, W.J., and P.J. Veshnock. 1987. Ankyrin binding to Na+,K+-ATPase and implications for the organization of membrane domains in polarized cells. *Nature (Lond.).* 328:533–536.

Orci, L., M. Tagaya, M. Amherdt, A. Perrelet, J. Donaldson, J. Lippincott-Schwartz, R. Klausner, and J. Rothman. 1991. Brefeldin A, a drug that blocks secretion, prevents the assembly of non-clathrin-coated buds on Golgi cisternae. *Cell.* 64:1183–1195.

Pearce, B.M.F., and M. Robinson. 1990. Clathrin, adaptors, and sorting. *Annu. Rev. Cell Biol.* 6:151–171.

Pfeffer, S.R., and J.E. Rothman. 1987. Biosynthetic protein transport and sorting by the endoplasmic reticulum and Golgi. *Annu. Rev. Biochem.* 56:829–852.

Piepenhagen, P., L. Peters, S. Lux, and W. Nelson. 1995a. Differential expression of Na/K-ATPase, ankyrin, fodrin, and E-cadherin along the kidney nephron. *Am. J. Physiol.* 269:C1417–C1432.

Piepenhagen, P., L.L. Peters, S.E. Lux, and W.J. Nelson. 1995b. Differential distribution of Na/K-ATPase, ankyrin, fodrin, and E-cadherin along the kidney nephron. *Am. J. Physiol.* 269:1433–1449.

Rimm, D.L., E. Koslov, P. Kebriaei, D. Cianci, and J.S. Morrow. 1995. Alpha1(E)-catenin is an actin binding and -bundling protein mediating the attachment of F-actin to the membrane adhesion complex. *Proc. Natl. Acad. Sci. USA.* 92:8813–8817.

Robinson, M. 1994. The role of clathrin, adaptors and dynamin in endocytosis. *Cur. Opin. Cell Biol.* 6:538–544.

Rodriguez-Boulan, E., and W.J. Nelson. 1989. Morphogenesis of the polarized epithelial cell phenotype. *Science.* 245:718–725.

Rodriguez-Boulan, E., and S.K. Powell. 1992. Polarity of epithelial and neuronal cells. *Annu. Rev. Cell Biol.* 8:395–427.

Rothman, J., and L. Orci. 1992. Molecular dissection of the secretory pathway. *Nature.* 355:409–415.

Rothman, J.E., and G. Warren. 1994. Implications of the SNARE hypothesis for intracellular membrane topology and dynamics. *Curr. Biol.* 4:220–233.

Sheetz, M.P., M. Schindler, and D. Koppel. 1980. Lateral mobility of integral membrane proteins is increased in spherocytic erythrocytes. *Nature.* 285:510–512.

Smith, P.R., G. Saccomani, E.H. Joe, K.J. Angelides, and D.J. Benos. 1991. Amiloride-sensitive sodium channel is linked to the cytoskeleton in renal epithelial cells. *Proc. Natl. Acad. Sci. USA.* 88:6971–6975.

Smith, P.R., A.L. Bradford, E.H. Joe, K.J. Angelides, D.J. Benos, and G. Saccomani. 1993. Gastric parietal cell H/K-ATPase microsomes are associated with isoforms of ankyrin and spectrin. *Am. J. Physiol.* 264:63–70.

Srinivasan, Y., L. Elmer, J. Davis, V. Bennett, and K. Angelides. 1988. Ankyrin and spectrin associate with voltage-dependent sodium channels in brain. *Nature.* 333:177–180.

Tsuji, A., and S. Ohnishi. 1986. Restriction of the lateral motion of band 3 in the erythrocyte membrane by the cytoskeletal network: dependence on spectrin association sites. *Biochemistry.* 25:6133–6139.

Warren, G. 1993. Membrane partitioning during cell division. *Annu. Rev. Biochem.* 62:323–348.

Role of Cytoplasmic Motors in Post-Golgi Vesicular Traffic

Anne Müsch and Enrique Rodriguez-Boulan

Margaret Dyson Vision Research Institute, Department of Ophthalmology, and Department of Cell Biology, Cornell University Medical College, New York, New York 10021

Transport of material in the secretory pathway and in the endocytic route is mediated by carrier vesicles specific for each transport step. Assembly of the vesicles from the donor membrane, their transport to specific destinations, as well as docking and fusion of these vesicles to their target membranes, is mediated by distinct sets of cytosolic components (Fig. 1) (Rothman, 1994; Rothman and Wieland, 1996). Understanding the molecular bases of these fundamental processes is a major goal in cell biology. In this review, we will focus primarily on the first two steps in vesicular transport: the assembly and transport of transport vesicles. We will briefly review the role of motors in vesicular transport and discuss a novel role for myosin in the production of transport vesicles from the TGN.

Coat and Adaptor Proteins that Control the Assembly of Transport Vesicles

The first step in vesicular transport, the assembly of the carrier vesicle, is believed to be controlled by two major families of coat proteins and related adaptor complexes (AP). The *clathrin* family includes the adaptor complexes *AP1* (*trans* Golgi network), *AP2* (plasma membrane), and neuronal*AP3*. (Pearse and Robinson, 1990). The *coatomer* (*COP*) family comprises three different classes, *COP I* (intraGolgi and Golgi-ER transport) (Orci et al., 1993), *COPII* (ER-Golgi transport) (Barlowe et al., 1994), and *endosomal COP* (Whitney et al., 1995). Coat proteins and adaptors are thought to mechanically promote the curvature of the donor membrane into a bud, leading to the formation of a vesicle with a characteristic diameter of 60–80 nm (Schekman and Orci, 1996). Attachment of coat proteins and adaptors to the donor membrane is mediated by small GTP binding proteins, either directly or via activation of phospholipase D, which generates negative charges in the membrane through the generation of lysophosphatidic acid (Ktistakis et al., 1996). Until recently, the dogma was that clathrin promotes concentration of the transported proteins, whereas coatomer is involved in bulk flow transport. However, a growing body of experimental evidence indicates that coatomers recognize signals in the transported proteins and may participate in concentrative processes as well.

Cytoskeletal Regulation of Membrane Function © 1997 by The Rockefeller University Press

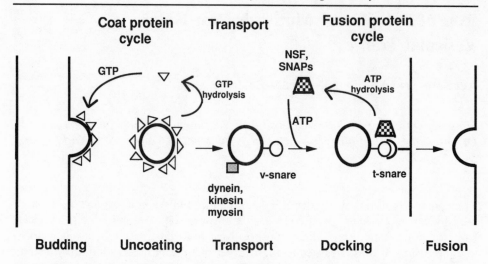

Figure 1. Successive steps in one cycle of vesicular transport. The first step is the bending of the donor membrane by coat/adaptor proteins, which bind in a GTP-dependent fashion. The hydrolysis of GTP leads to the dissociation of the coat/adaptor complex from the vesicle. Motor attachment sites are exposed and the vesicle travels along microfilaments or microtubules. Specific v- and t-SNAREs mediate the docking of the vesicle. NSF and SNAPs direct the fusion with the target membrane.

Cytoplasmic Motors Transport Vesicles in the Exocytic and Endocytic Routes

Cytoplasmic motors are presumed to be essential for vesicular transport, since the tight, gel-like, structure of the cytoplasmic matrix, prevents or greatly impairs the free diffusion of particles the size of transport vesicles (Luby-Phelps et al., 1987). Once the coated vesicle is assembled and released, the coat proteins are removed, a step that usually requires GTP hydrolysis; at this stage, cytoplasmic motors are believed to take over (Fig. 1). The motors may recognize newly exposed components of the transported vesicle, such as the recently identified putative receptor kinectin. After formation of the transport vesicle by the TGN, the sorting compartment of the Golgi complex, microtubule-associated motors are believed to transport carrier vesicles to the cortical region of the cytoplasm. At this point, the microtubules end and the vesicles shift to a different motor, myosin, in order to move across the thick cortical actin network (see below).

The microtubule-associated cytoplasmic motors kinesin and dynein were initially described as the generators of the force responsible for long distance vesicle translocations along the axon of neuronal cells (Craig and Banker, 1994; Mandell and Banker, 1995). Kinesin was identified as motor transporting vesicles forward towards the plus end of microtubules, whereas dynein was responsible for retrograde transport. Whereas in the axon all microtubules are oriented with the plus end pointing forward, in dendrites microtubules have both centrifugal and centripetal orientations, suggesting that both kinesin and dynein are involved in transport to the dendritic terminals. In recently plated, flat epithelial cells, microtubules have the same peripheral plus end distribution observed in fibroblasts or other "nonpolar" cells[1] (Cole and Lippincott-Schwartz, 1995). On the other hand, as epithelial cells form confluent monolayers and segregate their plasma membrane domains, microtubules orient along a vertical axis with the minus end under the apical membrane and the plus end above the basal surface (Nelson, 1991). Many drug-inhibition studies have described the participation of microtubules in apical transport and transcytosis, and a more selective participation of microtubules in only some forms of basolateral transport (Gilbert et al., 1991; Hunziker et al., 1990; Matter et al., 1990; Rindler et al., 1987; van Zeijl and Matlin, 1990). A recent study has suggested that the lack of effect of microtubule-disrupting agents on some forms of basolateral transport may be due to incomplete depolymerization of microtubules (Lafont et al., 1994). The authors report that in SLO-permeabilized MDCK cells, cytosol-dependent transport of the basolateral marker VSV G is decreased by inactivation of kinesin whereas the transport of the apical marker influenza hemagglutinin (HA) is inhibited by inactivation of both kinesin and dynein motors. Several reports have pointed out that microtubule disruption in epithelial cells affects the kinetics of delivery of plasma membrane proteins to the cell surface, but only slightly their steady-state polarized surface distribution, in agreement with the concept that specialized apical and basolateral vesicle docking and fusion systems are ultimately responsible for the correct alignment of vesicle and target membrane (Salas et al., 1986; Gilbert et al., 1991; Hunziker et al., 1990; Matter et al., 1990; van Zeijl and Matlin, 1990).

By comparison with microtubule-associated motors, the role of myosin in post-Golgi transport is less well characterized. Video and electron microscopy of in vitro systems has shown that cytoplasmic vesicles and organelles can bind and move along actin filaments, indicating that some form of myosin might be involved in this process (Fath and Burgess, 1993; Johnston et al., 1991; Kuznetsov et al., 1992; Smith et al., 1995). The myosin family has grown exponentially in the past few years, with approximately 12 myosins grouped in separate classes that differ in their structure and function (Mooseker and Cheney, 1995). Burgess and collaborators have reported the presence of one headed myosin I, as well as dynein and its activator, dynactin, on Golgi membranes, and of myosin I in apical vesicles isolated from intestinal cells (Fath et al., 1994). These authors suggested that myosin might be involved in the translocation of vesicles across the actin-rich terminal web under the apical surface.

A shift from microtubule-associated motor to actin-associated motors at the cell periphery is probably a general property of vesicular transport in all cells. A similar shift, but in reverse, may occur during endocytosis. Myosin may be involved in moving a recently formed endocytic vesicle across the cortical actin network. Exposure to cytochalasin D results in accumulation of clathrin-coated pits selectively at the apical membrane of MDCK cells (Gottlieb et al., 1993) and the accumulation of cytoplasmic vessels in yeast (Kuebler and Riezman, 1993). Microtubule-associated motors participate in vesicular transport between early and late endosomes, or alterna-

---

[1]The quotation marks indicate that we are referring here to the specific type of polarity observed in epithelial and neuronal cells, i.e., the possession of very well defined plasma membrane regions. Fibroblasts and other "nonpolar" cells can show very polarized movement and organization in response to specific stimuli.

tively, according to a different view, in the transport of mature peripheral endosomes into the deeper position characteristic of late endosomes (Matteoni and Kreis, 1987; Gruenberg et al., 1989; Oka and Weigel, 1983). Recently, Durrbach et al. showed the involvement of actin filaments in an early and a late stage of endocytosis and a role for microtubules in an intermediate step (Durrbach et al., 1996).

Role of Motors in the Production of Transport Vesicles

Whereas there is abundant evidence for the involvement of microtubule and microfilament-associated motors in vesicular transport across the cytoplasm, little attention has been paid to potential roles of motors in bud formation and vesicle fission from organelles. Elegant morphological and biochemical work supports the involvement of the GTPase motor dynamin in the fission of clathrin-coated endocytic vesicles from the plasma membrane of synaptic terminals. Dynamin filaments wind around the neck region of clathrin-coated buds, and the fission of the endocytic vesicle requires GTP hydrolysis (Damke et al., 1995; Hinshaw and Schmid, 1995; Takel et al., 1995). Dynamin has been recently found in association with the Golgi apparatus but no function has been assigned to it yet (Henley and McNiven, 1996). Although dynamin is a microtubule-associated motor, there are no reports on the involvement of microtubules in an early step of endocytosis.

On the other hand, there is experimental evidence available that supports a role of actin in an early endocytic step. It is well accepted that the disruption of the cortical actin cytoskeleton is a pre-condition for both an early step in endocytosis and a late step in exocytosis (Bretscher, 1991; Muallem et al., 1995; Vitale et al., 1995). This actin reorganization is thought to be achieved via actin binding proteins that either favor or oppose the polymerization of actin filaments. A role for the actin-based motor myosin in actin reorganization has not been established. There is, however, evidence for the involvement of myosin I in early stages of endocytosis in yeast (Geli and Riezman, 1996). Similarly, in *Dictyostelium*, myosin I is involved in fluid phase pinocytosis and phagocytosis (Jung et al., 1996). Myosin might function in pulling pre-formed vesicles through the cortical actin meshwork towards the cell center. Alternatively, myosin could be involved in an earlier step, such as bending the plasma membrane into a bud or redistributing transmembrane proteins that are required for vesicle formation into the bud.

Myosin II Participates in the Formation of Carrier Vesicles from the TGN

Whereas the participation of both actin and microtubule-associated motors in translocation of Golgi-derived vesicles across the cytoplasm is currently accepted, recent work by our laboratory suggests that motors may be involved in earlier steps in the transport process, such as the production or release of vesicles from the TGN. A peripheral Golgi protein, p200 (Narula et al., 1992), initially presumed to be a coat protein because of its sensitivity to Brefeldin A and regulation by trimeric G proteins, has turned out to be identical to myosin II. Using an in vitro assay that measures the cytosol-dependent release of vesicles carrying VSVG or influenza HA from the TGN of semi-intact MDCK cells or from the TGN of a Golgi-enriched membrane fraction, we found that the release of VSVG, but not HA transport vesicles, was dependent on myosin II and its actin-dependent ATPase activity. When cytosol was immuno-depleted with antibodies against either p200 or myosin II, its ability to promote VSVG vesicle budding was greatly reduced. The release of VSVG transport vesicles was reconstituted when the depleted cytosol was supplemented with purified platelet myosin II or a myosin-enriched fraction from rat liver cytosol. The S1 fragment of myosin II, the COOH-terminal fragment of caldesmon and the myosin ATPase inhibitor butane dione monoxime (BDM) all inhibited the release of VSVG containing transport vesicles from the TGN, as expected for a myosin ATPase activity involved in vesicle formation. The binding of myosin to purified Golgi fractions required actin and additional cytosolic proteins. Recruitment of both myosin and actin from the cytosol was stimulated by aluminum fluoride and GTPγS and inhibited by BFA, suggesting regulation by G-proteins.

Several mechanisms can be envisioned on how myosin could selectively promote the production of basolateral vesicles from the TGN. Basolateral proteins expose well defined basolateral signals on the cytoplasmic domain. These signals could be used to recruit specific adaptor proteins, in a process that may be regulated by trimeric Gαi, since basolateral transport is stimulated by pertussis toxin. Upon reaching a threshold density in the TGN, the basolateral protein may activate a signalling cascade that attracts a GTP-activated protein with affinity for myosin or actin or which activates local enzymes (e.g., phospholipase D) resulting in the generation of membrane binding sites for actin or myosin, a similar scheme as proposed for the ARF-mediated recruitment of coatomers or gamma adaptin to the Golgi complex. Myosin could participate in the generation of the vesicle in a conventional manner, by sliding on actin filaments (Fig. 2 *A*), or unconventionally, as a coat protein, bending the membrane in an ATP dependent manner (Fig. 2 *B*), or by carrying out the fission of the vesicle, as described for dynamin (Fig. 2 *C*). It is clear that other possibilities exist.

It is noteworthy that both spectrin and ankyrin have been detected at the Golgi membrane (Beck et al., 1994; Devarajan et al., 1996). Spectrin shows a dynamic interaction with the Golgi that is reminiscent of the membrane interactions of coat proteins and p200/myosin. This behavior would be expected if a spectrin-based cytoskeleton responsible for the stabilization of the Golgi membrane has to be locally disassembled to allow for vesicle budding. These cytoskeletal rearrangements could be controlled by small G proteins, such as, for example, rab 8 (shown to be involved in basolateral transport) in a similar manner as the cortical cytoskeleton is regulated by rho and rac under conditions of enhanced secretion or endocytosis.

A similar role as we are postulating for myosin in basolateral transport could be envisioned for a MT-based motor in apical vesicle formation. This is suggested by the specific inhibitory effect of tubulin on the release of HA (and not VSVG) transport vesicles in vitro (Mayer et al., 1996). This effect of tubulin might be rationalized as resulting from the removal of MT-binding motors from the cytosol, required for the release of HA containing vesicles at the membrane. It will be interesting to test whether depletion of kinesin and dynein, which have been postulated to mediate TGN to surface transport of VSVG and HA in vitro, has also an effect on vesicle release in the in vitro budding assay.

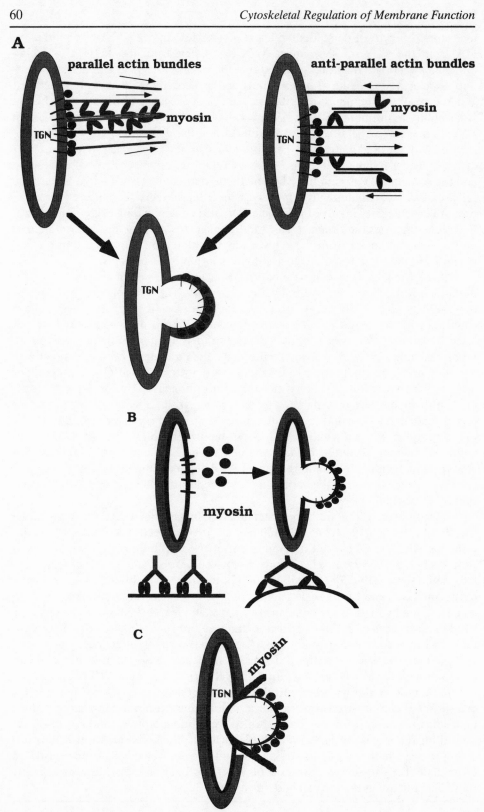

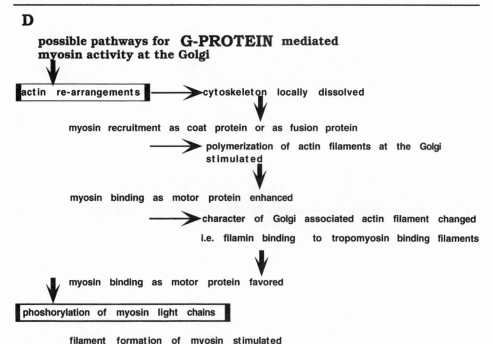

Figure 2. Possible roles for myosin in vesicle formation. (*A*) Classical role for myosin in mediating movements of Golgi-associated actin filaments. Myosin filaments are either tethered to the membrane and walk along parallel actin bundles that are locally associated with the Golgi, causing membrane protrusions (*left*), or small myosin filaments move two anti-parallel actin filaments against each other, causing the bending of the membrane (*right*). In this case, myosin is not associated with the Golgi membrane but with the actin filaments only. (*B*) Myosin as coat protein. Changes in the distance between the two head-groups of a myosin dimer that associate at adjacent sites with the Golgi membrane and/or structural alterations in short myosin filaments, consisting of two dimers with both double headgroups associated with the membrane, cause the curvature of the membrane into a bud. (*C*) Myosin in vesicle fission. In analogy to the MT motor dynamin, myosin ATPase activity might be responsible for vesicle fission. (*D*) Possible mechanisms for G-protein–induced enhancement of Golgi myosin activity. Small G-proteins might either signal polymerization of actin filaments at the site of vesicle formation or induce the rearrangement of existing actin structures at the Golgi, leading to an enhanced binding of myosin. Alternatively, G-proteins might induce a cascade of signal transduction events that lead to the local phosphorylation of myosin light chains and thereby activate myosin ATPase activity and stimulate the formation of myosin filaments at sites of vesicle formation.

Vesicular or Tubular Post-Golgi Traffic?

Two main modes of interorganellar membrane transport have been described: vesicular and tubular. Different views have been offered on the significance of these two transport mechanisms, which propose that either one or the other is the preferred mode of transportation between membrane-bound compartments by the cell. The tubular transport is best understood as a result of the effect of the fungal metabolite Brefeldin A (BFA) (Lippincott-Schwartz et al., 1991). Within minutes of addition to cell cultures BFA causes a dramatic tubulation of the Golgi apparatus and the TGN. The Golgi-derived tubules fuse specifically with ER membranes whereas the TGN tubules show specificity for the endosomal system. The final effect is the absorption of the Golgi complex into the ER and of the TGN into the endosomal system. Tubulation of the Golgi complex observed in the presence of BFA is completely dependent on the presence of intact microtubules (Lippincott-Schwartz et al., 1990). Since BFA prevents the binding of coat proteins to the Golgi or the *trans* Golgi network (TGN), binding sites are likely exposed for microtubule-associated motors, which are then able to pull membrane tubules along MT. Tubulation has been also described in the TGN, by video recording of live cells in which the TGN was labeled by fluorescent ceramides (Cooper et al., 1990). The tubules showed a tendency to fuse with each other, as expected from the behavior of Golgi tubules generated by BFA treatment. As is the case for Golgi tubules, the formation of TGN tubules is completely dependent on the presence of intact microtubules. Finally, similar phenomena have been described upon destabilization of the actin cytoskeleton at the plasma membrane; actin depolymerization promotes tubulation that is dependent on the presence of intact MT (van Deurs et al., 1996).

Could tubular transport be a major mode of transport in the cell? Although the BFA effects on Golgi and TGN demonstrate that tubular transport shows some targeting specificity, large targeting errors of tubular transport have been documented (Apodaca et al., 1993; Low et al., 1992; Wagner et al., 1994). In epithelial cells, anterograde transport of secretory proteins from the TGN to the plasma membrane is not abolished but may be delayed. In addition, there is considerable missorting of apical proteins to the basolateral surface, and polarized recycling of transferrin to the basolateral membrane is completely disrupted (Wan et al., 1992). Tubular transport (at least, as promoted by drug treatment) appears not to be an acceptable mechanism for interorganellar transport. However, it could be an appropriate mechanism to move material when sorting is not required, e.g., within subcompartments of a given organelle. Taylor et al. provided evidence that tubules derived from the *cis* Golgi network fused with the *cis* Golgi cisternae in a cell-free assay designed to study the anterograde transport of VSVG protein in CHO cells (Taylor et al., 1994). Recent biophysical studies on Golgi enzymes coupled to green fluorescent protein, in which a fraction of the cisternae in the Golgi stack were analyzed by photo-bleaching recovery, resulted in the surprising observation that Golgi enzymes appear to diffuse quite rapidly between different cisternae, which would be incompatible with vesicular transport but would support the existence of tubular connections between adjacent cisternae (Cole et al., 1996). Theoretical considerations suggest that a sequential addition of Golgi sugars can proceed without compartmentation of the glycosyl transferases to different Golgi cisternae. Thus, tubular connections between Golgi cisternae may constitute a viable alternative for intra Golgi transport to the more widely publicized coatomer mediated vesicular transport.

Fig. 3 rationalizes the need for coat protein-mediated vesicular transport whenever molecular sorting is required. Coat proteins appear to have the ability to bind to sorting signals in the transported proteins and thus provide a mechanism to concentrate proteins that will be transported and exclude proteins that must remain

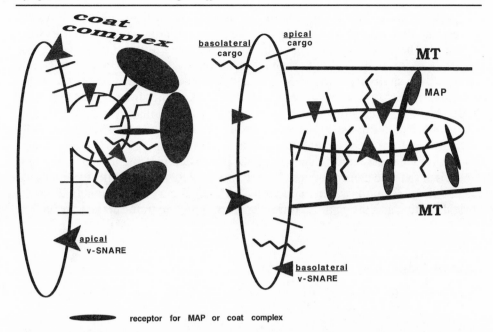

Figure 3. Protein transport via coated vesicles or MT-dependent tubule formation. (*left*) Coat proteins are recruited to membrane receptors that have also affinity for MT associated proteins. Coat assembly results in the sorting and concentration of cargo proteins and their specific v-SNAREs that excludes proteins with other destinations or resident proteins. The vesicle fuses with the target organelle that recognizes its v-SNAREs. (*right*) In the absence of coat proteins, the coat receptor binds to MT-associated motors (MAPs) which pull tubules out of the membrane plane of the organelle. The motors are not able to concentrate and sort proteins and hence the emerging tubule contains undiscriminated cargo protein and v-SNAREs that were present in the donor organelle.

behind in the donor organelle. The process is best understood for chathrin/AP2-mediated concentration of receptor in plasma membrane–coated pits. Recent work has provided evidence for the ability of coatomers to bind to ER recycling signals (KDEL) (Letourneur et al., 1994) and for the concentration of transported proteins during transport between ER and Golgi (Balch et al., 1994). On the other hand, as discussed above, tubules appear to be relatively promiscuous in the selection of a target membrane for fusion. The promiscuous behavior of tubules might be the consequence of a mixed population of v-SNARES (Fig. 3), which can only be segregated by coat proteins.

The Involvement of Motors in Post-Golgi Transport

Whether interorganellar transport is mediated by vesicles or tubules, cytoplasmic motors may play a larger role than envisioned at present. In addition to the well established roles of motors in transporting material between two separate locales in the cell, the evidence discussed above suggests that motors could play a crucial role in the removal of material, tubular or vesicular, from the donor compartment. Motor participation is well documented for the formation of tubules, but recent results suggest that they may also be involved in the generation of vesicles from the Golgi apparatus. The vesicle generating ability of motors may reflect hitherto unknown properties of these molecules. Crystallographic analysis of motors has revealed not only a striking similarity in the tridimensional structure of different motors (e.g., kinesin and dynein), but, also, a remarkable similarity to the structure of G proteins (Vale, 1996). It appears that evolution has utilized very similar strategies in the development of molecular switches and molecular motors. Given this structural similarity, it would not be surprising to find out that motors may perform novel functions in vesicular assembly which could not have been imagined a few years ago. Analysis of the mechanisms of post-TGN transport with the sophisticated biochemical and high resolution video microscopic techniques currently available is bound to yield exciting insights on novel roles of motors in protein traffic in the near future.

Acknowledgments

This work was supported by NIH grant GM34107, by a Revson-Wilson Fellowship to A. Müsch and by an award from the Research to Prevent Blindness to E. Rodriguez-Boulan.

References

Apodaca, G., B. Aroeti, K. Tang, and K.E. Mostov. 1993. Brefeldin-A inhibits the delivery of the polymeric immunoglobulin receptor to the basolateral surface of MDCK cells. *J. Biol. Chem.* 268:20380–20385.

Balch, W.E., J.M. McCaffery, H. Plutner, and M.G. Farquhar. 1994. Vesicular stomatitis virus glycoprotein is sorted and concentrated during export from the endoplasmic reticulum. *Cell.* 76:841–852.

Barlowe, C., L. Orci, T. Yeung, M. Hosobuchi, S. Hamamoto, N. Salama, M.F. Rexach, M. Ravazzola, M. Amherdt, and R. Schekman. 1994. COP II: a membrane coat formed by Sec proteins that drive vesicle budding from the endoplasmic reticulum. *Cell.* 77:895–907.

Beck, K.A., J.A. Buchanan, V. Malhotra, and W.J. Nelson. 1994. Golgi spectrin: Identification of an erythroid beta-spectrin homolog associated with the Golgi complex. *J. Cell Biol.* 127:707–723.

Bretscher, A. 1991. Microfilament structure and function in the cortical cytoskeleton. *Annu. Rev. Cell Biol.* 7:337–374.

Cole, N.B., and J. Lippincott-Schwartz. 1995. Organization of organelles and membrane traffic by microtubules. *Curr. Opin. Cell Biol.* 7:55–64.

Cole, N.B., C.L. Smith, N. Sciaky, M. Terasaki, M. Edidin, and J. Lippincott-Schwartz. 1996. Diffusional mobility of Golgi proteins in membranes of living cells. *Science.* 273:797–801.

Cooper, M.S., A.H. Cornell-Bell, A. Chernjavsky, J.W. Dani, and S.J. Smith. 1990. Tubulovesicular processes emerge from Trans-Golgi cisternae, extend along microtubules, and interlink adjacent Trans-Golgi elements into a reticulum. *Cell.* 61:135–145.

Craig, A.M., and G. Banker. 1994. Neuronal polarity. *Annu. Rev. Neurosci.* 17:267–310.

Damke, H., T. Baba, A.M. van der Bliek, and S.L. Schmid. 1995. Clathrin-independent pinocytosis is induced in cells overexpressing a temperature-sensitive mutant of dynamin. *J. Cell Biol.* 131:69–80.

Devarajan, P., P.R. Stabach, A.S. Mann, T. Ardito, M. Kashgarian, and J.S. Morrow. 1996. Identification of a small cytoplasmic ankyrin (Ank G119) in the kidney and muscle that binds βIΣ* spectrin and associates with the Golgi apparatus. *J. Cell Biol.* 133:819–830.

Durrbach, A., D. Louvard, and E. Coudrier. 1996. Actin filaments facilitate two steps of endocytosis. *J. Cell Sci.* 109:457–465.

Fath, K.R., and D.R. Burgess. 1993. Golgi-derived vesicles from developing epithelial cells bind actin filaments and possess myosin-I as a cytoplasmically oriented peripheral membrane protein. *J. Cell Biol.* 120:117–127.

Fath, K.R., G.M. Timbur, and D.R. Burgess. 1994. Molecular motors are differentially distributed on Golgi membranes from polarized epithelial cells. *J. Cell Biol.* 126:661–675.

Geli, M.I., and H. Riezman. 1996. Role of type I myosins in receptor-mediated endocytosis in yeast. *Science.* 272:533–535.

Gilbert, T., A. Le Bivic, A. Quaroni, and E. Rodriguez-Boulan. 1991. Microtubular organization and its involvement in the biogenetic pathways of plasma membrane proteins in Caco-2 intestinal epithelial cells. *J. Cell Biol.* 113:275–288.

Gottlieb, T.A., I.E. Ivanov, M. Adesnik, and D.D. Sabatini. 1993. Actin filaments play a critical role in endocytosis at the apical but not the basolateral surface of polarized epithelial cells. *J. Cell Biol.* 120:695–710.

Gruenberg, J., G. Griffiths, and K.E. Howell. 1989. Characterization of the early endosome and putative endocytic carrier vesicles in vivo and with an assay of vesicle fusion in vitro. *J. Cell Biol.* 108:1301–1316.

Henley, J.R., and M.A. McNiven. 1996. Association of a dynamin-like protein with the Golgi apparatus in mammalian cells. *J. Cell Biol.* 133:761–775.

Hinshaw, J.E., and S.L. Schmid. 1995. Dynamin self-assembles into rings suggesting a mechanism for coated vesicle budding. *Nature.* 274:190–192.

Hunziker, W., P. Male, and I. Mellman. 1990. Differential microtubule requirements for transcytosis in MDCK cells. *EMBO J.* 9:3515–3525.

Johnston, G.C., J.A. Prendergast, and R.A. Singer. 1991. The Saccharomyces cerevisiae MYO2 gene encodes an essential myosin for vectorial transport of vesicles. *J. Cell Biol.* 113:539–551.

Jung, G., X. Wu, and J.A. Hammer III. 1996. Dictyostelium mutants lacking multiple classic myosin I isoforms reveal combinations of shared and distinct functions. *J. Cell Biol.* 133:305–323.

Ktistakis, N.T., H.A. Brown, M.G. Waters, P.C. Sternweis, and M.G. Roth. 1996. Evidence that phospholipase D mediates ADP Ribosylation factor-dependent formation of Golgi coated vesicles. *J. Cell Biol.* 134:295–306.

Kuebler, E., and H. Riezman. 1993. Actin and fimbrin are required for the internalization step of endocytosis in yeast. *EMBO J.* 12:2855–2862.

Kuznetsov, S.A., G.M. Langford, and D.G. Weiss. 1992. Actin-dependent organelle movement in squid axoplasm. *Nature.* 356:722–725.

Lafont, F., J. Burkhardt, and K. Simons. 1994. Involvement of microtubule motors in basolateral and apical transport in kidney cells. *Nature.* 372:801–803.

Letourneur, F., E.C. Gaynor, S. Hennecke, C. Demolliere, R. Duden, S. Emr, H. Riezman, and P. Cosson. 1994. Coatomer is essential for retrieval of dilysine-tagged proteins to the endoplasmic reticulum. *Cell.* 79:1199–1207.

Lippincott-Schwartz, J., J.G. Donaldson, A. Schweizer, E.G. Berger, H.P. Hauri, L.C. Yuan, and R.D. Klausner. 1990. Microtubule-dependent retrograde transport of proteins into the ER in the presence of Brefeldin A suggests an ER recycling pathway. *Cell.* 60:821–836.

Lippincott-Schwartz, J., L. Yuan, C. Tipper, M. Amherdt, L. Orci, and R.D. Klausner. 1991. Brefeldin A's effects on endosomes, lysosomes, and the TGN suggest a general mechanism for regulating organelle structure and membrane trafficking. *Cell.* 67:601–616.

Low, S.H., B.L. Tang, S.H. Wong, and W. Hong. 1992. Selective inhibition of protein targeting to the apical domain of MDCK cells by brefeldin A. *J. Cell Biol.* 118:51–62.

Luby-Phelps, K., P.E. Castle, D.L. Taylor, and F. Lanni. 1987. Hindered diffusion of inert tracer particles in the cytoplasm of mouse 3T3 cells. *Proc. Natl. Acad. Sci. USA.* 84:4910–4913.

Mandell, J.W., and G.A. Banker. 1995. The microtubule cytoskeleton and the development of neuronal polarity. *Neurobiol. Aging.* 16:229–238.

Matteoni, R., and T.E. Kreis. 1987. Translocation and clustering of endosomes and lysosomes depends on microtubules. *J. Cell Biol.* 105:1253–1265.

Matter, K., K. Bucher, and H.P. Hauri. 1990. Microtubule perturbation retards both the direct and the indirect apical pathway but does not affect sorting of plasma membrane proteins in intestinal epithelial cells (Caco-2). *EMBO J.* 9:3163–3170.

Mayer, A., I.E. Ivanov, D. Gravotta, M. Adesnik, and D.D. Sabatini. 1996. Cell-free reconstitution of the transport of viral glycoproteins from the TGN to the basolateral plasma membrane of MDCK cells. *J. Cell Sci.* 109:1667–1676.

Mooseker, M.S., and R.E. Cheney. 1995. Unconventional myosins. *Annu. Rev. Cell. Dev. Biol.* 11:633–675.

Muallem, S., K. Kwiatkowska, X. Xu, and H.L. Yin. 1995. Actin filament disassembly is a sufficient final trigger for exocytosis in nonexcitable cells. *J. Cell Biol.* 128:589–598.

Narula, N., I. McMorrow, G. Plopper, J. Doherty, K.S. Matlin, B. Burke, and J.L. Stow. 1992. Identification of a 200kD, Brefeldin-sensitive protein on Golgi membranes. *J. Cell Biol.* 114:1113–1124.

Nelson, W.J. 1991. Cytoskeleton functions in membrane traffic in polarized epithelial cells. *Semin. Cell Biol.* 2:375–385.

Oka, J.A., and P.H. Weigel. 1983. Microtubule-depolymerizing agents inhibit asialo-orosomucoid delivery to lysosomes but not its endocytosis or degradation in isolated rat hepatocytes. *Biochim. Biophys. Acta.* 763:368–376.

Orci, L., D.J. Palmer, M. Ravazzola, A. Perrelet, M. Amherd, and J.E. Rothman. 1993. Budding from Golgi membranes requires coatomer complex of non-clathrin coated vesicles. *Nature.* 362:648–652.

Pearse, B.M., and M.S. Robinson. 1990. Clathrin, adaptors, and sorting. *Annu. Rev. Cell Biol.* 6:151–171.

Rindler, M.J., I.E. Ivanov, and D.D. Sabatini. 1987. Microtubule-acting drugs lead to the nonpolarized delivery of the influenza hemagglutinin to the cell surface of polarized Madin-Darby canine kidney cells. *J. Cell Biol.* 104:231–241.

Rothman, J.E. 1994. Mechanisms of intracellular protein transport. *Nature.* 372:55–63.

Rothman, J.E., and F.T. Wieland. 1996. Protein sorting by transport vesicles. *Science.* 272:227–234.

Salas, P.J., D.E. Misek, D.E. Vega Salas, D. Gundersen, M. Cereijido, and E. Rodriguez-Boulan. 1986. Microtubules and actin filaments are not critically involved in the biogenesis of epithelial cell surface polarity. *J. Cell Biol.* 102:1853–1867.

Schekman, R., and L. Orci. 1996. Coat proteins and vesicle budding. *Science.* 271:1526–1533.

Smith, M.G., V.R. Simon, H. O'Sullivan, and L.A. Pon. 1995. Organelle-cytoskeletal interactions: actin mutations inhibit meiosis-dependent mitochondrial rearrangement in the budding yeast Saccharomyces cerevisiae. *Mol. Biol. Cell.* 6:1381–1396.

Takel, K., P.S. McPherson, S.L. Schmid, and P. De Camilli. 1995. Tubular membrane invaginations coated by dynamin rings are induced by GTP-gamma S in nerve terminals. *Nature.* 374:186–190.

Taylor, T.C., M. Kanstein, P. Weidman, and P. Melancon. 1994. Cytosolic ARFs are required for vesicle formation but not for cell-free intra-Golgi transport: evidence for coated vesicle-independent transport. *Mol. Biol. Cell.* 5:237–252.

Vale, R.D. 1996. Switches, latches and amplifiers: common themes of G-proteins and molecular motors. *J. Cell Biol.* 135:291–302.

van Deurs, B., F. von Buelow, P.K. Holm, and K. Sandvig. 1996. Destabilization of plasma membrane structure by prevention of actin polymerization. *J. Cell Sci.* 109:1655–1665.

van Zeijl, M.J.A.H., and K.S. Matlin. 1990. Microtubule perturbation inhibits intracellular transport of an apical membrane glycoprotein in a substrate-dependent manner in polarized Madin-Darby canine kidney epithelial cells. *Cell Regul.* 1:921–936.

Vitale, M.L., E.P. Seward, and J.M. Trifaro. 1995. Chromaffin cell cortical actin network dynamics control the size of the release-ready vesicle pool and the initial rate of exocytosis. *Neuron.* 14:353–363.

Wagner, M., A.K. Rajasekaran, D.K. Hanzel, S. Mayor, and E. Rodriguez-Boulan. 1994. Brefeldin A causes structural and functional alterations of the trans-Golgi network of MDCK cells. *J. Cell Sci.* 107:933–943.

Wan, J., M.E. Taub, D. Shah, and W.C. Shen. 1992. Brefeldin A enhances receptor-mediated transcytosis of transferrin in filter-grown Madin-Darby canine kidney cells. *J. Biol. Chem.* 267:13446–13450.

Whitney, J.A., M. Gomez, D. Sheff, T.E. Kreis, and I. Mellman. 1995. Cytoplasmic coat proteins involved in endosome fusion. *Cell.* 83:703–713.

Molecular Architecture of Tight Junctions: Occludin and ZO-1

Shoichiro Tsukita, Mikio Furuse, and Masahiko Itoh

Department of Cell Biology, Faculty of Medicine, Kyoto University, Yoshida-Konoe, Sakyo-ku, Kyoto 606, Japan

Dual Roles of Tight Junctions

In epithelial and endothelial cells, the tight junction seals cells to create a primary barrier to the diffusion of solutes across the cell sheet, and it also works as a boundary between the apical and basolateral membrane domains to create their polarization (Schneeberger and Lynch, 1992; Gumbiner, 1987, 1993). The occurrence of tight junctions is thus essential for epithelial and endothelial cells to exert their various physiological functions.

In thin-section electron microscopy, tight junctions appear as a series of discrete sites of apparent fusion, involving the outer leaflet of the plasma membrane of adjacent cells (Farquhar and Palade, 1963). In freeze-fracture electron microscopy, this junction appears as a set of continuous, anastomosing intramembrane strands or fibrils in the P-face (the outwardly facing cytoplasmic leaflet) with complementary grooves in the E-face (the inwardly facing extracytoplasmic leaflets) (Staehelin, 1974). It has remained controversial whether the strands are predominantly lipid in nature, that is, cylindrical lipid micelles, or represent linearly aggregated integral membrane proteins (Pinto da Silva and Kachar, 1982; Kachar and Reese, 1982). However, given the detergent stability of tight junction strands visualized by negative staining (Stevenson and Goodenough, 1984) and freeze fracture (Stevenson et al., 1988), it was thought to be unlikely that these elements are composed solely of lipids.

ZO-1: A Peripheral Membrane Protein Localizing at Tight and Adherens Junctions

ZO-1 was defined as an antigen for monoclonal antibodies that recognize tight junctions in epithelial cells (Stevenson et al., 1986). It is a peripheral membrane protein with a molecular mass of 220 kD underlying the cytoplasmic surface of plasma membranes. In epithelial cells, ZO-1 is exclusively localized at tight junctions, with the exception that it is highly concentrated at the undercoat of plasma membranes of the slit diaphragm in kidney glomeruli. In nonepithelial cells such as cardiac muscle cells and fibroblasts, ZO-1 is precisely colocalized with cadherins to form adherens junctions (Itoh et al., 1991, 1993; Howarth et al., 1992). The molecular mechanism of this peculiar behavior of ZO-1 remains elusive.

The structure of ZO-1 has been analyzed by cDNA cloning and sequencing (Itoh et al., 1993; Tsukita et al., 1993; Willott et al., 1993). The amino-terminal half of this molecule displays significant similarity to the product of the lethal(1)discs large-1 (dlg) gene in *Drosophila*. The dlg gene product (and also the amino-terminal half of ZO-1) contains putative functional domains. From the amino terminus, there are filamentous, SH3, and guanylate kinase domains. The amino-terminal filamentous domain contains three internal repeats, which are called PDZ repeats. Various proteins localizing just beneath plasma membranes are now found to contain PDZ repeats, indicating a gene family called MAGUK (membrane-associated guanylate kinase homologues) family.

In addition to ZO-1, other peripheral proteins have been identified to be localized at tight junctions. ZO-2 with a molecular mass of 160 kD was identified as a ZO-1–binding protein by immunoprecipitation (Gumbiner et al., 1991). This molecule also shows sequence similarity to the dlg product, indicating that it is a member of the MAGUK family (Jesaitis and Goodenough, 1994). Furthermore, monoclonal antibodies identified two other tight junction-specific peripheral membrane proteins named cingulin and 7H6 antigen (Citi et al., 1988; Zhong et al., 1993). Despite intensive studies, the functions of these tight junction-associated peripheral membrane proteins totally remain elusive, although it is presumed that they are involved in the formation, maintenance, and regulation of tight junctions.

Identification of Occludin: An Integral Membrane Protein Localizing at Tight Junctions

To clarify the structure and functions of tight junctions at the molecular level, an integral membrane protein working at tight junctions had to be identified. However, this integral membrane component remained elusive for quite some time. We established a procedure for isolating cadherin-based cell-to-cell adherens junctions from the rat liver (Tsukita and Tsukita, 1989). During the course of identification of novel proteins enriched in this fraction, we found that ZO-1 was also concentrated, indicating that not only adherens, but also tight junctions are enriched in this fraction (Itoh et al., 1993). We then attempted to identify the putative integral membrane protein localizing at tight junctions using this isolated junctional fraction. To obtain a powerful antigen, we isolated the junctional fraction from chicken liver and injected it into rats to generate mAbs.

Three mAbs which are specific for an ~65-kD protein were obtained (Furuse et al., 1993). This antigen was not extractable from plasma membranes without detergent, suggesting that it is an integral membrane protein. Immunofluorescence and immunoelectron microscopy with these mAbs showed that this ~65-kD membrane protein is exclusively localized at tight junctions of both epithelial and endothelial cells (Fig. 1). Labels were detected directly over the points of membrane contact in tight junctions by ultra-thin section electron microscopy, and directly over the intramembranous particle strands of tight junctions by immuno-freeze fracture electron microscopy (Fujimoto, 1995). We therefore concluded that these mAbs recognize an integral membrane protein localizing at tight junctions. This antigen was then designated occludin from the Latin word "occlude."

To further clarify the nature and structure of chick occludin, we cloned and sequenced its cDNA. We found that it encoded a 504-amino acid polypeptide with a calculated molecular mass of 56 kD. A search of the data base identified no proteins with significant homology to occludin. A most striking feature of its primary structure was revealed by a hydrophilicity plot (see Fig. 3). (*1*) In the amino-terminal half, occludin contains four transmembrane domains. (*2*) A carboxyl-terminal half consisting of ~250 amino acid residues resides in the cytoplasm. (*3*) Charged amino acids mostly locate in the cytoplasm. (*4*) The content of tyrosine and glycine residues is very high in the extracellular domains.

Compared with adhesion molecules working at other intercellular junctions such as adherens junctions and desmosomes, those at tight junctions should be structurally and functionally unique. They must tightly obliterate the intercellular space for the barrier function in epithelial and endothelial cell sheets, and form a continuous strand within the membrane to form a fence against membranous lipids and proteins. To determine whether or not occludin fulfills these criteria for tight junction adhesion molecules, chicken occludin was overexpressed in insect Sf9 cells by recombinant baculovirus infection (Furuse et al., 1996). Most of the overexpressed occludin molecules did not appear on the cell surface but were concentrated in peculiar multilamellar structures in the cytoplasm. Thin section electron

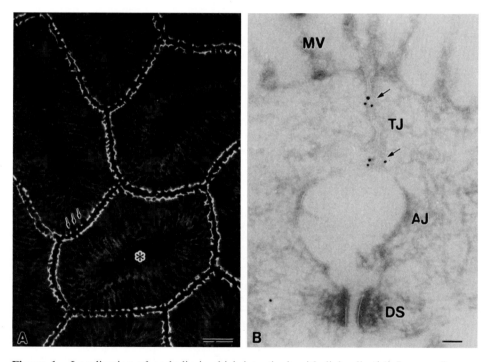

Figure 1. Localization of occludin in chick intestinal epithelial cells. (*A*) Immunofluorescence staining of a frozen section with anti–chick occludin mAb. Intestinal villi are cut transversely, and the connective tissue core of each villus (*) is covered with simple columnar epithelium. Occludin is concentrated at the most apical region of lateral membranes of epithelial cells (*arrows*). Bar, 20 μm. (*B*) Ultrathin cryosections of formalin-fixed intestinal epithelial cells were labeled with anti–chick occludin mAb. Gold particles accumulated at the TJ region (*TJ*) and are hardly detected in the AJ (*AJ*) and DS (*DS*) regions. *MV*, microvilli. Bar, 100 nm.

microscopy revealed that each lamella was transformed from intracellular membranous cisternae in which the luminal space was completely collapsed. Furthermore, the outer leaflets of opposing membranes in each lamella appeared to be fused with no gaps, like tight junctions. Short tight junction-like intramembranous particle strands were occasionally observed in freeze-fracture replicas of these multilamellar structures, which were specifically labeled by an anti-occludin mAb. These findings favor the notion that occludin is one of the major cell adhesion molecules in tight junctions.

Considering that tight junctions play a key role as a barrier in endothelial cells (and also in epithelial cells), it would be important in cell biological as well as in medical research to analyze the expression and localization of occludin in various pathological states of human samples and to modulate the functions of occludin not only at the cell culture but also at the whole body levels. As described above, occludin was identified in the chicken, and none of our mAbs and pAbs raised against chicken occludin crossreacted with the murine and human homologues (Furuse et al., 1993). Several investigators, including ourselves, have attempted to isolate cDNA encoding mammalian homologues, based upon the assumption that the occludin amino acid sequence is rather evolutionarily conserved due to its functional importance. However, these efforts have only recently been successful.

During our attempts to identify the mammalian homologues of occludin, we learned from the GenBank database, using the biological sequence search program, Mpsrch, that a 675-nucleotide sequence showing similarity to part of the carboxyl-terminal domain of chicken occludin had been found in close proximity to the human neuronal apoptosis inhibitory protein gene (Roy et al., 1995). To determine whether or not this sequence really encodes part of the human homologue of occludin, we performed PCR with two oligonucleotides as primers, using an expression cDNA library of human cultured cells. We obtained a DNA fragment that allowed us to isolate a full-length cDNA encoding human occludin (Ando-Akatsuka et al., 1996). The cDNAs encoding murine and canine occludin homologues were also isolated and sequenced by the same procedure. The amino acid sequences of the three mammalian (human, murine, and canine) occludins are closely related (~90% identity), whereas they diverged considerably from those of chicken (~50% identity). However, the hydrophylicity profile of chicken occludin is highly conserved in human occludin (Fig. 2).

Molecular Architecture of Tight Junctions

The identification of occludin allowed us to analyze the detailed molecular architecture of tight junctions, in terms of the interaction of occludin with tight junction peripheral proteins at the molecular level. Based on distance from the plasma membrane, tight junction peripheral proteins can be subclassified into two categories. The first class includes ZO-1 and ZO-2, which are localized in the immediate vicinity of plasma membranes. The second class includes cingulin and the 7H6 antigen, which are localized more than 40 nm from the plasma membranes. We showed that the GST-fusion protein of the carboxyl-terminal half cytoplasmic domain of occludin specifically associates with 220- and 160-kD bands among the various membrane peripheral proteins in the extract from the isolated junction fraction

(Furuse et al., 1994). These two bands were identified as ZO-1 and ZO-2, respectively, by immunoblotting and the domain responsible for this binding was narrowed down to the carboxyl-terminal ~150 amino acid sequence of occludin. Furthermore, using the various recombinant ZO-1 and ZO-2 proteins produced by recombinant baculovirus infection, we found that the amino-terminal halves of ZO-1 and ZO-2 directly and independently bound to the cytoplasmic domain of occludin (Itoh et al., manuscript in preparation). Spectrin tetramers are associated with ZO-1 at ~10–20 nm from their midpoint (Itoh et al., 1991), and spectrin was specifically trapped in a column containing the GST-occludin fusion protein (Furuse et al., 1994). Thus there may be a molecular linkage between occludin and actin filaments as shown in Fig. 3, since an intimate spatial relationship between tight junctions and actin-based cytoskeletons has been observed (Madara, 1987). It remains elusive

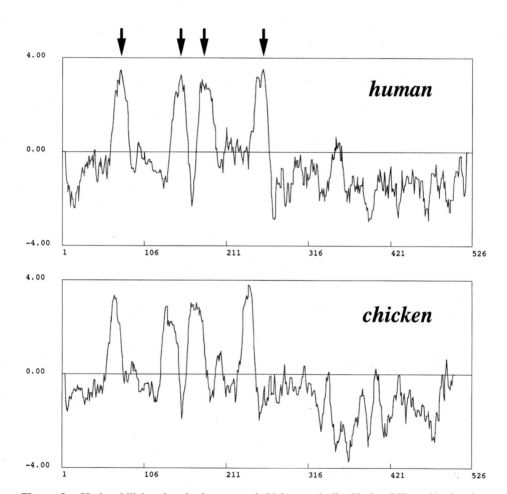

Figure 2. Hydrophilicity plots for human and chicken occludin. Hydrophilic and hydrophobic residues are in the lower and upper part of the frames, respectively. The axis is numbered in amino acid residues. At the amino-terminal half of each occludin, there are four major hydrophobic, potentially membrane-spanning regions (*arrows*).

where other tight junction peripheral proteins such as cingulin and the 7H6 antigen should be placed in this scheme, how the integrity of this molecular organization is regulated, and how this molecular organization is important for the functions of tight junctions.

Perspective

Now that the mammalian occludins have been identified, the organization and function of tight junctions can be structurally and functionally examined at the molecular level. Using various types of cultured human, murine and canine (MDCK) cells, the barrier function of tight junctions and the regulation mechanisms involved can be experimentally analyzed by modulating occludin gene expression or by blocking it with anti-sense probes or with antibodies. For example, it can now be determined whether upon the overexpression of occludin cDNA, the number of tight junctions strands, as seen in freeze-fracture replicas, will increase, with concomitant upregulation of the barrier function. Through the production of various transgenic and occludin gene knock-out mice, we will learn how tight junction formation is involved in the morphogenesis of various organs and whether or not tight junction dysfunction is related to various pathological states such as inflammation and tumor metastasis. The modulation of tight junction functions, especially its barrier function, is also interesting in relation to drug delivery.

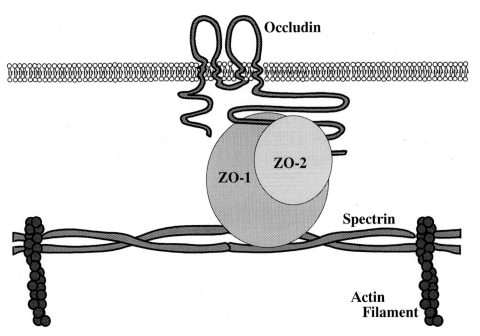

Figure 3. Schematic drawing of the possible molecular architecture of tight junctions. Both ZO-1 and ZO-2 are directly associated with occludin.

References

Ando-Akatsuka, Y., M. Saitou, T. Hirase, M. Kishi, A. Sakakibara, M. Itoh, S. Yonemura, M. Furuse, and S. Tsukita. 1996. Iterspecies diversity of the occludin sequence: cDNA cloning of human, mouse, dog, and rat-kangaroo homologues. *J. Cell Biol.* 133:43–47.

Citi, S., H. Sabanay, R. Jakes, B. Geiger, and J. Kendrick-Jones. 1988. Cingulin, a new peripheral component of tight junctions. *Nature.* 33:272–276.

Farquhar, M.G., and G.E. Palade. 1963. Junctional complexes in various epithelia. *J. Cell Biol.* 17:375–409.

Fujimoto, K. 1995. Freeze-fracture replica electron microscopy combined with SDS digestion for cytochemical labeling of integral membrane proteins. Application to the immunogold labeling of intercellular junctional complexes. *J. Cell Sci.* 108:3443–3449.

Furuse, M., K. Fujimoto, N. Sato, T. Hirase, S. Tsukita, and S. Tsukita. 1996. Overexpression of occludin, a tight junction-associated integral membrane protein, induces the formation of intracellular multilamellar bodies bearing tight junction-like structures. *J. Cell Sci.* 109:429–435.

Furuse, M., T. Hirase, M. Itoh, A. Nagafuchi, S. Yonemura, S. Tsukita, and S. Tsukita. 1993. Occludin: a novel integral membrane protein localizing at tight junctions. *J. Cell Biol.* 123:1777–1788.

Furuse, M., M. Itoh, T. Hirase, A. Nagafuchi, S. Yonemura, S. Tsukita, and S. Tsukita. 1994. Direct association of occludin with ZO-1 and its possible involvement in the localization of occludin at tight junctions. *J. Cell Biol.* 127:1617–1626.

Gumbiner, B. 1987. Structure, biochemistry, and assembly of epithelial tight junctions. *Am. J. Physiol.* 253:C749–C758.

Gumbiner, B. 1993. Breaking through the tight junction barrier. *J. Cell Biol.* 123:1631–1633.

Gumbiner, B., T. Lowenkopf, and D. Apatira. 1991. Identification of a 160kDa polypeptide that binds to the tight junction protein ZO-1. *Proc. Natl. Acad. Sci. USA.* 88:3460–3464.

Howarth, A.G., M.R. Hughes, and B.R. Stevenson. 1992. Detection of the tight junction-associated protein ZO-1 in astrocytes and other nonepithelial cell types. *Am. J. Physiol.* 262:C461–C469.

Itoh, M., A. Nagafuchi, S. Yonemura, T. Kitani-Yasuda, S. Tsukita, and S. Tsukita. 1993. The 220-kD protein colocalizing with cadherins in non-epithelial cells is identical to ZO-1, a tight junction-associated protein in epithelial cells: cDNA cloning and immunoelectron microscopy. *J. Cell Biol.* 121:491–502.

Itoh, M., S. Yonemura, A. Nagafuchi, S. Tsukita, and S. Tsukita. 1991. A 220-kD undercoat-constitutive protein: Its specific localization at cadherin-based cell-cell adhesion sites. *J. Cell Biol.* 115:1449–1462.

Jesaitis, L.A., and D.A. Goodenough. 1994. Molecular characterization and tissue distribution of ZO-2, a tight junction protein homologous to ZO-1 and the *Drosophila* discs-large tumor suppresser protein. *J. Cell Biol.* 124:949–961.

Kachar, B., and T.S. Reese. 1982. Evidence for the lipidic nature of tight junction strands. *Nature.* 296:464–466.

Madara, J.L. 1987. Tight junction dynamics: is paracellular transport regulated? *Cell.* 53:497–498.

Pinto da Silva, P., and B. Kachar. 1982. On tight junction structure. *Cell.* 28:441–450.

Roy, N., M.S. Mahadevan, M. McLean, G. Shutler, Z. Yaraghi, R. Farahani, S. Baird, A. Besner-Jonston, C. Lefebvre, X. Kang et al. 1995. The gene for neuronal apoptosis inhibitory protein is partially deleted in individuals with spinal muscular atrophy. *Cell.* 80:167–178.

Schneeberger, E.E., and R.D. Lynch. 1992. Structure, function, and regulation of cellular tight junctions. *Am. J. Physiol.* 262:L647–L661.

Staehelin, L.A. 1974. Structure and function of intercellular junctions. *Int. Rev. Cytol.* 39:191–282.

Stevenson, B.R., J.M. Anderson, and S. Bullivant. 1988. The epithelial tight junction: structure, function and preliminary biochemical characterization. *Mol. Cell. Biochem.* 83:129–145.

Stevenson, B.R., and D.A. Goodenough 1984. Zonula occludentes in junctional complex-enriched fractions from mouse liver: preliminary morphological and biochemical characterization. *J. Cell Biol.* 98:1209–1221.

Stevenson, B.R., J.D. Siliciano, M.S. Mooseker, and D.A. Goodenough. 1986. Identification of ZO-1: a high molecular weight polypeptide associated with the tight junction (zonula occludens) in a variety of epithelia. *J. Cell Biol.* 103:755–766.

Tsukita, S., M. Itoh, A. Nagafuchi, S. Yonemura, and S. Tsukita. 1993. Submembranous junctional plaque proteins include potential tumor suppresser molecules. *J. Cell Biol.* 123:1049–1053.

Tsukita, S., and S. Tsukita. 1989. Isolation of cell-to-cell adherens junctions from rat liver. *J. Cell Biol.* 108:31–41.

Willott, E., M.S. Balda, A.S. Fanning, B. Jameson, C. Van Itallie, and J.M. Anderson. 1993. The tight junction protein ZO-1 is homologous to the *Drosophila* discs-large tumor suppresser protein of septate junctions. *Proc. Natl. Acad. Sci. USA.* 90:7834–7838.

Zhong, Y., T. Saitoh, T. Minase, N. Sawada, K. Enomoto, and M. Mori. 1993. Monoclonal antibody 7H6 reacts with a novel tight junction-associated protein distinct from ZO-1, cingulin and ZO-2. *J. Cell Biol.* 120:477–483.

Chapter 3

Spectrin and Associated Proteins

Capping Actin Filament Growth:
Tropomodulin in Muscle and Nonmuscle Cells

Velia M. Fowler

Department of Cell Biology, The Scripps Research Institute,
La Jolla, California 92037

Actin filament lengths are precisely regulated and very stable in the sarcomeres of striated muscle, in the erythrocyte membrane skeleton, and in cell protrusions such as microvilli in intestinal epithelial cells and stereocilia in hair cells of the inner ear. In contrast, in motile cells, actin filament lengths are dynamically regulated when cells extend lamellipodia. Control of actin filament lengths and dynamics in cells is expected to be achieved in part by capping proteins that prevent filament growth or shrinkage by blocking subunit exchange at both the fast growing (barbed) and slow growing (pointed) filament ends. Much is known about how barbed end capping proteins control actin filament assembly and length in many cells, but little is known about the significance of regulating actin filament assembly at the pointed end. Tropomodulin is the only known capping protein for the pointed ends of actin filaments and is a ~40-kD protein that is expressed in erythrocytes, striated muscle, lens fiber cells, some regions of the adult brain, as well as sensory neurons and epithelial cells of the inner ear. A related isoform (59% identical at the protein level) is expressed principally in neurons of both embryonic and adult brain. In striated muscle and in erythrocytes, tropomodulin is tightly associated with the actin filament pointed ends where it functions to maintain actin filament length in vivo (for recent reviews, see Fowler, 1996; Gregorio and Fowler, 1996). Unlike proteins that cap actin filament barbed ends, tropomodulin also binds tropomyosin and requires tropomyosin for tight capping of actin filament pointed ends. Mapping of functional domains on tropomodulin shows that the COOH-terminal end of tropomodulin is important for actin filament pointed end capping activity while the NH_2-terminal portion of tropomodulin contains the tropomyosin binding domain. Searches of protein and EST databases for tropomodulin-like sequences reveal a number of proteins with homologies to both the tropomyosin binding and the actin filament capping portions of tropomodulin. In particular, we have identified a tropomodulin-like 64-kD protein that is principally expressed in smooth muscle cells. We anticipate that tropomodulin and this 64-kD protein are members of a larger family of tropomyosin and actin binding proteins that are responsible for capping actin filament pointed ends and regulating actin filament lengths in muscle and nonmuscle cells.

Introduction

Actin filament lengths are precisely determined and stably maintained in a variety of elaborate supramolecular structures in differentiated cells. Some striking exam-

ples of these specialized actin filament cytoskeletons include the sarcomeres of striated muscle (Huxley, 1963), the membrane skeleton of erythrocytes (Lux and
Palek, 1995), and cell protrusions such as microvilli in intestinal epithelial cells (Hirokawa et al., 1982), stereocilia in the hair cells of the inner ear (Tilney et al.,
1992a), and cuticular bristles in *Drosophila melanogaster* (Tilney et al., 1996). Measurements of actin filament lengths in these specialized cytoskeletons reveal both
the incredible precision with which the lengths of individual filaments are regulated
in each structure, as well as the diversity of filament lengths among different structures. For example, the long actin filaments in the sarcomeres of skeletal muscle are
all 1.1 ± 0.025 μm long (Sosa et al., 1994) whereas the short actin filaments in the
red blood cell membrane skeleton are less than 1/20th as long (33 ± 5 nm) (Shen et
al., 1986; but see Fowler, 1996) (Fig. 1). These narrow filament length distributions
contrast dramatically with the exponential length distribution of pure actin filaments polymerized in vitro, which is a consequence of stochastic monomer associa-

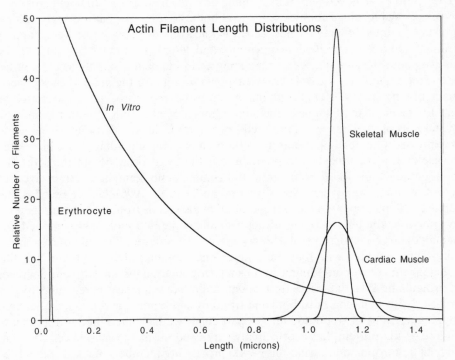

Figure 1. Actin filament length distributions for pure actin filaments polymerized in vitro
compared to actin filament length distributions for spread human erythrocyte membrane
skeletons, thin filaments in glycerinated rabbit psoas muscle, and thin filaments in frog cardiac (atrial) muscle. The length distributions were plotted with the Microsoft® Excel soft-
ware, using an exponential distribution function for pure actin filaments: $F(x) = n\lambda e^{-\lambda x}$,
where $n = 25,000$ and $\lambda = 0.0025$ (Oosawa, 1975) and using a normal gaussian distribution
function for actin filaments in erythrocytes, skeletal and cardiac muscle: $(F(x) = [n/(\sqrt{2\pi}\sigma)]e -
[(x - \mu)^2/2\sigma^2]$, where $n = 150$, $\mu = 33$ nm, $\sigma = 5$ nm for erythrocytes (Shen et al., 1986), $n =
3000$, $\mu = 1.11$ μm, $\sigma = 0.025$ μm for skeletal muscle (Sosa et al., 1994), and $n = 3000$, $\mu =
1.11$ μm, $\sigma = 0.075$ μm for cardiac muscle (see Fig. 2A in Robinson and Winegrad, 1979).

tion and dissociation events at both filament ends at steady state (Oosawa, 1975) (Fig. 1). Even in cardiac muscle where the filament lengths are not as tightly regulated as in skeletal muscle, the range of lengths is still relatively narrow and the shape of the filament length distribution does not at all resemble that of pure actin filaments (e.g., $\sim$1.1 $\pm$ 0.075 μm for frog atrial muscle; Robinson and Winegrad, 1979) (Fig. 1).

In both skeletal and cardiac muscle, maintenance of actin filament length is critical for efficient interaction of thin and thick filaments and is an important determinant of the length-tension relationship during contraction, thus influencing the physiological properties of the muscle (Robinson and Winegrad, 1979). In erythrocytes, actin filament length is likely to influence the connectivity of the spectrin-actin network and to be critical for maintaining cell deformability and survival in the circulation (Lux and Palek, 1995). In microvilli, stereocilia, and *Drosophila* bristles, the actin filament bundles appear to provide structural support for the cell protrusions and are also important for their morphogenesis during development (Mooseker, 1985; Tilney et al., 1992a; Tilney et al., 1995; Tilney et al., 1997).

These diverse yet highly restricted filament length distributions are thought to be achieved in part by the regulated action of actin filament capping proteins that bind to both ends of the actin filaments and block monomer association and dissociation, thus preventing filaments from stochastically growing or shrinking. Additionally, in the cases in which actin filament lengths are extremely tightly regulated (e.g., erythrocytes and skeletal muscle; see Fig. 1), template or vernier proteins might exist to specify the characteristic filament lengths in each cell type (for a discussion of this topic, see Fowler, 1996). Many proteins that cap the fast growing (barbed) filament ends have been described in both muscle and nonmuscle cells (Weeds and MacIver, 1993; Schafer and Cooper, 1995). For example, capZ ($\alpha\beta1$) is the capping protein for the barbed ends of the thin filaments in striated muscle, and a related isoform ($\alpha\beta2$) caps some of the actin filaments in nonmuscle cells (for a review, see Schafer and Cooper, 1995). However, capping only one end of the filament is not in itself sufficient to alter the steady-state filament length distribution from the exponential distribution observed for uncapped filaments, since monomers can still associate and dissociate stochastically from the remaining uncapped end (e.g., Burlacu et al., 1992). Thus, a mechanism to block monomer exchange at the slow growing (pointed) filament is required. In this review, I will focus on the domain structure and isoforms of tropomodulin, a highly conserved, $\sim$40-kD tropomyosin and actin binding protein which caps the pointed ends of the actin filaments in erythrocytes and striated muscle (Fig. 2) (for recent reviews, see Fowler, 1996; Gregorio and Fowler, 1996). Tropomodulin is the only pointed end capping protein yet identified and is therefore likely to play an important role in regulating actin filament length and assembly (Weber et al., 1994; Coluccio, 1994).

Tropomodulin Functional Domains

Tropomodulin was originally discovered as a tropomyosin binding protein in the erythrocyte membrane skeleton (Fowler, 1987) and only much later was shown to have actin filament pointed end capping activity (Weber et al., 1994). The tropomyosin binding region of tropomodulin has been mapped to the NH$_2$-terminal portion of the molecule using competition binding assays with recombinant tropomodulin fragments (Babcock and Fowler, 1994). An NH$_2$-terminal fragment containing amino acids 6–184 binds as well to tropomodulin as does the full length molecule. Curiously, human erythrocyte and chicken skeletal muscle tropomodulins (see below) appear to bind equally well to both nonmuscle (erythrocyte) and muscle (skeletal) tropomyosins (Babcock and Fowler, 1994). This was initially puzzling since erythrocyte tropomodulin binds to the NH$_2$-terminal end of nonmuscle tropomyosin (human TM5) (Sung and Lin, 1994) but the NH$_2$-terminal sequences of muscle and nonmuscle tropomyosins are completely different (Pittenger et al., 1994). However, further dissection of the tropomyosin binding region of tropomodulin demonstrated that the NH$_2$-terminal half of tropomodulin contains two distinct tropomyosin binding regions. Residues 6-94 contain the binding site for skeletal muscle tropomyosin while residues 90-184 contain the binding site for erythrocyte tropomyosin (Babcock and Fowler, 1994). This provides a novel mechanism by which one tropomodulin isoform can recognize either skeletal or nonmuscle tropomyosin.

Although considerably less is known about the regions of tropomodulin involved in actin filament capping, results so far indicate that the COOH-terminal end of tropomodulin is important. A monoclonal antibody (mAb9) that binds within the COOH-terminal portion of tropomodulin blocks tropomodulin's ability to cap actin filaments both in vitro and in vivo (Gregorio et al., 1995). This antibody has no effect on the ability of tropomodulin to bind tropomyosin in vitro and in vivo (Gregorio et al., 1995; also see below). Thus, tropomodulin appears to have two, at least partially independent, protein binding domains: an NH$_2$-terminal one involved in tropomyosin binding and a COOH-terminal one involved in actin binding and pointed end capping. Further dissection of the functional domains of tropomodulin using recombinant tropomodulin fragments as well as other monoclonal antibodies which bind to epitopes along the entire length of tropomodulin is in progress.

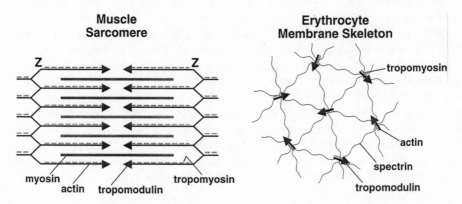

Figure 2. Schematic diagram of the location of tropomodulin at the free (pointed) ends of the thin filaments in striated muscle (*left*) and at the pointed ends of the short actin filaments in the erythrocyte membrane skeleton (*right*). For simplicity, not all of the components of the sarcomere or of the membrane skeleton are included. The relative sizes of the components in the diagrams are not to scale; for example, the erythrocyte actin filaments are actually about 1/5 as long as the spectrin tetramers and less than 1/20 as long as the muscle thin filaments (see Fowler, 1996).

However, the actin capping activity of tropomodulin is by no means functionally independent from its ability to bind tropomyosin; the ability of tropomodulin to cap actin filaments is greatly enhanced by tropomyosin. Thus, tropomodulin completely blocks elongation and depolymerization from the pointed ends of tropomyosin-actin filaments in vitro ($K_d < 1$ nM), but in the absence of tropomyosin, tropomodulin is a "leaky" cap and only partially inhibits monomer association and dissociation at the pointed filament end ($K_d \sim 0.1–0.4$ μM) (Weber et al., 1994). One possibility is that tropomyosin may provide a second binding site for tropomodulin at the pointed filament end, thus increasing the affinity of tropomodulin for the filament end. This is also consistent with the relatively low affinity of tropomodulin for tropomyosin in the absence of actin (0.2 μM; Babcock and Fowler, 1994). The tropomyosin requirement for tight actin filament capping by tropomodulin is unusual in that a role for tropomyosin has not been observed for any previously characterized barbed end capping proteins. For example, gelsolin and capZ have picomolar and nanomolar affinities, respectively, for capping actin filament barbed ends (Weeds and MacIver, 1993).

These observations suggest that in vivo, tropomyosin may serve to target tropomodulin only to the pointed ends of stable tropomyosin-coated actin filaments in cells. Indeed, these are the filaments with which tropomodulin has so far been observed to be associated (Fowler, 1996). A targeting or anchoring function for tropomyosin is supported by experiments with a saponin-permeablized cardiac myocyte cell model. Reconstitution of tropomodulin onto the pointed ends of thin filaments after high salt extraction was achieved only onto those filaments to which tropomyosin had first been bound (Gregorio and Fowler, 1995). A role for tropomyosin in anchoring tropomodulin to thin filaments in vivo is also suggested by the results of antibody microinjection experiments. When the actin capping activity of tropomodulin is inhibited by microinjection of mAb9 into chick cardiac myocytes, tropomodulin remains bound to the terminal tropomyosin molecule on the (former) ends of the thin filaments where it continues to cap the tropomyosin polymer and prevent addition of tropomyosin to the newly elongated actin filament extensions extending across the H zone (Gregorio et al., 1995). Nonetheless, it is likely that binding of tropomodulin to thin filament pointed ends in vivo may be strengthened by tropomodulin's interaction with additional, as yet unidentified, thin filament components, in view of the relatively low affinity of tropomodulin for tropomyosin ($K_d \sim 0.2$ μM) (Babcock and Fowler, 1994) as compared to its high affinity for tropomyosin-actin filaments ($K_d < 1$ nM) in vitro (Weber et al., 1994), (for more discussion of this point, see Gregorio and Fowler, 1995; Gregorio and Fowler, 1996).

Tropomodulin Isoforms

In mammals, the same isoform of tropomodulin appears to be expressed in erythrocytes and skeletal and cardiac muscle, based on comparison of amino acid sequences (Sung et al., 1992; Sussman et al., 1994*a*; Ito et al., 1995; Watakabe et al., 1996) and comigration of immunoreactive polypeptides on 2-dimensional gels (M. Bondad and V.M. Fowler, unpublished observations). This isoform of tropomodulin is also expressed in lens fiber cells of the eye, in sensory neurons and epithelial cells of the inner ear as well as in some regions of the adult rat brain (Ito et al., 1995;

Sussman et al., 1994*b*; Woo and Fowler, 1994). Expression of tropomodulin in the epithelial cells of the inner ear is interesting in light of previous observations that actin filament pointed ends are capped in the mature hair cells of the cochlea of birds (see Discussion in Tilney et al., 1992*c*). Southern analysis of mouse genomic DNA indicates that this tropomodulin is a product of a single copy gene (Ito et al., 1995); it has been mapped to chromosome 9q22 in the human (Sung et al., 1996) and to the homologous region of chromosome 4 in the mouse (White et al., 1995). Only one tropomodulin isoform has been identified in birds; it is expressed in cardiac and skeletal muscle and is 86% identical to the human erythrocyte isoform (Babcock and Fowler, 1994).

Recently, a new mammalian isoform of tropomodulin has been identified which is 59% identical at the amino acid level to the erythrocyte/muscle tropomodulin isoform (Watakabe et al., 1996). This new isoform is predominantly expressed in neurons of both embryonic and adult brain (Watakabe et al., 1996), unlike the erythrocyte/muscle isoform which is expressed prenatally at very low levels in the rat brain (Sussman et al., 1994*b*). In addition, low level expression of the new brain isoform is also detected in uterus (smooth muscle?). Sequence comparisons indicate that the new brain isoform is a product of a distinct gene; furthermore, RNAse protection assays do not reveal any alternatively spliced regions within the coding sequence. This brain tropomodulin isoform preferentially recognizes tropomyosins with the nonmuscle NH_2-terminal sequence in comparison to isoforms containing the skeletal muscle type NH_2-terminal sequence (Watakabe et al., 1996), suggesting it could be associated specifically with nonmuscle tropomyosin-actin filaments. The identification of a tropomodulin isoform specialized for interaction with nonmuscle tropomyosins raises the intriguing possibility that yet a third variety of tropomodulin might exist that is specialized for interaction with muscle-type tropomyosins. In such a case, different tropomodulins in the same cell could be selectively targeted to actin filaments coated with one or another tropomyosin isoform (see below).

Tropomodulin-related Proteins

A BLAST search of the combined protein sequence data bases with the human erythrocyte tropomodulin amino acid sequence has identified a 64-kD human protein with significant sequence homology to tropomodulin (Dong et al., 1991; Wall et al., 1993; C.A. Conley and V.M. Fowler, manuscript in preparation). A detailed comparison of the sequence of the 64-kD protein with tropomodulin reveals that the 64-kD protein contains two separate regions of homology to tropomodulin: the NH_2-terminal 55 amino acids of the 64-kD protein are 47% identical and 56% similar to residues 27-82 of erythrocyte tropomodulin, while residues 273-444 of the 64-kD protein are 47% identical and 64% similar to almost the entire COOH-terminal half of erythrocyte tropomodulin (residues 166-335) (Fig. 3). The presence of significant regions of sequence homology with regions of tropomodulin shown to be involved in both tropomyosin binding and actin filament capping (see above) strongly suggests that this protein is likely to be a functional analog of tropomodulin. We have also identified several classes of ESTs encoding tropomodulin-like sequences that are present in humans, mice and *Caenorhabditis elegans* (C.A. Conley and V.M. Fowler, unpublished observations), indicating that the 40-kD tropomodulin isoforms and the 64-kD tropomodulin-related protein may turn out to be members of a larger family of actin filament pointed end capping proteins.

Western blots of a variety of rat tissues with affinity purified rabbit antibodies to recombinant 64-kD protein from *E. coli* reveal that this protein is principally expressed in a variety of smooth muscle containing tissues including small intestine, stomach, and uterus, but is not detected in erythrocytes, striated muscle, or brain (C.A. Conley and V.M. Fowler, manuscript in preparation). Consistent with the presence of a region in the 64-kD protein (amino acids 1-55) which is homologous to part of the skeletal muscle tropomyosin binding region in the erythrocyte tropomodulin isoform (amino acids 27-82; see Fig. 3), the 64-kD protein binds to rabbit skeletal muscle tropomyosin but not to nonmuscle (rat brain) tropomyosin. This suggests that it will also bind smooth muscle tropomyosin, which contains a muscle NH_2-terminal type sequence (Pittenger et al., 1994). Since the new brain 40-kD isoform of tropomodulin identified by Watakabe et al. (1996) may also be expressed in smooth muscle tissue, we speculate that this 40-kD tropomodulin isoform and the 64-kD tropomodulin-related protein may each function to cap different populations of nonmuscle and muscle tropomyosin-actin filaments, respectively, in smooth muscle cells.

The cDNA clone for the 64-kD tropomodulin-related protein was originally isolated by screening a thyroid cDNA library with autoimmune serum from patients with thyroid-associated opthalmopathy (Dong et al., 1991). Although the autoimmune sera used to isolate the cDNAs for this clone recognized a 64-kD protein on Western blots of extraocular muscle and thyroid tissue, at the present time the relationship between the 64-kD autoantigen and the cloned 64-kD tropomodulin-related protein is not clear.

Tropomodulins in Motile Cells?

The 40-kD tropomodulin isoforms and the 64-kD tropomodulin-related protein described above are all expressed in highly differentiated and generally nonmotile cells which tend to be characterized by the presence of populations of relatively stable actin filaments (muscle, erythrocytes, neurons, lens fiber cells, etc.). What about highly motile cells with dynamic actin filaments? Do these cells also contain tropomodulins or related proteins to cap their pointed filament ends? Preliminary evidence for this is that affinity-purified polyclonal antibodies to human erythrocyte tropomodulin were shown to cross-react with polypeptides of about 70- and 40-kD

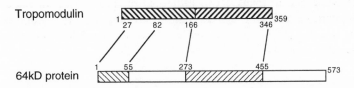

Figure 3. Schematic alignment of the amino acid sequences of erythrocyte tropomodulin and the 64-kD tropomodulin-related protein. Amino acids 1-55 and 273-444 of the 64-kD protein are homologous to amino acids 27-82 and 166-335, respectively, of tropomodulin (see text).

on Western blots of human neutrophils and bovine aorta endothelial cells (Fowler, 1990), suggesting that both tropomodulin and possibly a(nother) 64-kD tropomodulin-related protein may be present in these cells.

Certainly, although actin filaments are highly dynamic in motile cells, actin filament length distributions are considerably restricted. Actin filament lengths are reported to average $\sim$0.5 μm long in macrophages (Hartwig and Shevlin, 1986), $\sim$0.2 μm long in the comet tails of *Listeria monocytogenes* (Tilney et al., 1992*b*), $\sim$0.75 μm long in the comet tails of *Vaccinia* virus (Cudmore et al., 1996), and to range from $\sim$0.18–2.1 μm long in neutrophils (Cano et al., 1991). In the leading lamellipodia of nerve growth cones and fish keratocytes, there is evidence for two populations of actin filaments: one consisting of very long parallel filaments about 3–6 μm long and another that may be considerably shorter and possibly more labile (Lewis and Bridgman, 1992; Small et al., 1995; for a review, see Mitchison and Cramer, 1996). Evidence that the pointed ends of such dynamic actin filaments are capped in at least some of these situations is that exogenous actin has been observed to elongate only from the barbed but not the pointed ends of the filaments in the comet-tail of *Listeria* (Tilney et al., 1992*c*) and of *Vaccinia* virus (Cudmore et al., 1996), and in the leading lamellipodia in permeabilized cells (Symons and Mitchison, 1991). Analysis of actin depolymerization rates in neutrophil lysates is also consistent with the presence of a pointed end capping protein, although actin filament stabilization due to filament side-binding proteins is another possible explanation (Cano et al., 1992). In these types of motile cells, if pointed end capping proteins are present, their interactions with actin filaments are likely to be dynamic, with their activity being regulated both temporally and spatially in response to extracellular signals (for a review, see Schafer and Cooper, 1995).

Finally, the action of pointed end capping proteins may not be limited to conventional actin filaments. The short filament in the dynactin complex is comprised of a polymer of the actin-related protein, Arp1, and is likely to be capped at both ends based on the restricted length distribution of filaments in dynactin complexes (37.2 ± 2.4 nm) (Schafer et al., 1994). The β2 isoform of nonmuscle actin capping protein has been localized to one end of this short filament suggesting that β2 capping protein caps the equivalent of the Arp1 filament's barbed end, while a 62-kD protein has been localized to the other end, suggesting that it may be a cap for the Arp1 filament's pointed end (Schafer et al., 1994).

Acknowledgments

I would especially like to thank Ryan Littlefield for preparation of Figure 1, Peggy Meyer for preparation of Figure 2, and Catharine A. Conley for preparation of Figure 3.

This research was supported by grants from the NIH to V.M. Fowler (GM34225 and EY10814).

References

Babcock, G.G., and V.M. Fowler. 1994. Isoform specific interaction of tropomodulin with skeletal muscle and erythrocyte tropomyosins. *J. Biol. Chem.* 269:27510–27518.

Burlacu, S., P.A. Janmey, and J. Borejdo. 1992. Distribution of actin filament lengths measured by fluorescence microscopy. *Am. J. Physiol.* 262:C569–C577.

Cano, M.L., L. Cassimeris, M. Fechheimer, and S.H. Zigmond. 1992. Mechanisms responsible for F-actin stabilization after lysis of polymorphonuclear leukocytes. *J. Cell Biol.* 116:1123–1134.

Cano, M.L., D.A. Lauffenburger, and S.H. Zigmond. 1991. Kinetic analysis of F-actin depolymerization in polymorphonuclear leukocyte lysates indicates that chemoattractant stimulation increases actin filament number without altering the filament length distribution. *J. Cell Biol.* 115:677–687.

Coluccio, L.M. 1994. An end in sight: tropomodulin. *J. Cell Biol.* 127:1497–1499.

Cudmore, S., I. Reckmann, G. Griffiths, and M. Way. 1996. Vaccinia virus: a model system for actin-membrane interactions. *J. Cell Sci.* 109:1739–1747.

Dong, Q., M. Ludgate, and G. Vassart. 1991. Cloning and sequencing of a novel 64-kDa autoantigen recognized by patients with autoimmune thyroid disease. *J. Clin. Endocrin. Metabol.* 72:1375–1381.

Fowler, V.M. 1987. Identification and purification of a novel M_r 43,000 tropomyosin-binding protein from human erythrocyte membranes. *J. Biol. Chem.* 262:12792–12800.

Fowler, V.M. 1990. Tropomodulin: a cytoskeletal protein that binds to the end of erythrocyte tropomyosin and inhibits tropomyosin binding to F-actin. *J. Cell Biol.* 111:471–482.

Fowler, V.M. 1996. Regulation of actin filament length in erythrocytes and striated muscle. *Curr. Opin. Cell Biol.* 8:86–96.

Gregorio, C.C., and V.M. Fowler. 1995. Mechanisms of thin filament assembly in embryonic chick cardiac myocytes: tropomodulin requires tropomyosin for assembly. *J. Cell Biol.* 129:683–695.

Gregorio, C.C., and V.M. Fowler. 1996. Tropomodulin function and thin filament assembly in cardiac myocytes. *Trends Cardiovas. Med.* 6:136–141.

Gregorio, C.C., A. Weber, M. Bondad, C.R. Pennise, and V.M. Fowler. 1995. Requirement of pointed end capping by tropomodulin to maintain actin filament length in embryonic chick cardiac myocytes. *Nature.* 377:83–86.

Hartwig, J.H., and P. Shelvin. 1986. The architecture of actin filaments and the ultrastructural location of actin-binding protein in the periphery of lung macrophages. *J. Cell Biol.* 103:1007–1020.

Hirokawa, N., L.G. Tilney, K. Fujiwara, and J.E. Heuser. 1982. Organization of actin, myosin, and intermediate filaments in the brush border of intestinal epithelial cells. *J. Cell Biol.* 94:425–443.

Huxley, H.E. 1963. Electron microscope studies on the structure of natural and synthetic protein filaments from striated muscle. *J. Mol. Biol.* 7:281–308.

Ito, M., B. Swanson, M.A. Sussman, L. Kedes, and G. Lyons. 1995. Cloning of tropomodulin cDNA and localization of gene transcripts during mouse embryogenesis. *Dev. Biol.* 167:317–328.

Lewis, A.K., and P.C. Bridgman. 1992. Nerve growth cone lamellipodia contain two populations of actin filaments that differ in organization and polarity. *J. Cell Biol.* 119:1219–1243.

Lux, S.E., and J. Palek. 1995. Disorders of the red cell membrane. *In* Blood: Principles and Practice of Hematology. R. Handin, S.E. Lux, and T.P. Stossel, editors. J.B. Lippincott Co., Philadelphia. 1701–1818.

Mitchison, T.J., and L.P. Cramer. 1996. Actin-based cell motility and cell locomotion. *Cell.* 84:371–379.

Mooseker, M.S. 1985. Organization, chemistry, and assembly of the cytoskeletal apparatus of the intestinal brush border. *Annu. Rev. Cell Biol.* 1:209–241.

Oosawa, F., and S. Asakura. 1975. Thermodynamics of the Polymerization of Protein. Academic Press, New York. 194 pp.

Pittenger, M.F., J.A. Kazzazz, and D.M. Helfman. 1994. Functional properties of non-muscle tropomyosin isoforms. *Curr. Opin. Cell Biol.* 6:96–104.

Robinson, T.F., and S. Winegrad. 1979. The measurement and dynamic implications of thin filament lengths in heart muscle. *J. Physiol.* 286:607–619.

Schafer, D.A., S.R. Gill, J.A. Cooper, J.E. Heuser, and T.A. Schroer. 1994. Ultrastructural analysis of the dynactin complex: an actin-related protein is a component of a filament that resembles F-actin. *J. Cell Biol.* 126:403–412.

Schafer, D.A., and J.A. Cooper. 1995. Control of actin assembly at filament ends. *Annu. Rev. Cell Dev. Biol.* 11:497–518.

Shen, B.W., R. Josephs, and T.L. Steck. 1986. Ultrastructure of the intact skeleton of the human erythrocyte membrane. *J. Cell Biol.* 102:997–1006.

Small, J.V., M. Herzog, and K. Anderson. 1995. Actin filament organization in the fish keratocyte lamellipodium. *J. Cell Biol.* 129:1275–1286.

Sosa, H., D. Popp, G. Ouyang, and H.E. Huxley. 1994. Ultrastructure of skeletal muscle fibers studied by a plunge quick freezing method: myofilament lengths. *Biophys. J.* 67:283–292.

Sung, L.A., V.M. Fowler, K. Lambert, M.A. Sussman, D. Karr, and S. Chien. 1992. Molecular cloning and characterization of human fetal liver tropomodulin: a tropomyosin-binding protein. *J. Biol. Chem.* 267:2616–2621.

Sung, L.A., Y.S. Fan, and C.C. Lin. 1996. Gene assignment, expression, and homology of human tropomodulin. *Genomics.* 34:92–96.

Sung, L.A., and J.J.-C. Lin. 1994. Erythrocyte tropomodulin binds to the N-terminus of hTM5, a tropomyosin isoform encoded by the γ-tropomyosin gene. *Biochem. Biophys. Res. Comm.* 201:627–634.

Sussman, M.A., S. Sakhi, P. Barrientos, M. Ito, and L. Kedes. 1994*a*. Tropomodulin in rat cardiac muscle. Localization of protein is independent of messenger RNA distribution during myofibrillar development. *Circ. Res.* 75:221–232.

Sussman, M.A., S. Sakhi, G. Tocco, I. Najm, M. Baudry, L. Kedes, and S.S. Schreiber. 1994*b*. Neural tropomodulin: developmental expression and effect of seizure activity. *Dev. Brain Res.* 80:45–53.

Symons, M.H., and T.J. Mitchison. 1991. Control of actin polymerization in live and permeabilized fibroblasts. *J. Cell Biol.* 114:503–513.

Tilney, L.G., M.S. Tilney, and D.J. DeRosier. 1992*a*. Actin filaments, stereocilia, and hair cells: how cells count and measure. *Annu. Rev. Cell Biol.* 8:257–274.

Tilney, L.G., D.J. DeRosier, and M.S. Tilney. 1992*b*. How *Listeria* exploits host cell actin to form its own cytoskeleton. I. Formation of a tail and how that tail might be involved in movement. *J. Cell Biol.* 118:71–81.

Tilney, L.G., D.J. DeRosier, A. Weber, and M.S. Tilney. 1992*c*. How *Listeria* exploits host cell actin to form its own cytoskeleton. II. Nucleation, actin filament polarity, filament assembly, and evidence for a pointed end capper. *J. Cell Biol.* 118:83–93.

Tilney, L.G., M.S. Tilney, and G.M. Guild. 1995. F actin bundles in *Drosophila* bristles I. Two filament cross-links are involved in bundling. *J. Cell Biol.* 130:629–638.

Tilney, J.G., P. Connelly, S. Smith, and G.M. Guild. 1996. F-actin bundles in *Drosophila* bristles are assembled from modules composed of short filaments. *J. Cell Biol.* 135:1291–1308.

Wall, J.R., N. Bernard, A. Boucher, M. Salvi, Z.G. Zhang, J. Kennerdell, A. Tyutyunikov, and C. Genovese. 1993. Pathogenesis of thyroid-associated ophthalmopathy: an autoimmune disorder of the eye muscle associated with Graves' hyperthyroidism and Hashimoto's thyroiditis. *Clin. Immunol. Immunopathol.* 68:1–8.

Watakabe, A., R. Kobayashi, and D.M. Helfman. 1996. N-tropomodulin: a novel isoform of tropomodulin identified as the major binding protein to brain tropomyosin. *J. Cell Sci.* 109:2299–2310.

Weber, A., C.R. Pennise, G.G. Babcock, and V.M. Fowler. 1994. Tropomodulin caps the pointed ends of actin filaments. *J. Cell Biol.* 127:1627–1635.

Weeds, A., and S. MacIver. 1993. F-actin capping proteins. *Curr. Opin. Cell Biol.* 5:63–69.

White, R.A., L.L. Dowler, M. Woo, L.R. Adkison, S. Pal, D. Gershon, and V.M. Fowler. 1995. The tropomodulin (Tmod) gene maps to chromosome 4, closely linked to Mup1. *Mammalian Genome.* 6:332–333.

Woo, M.K., and V.M. Fowler. 1994. Identification and characterization of tropomodulin and tropomyosin in the adult rat lens. *J. Cell Sci.* 107:1359–1367.

Functional Studies of the Membrane Skeleton in *Drosophila*: Identification of a Positional Cue that Targets Polarized Membrane Skeleton Assembly

Ronald R. Dubreuil, Gary R. MacVicar, and
Pratumtip Boontrakulpoontawee Maddux

Department of Pharmacological & Physiological Sciences and the Committees on Cell Physiology and Developmental Biology, University of Chicago, Chicago, Illinois 60637

Introduction

Members of the spectrin family of proteins are widely distributed among eukaryotic organisms ranging from amoebas to humans. Conserved features among family members include their unusually large size, their direct or indirect interaction with integral plasma membrane proteins (IMPs), and their interaction with cytoplasmic actin filaments. Spectrins are related to a larger superfamily of cytoplasmic actin-crosslinking proteins that includes α-actinin, filamin, fimbrin, and the Dictyostelium actin binding proteins (Dubreuil, 1991; Hartwig and Kwiatkowski, 1991; Matsudaira, 1991). However, spectrin family members are distinguished by their formation of a subplasma membrane protein network known as the membrane skeleton.

Genetic studies have established important structural roles for the membrane skeleton in cell shape and tissue integrity. First, defects in human erythrocyte spectrin cause a shortened lifespan of circulating erythrocytes and consequent anemia (reviewed by Lux and Palek, 1995). The affected red blood cells are abnormal in shape and have an unusually fragile plasma membrane. Second, DuChenne muscular dystrophy is caused by a defect in the spectrin-like molecule dystrophin (Koenig et al., 1988). The disease is characterized by muscle wasting and the replacement of functioning muscle fibers with fatty tissue and fibrosis. In view of its similarity to spectrin, it has been proposed that dystrophin normally functions to stabilize the sarcolemma during muscle contraction (Ahn and Kunkel, 1993; Campbell, 1995). Third, mutations in *Drosophila* α spectrin are lethal and cause defects in the shape and adhesion of epithelial cells in the larval digestive tract (Lee et al., 1993). In each case, a defect in the major structural component of the membrane skeleton leads to a corresponding defect in cell and/or tissue structure.

In addition to their proposed structural roles, spectrin family members are thought to contribute to plasma membrane organization. One contribution is the stabilization of interacting plasma membrane proteins at the cell surface, thereby influencing the composition of the membrane. Another contribution is the formation of polarized membrane skeleton domains that confer unique properties on specialized regions of the plasma membrane. A third contribution arises from the sheer complexity of protein interactions between the plasma membrane and the membrane skeleton. The membrane skeleton may act as an organizational hub that physically links together many different components of the plasma membrane.

Interactions with the membrane skeleton can stabilize integral membrane proteins that otherwise have a short half-life at the cell surface. For example, dystrophin normally interacts with a complex of integral sarcolemmal glycoproteins (Campbell, 1995). Interestingly, when dystrophin is absent in muscular dystrophy patients, there is a concomitant loss of the glycoprotein complex as well (Ervasti et al., 1990). Therefore, it appears that the stable expression of the dystrophin associated glycoproteins is dependent on the presence of dystrophin. Another example is the sodium pump, which has been shown to directly interact with the membrane skeleton (Nelson and Veshnock, 1987). There is a dramatic increase in the half-life of sodium pump molecules that form interactions with the membrane skeleton in comparison to molecules that fail to associate with the membrane skeleton (Hammerton et al., 1991). These examples suggest a general mechanism through which the membrane skeleton can influence the composition of the plasma membrane.

By stabilizing integral membrane proteins within specialized plasma membrane domains, the membrane skeleton may also contribute to the establishment and maintenance of cell surface polarity. The membrane skeleton is selectively associated with subdomains of the plasma membrane in many cell types including neurons (Lazarides and Nelson, 1983), kidney (Drenckhahn et al., 1985; Nelson and Veshnock, 1986), and muscle (Craig and Pardo, 1983). Drubin and Nelson (1996) proposed that the assembly of the membrane skeleton in response to positional cues at the cell surface may be an important step in the establishment of cell polarity. Immunofluorescent studies of transfected fibroblasts revealed that membrane skeleton assembly occurs at sites of cell–cell contact in response to the positional cue of E-cadherin–mediated cell adhesion (McNeill et al., 1990). The sodium pump also accumulates at sites of E-cadherin–mediated adhesion, presumably through its direct interaction with the membrane skeleton. Thus it appears that a polarized membrane skeleton can transmit a positional cue at the cell surface from one membrane protein to another. The same mechanism is thought to explain the polarized distribution of the sodium pump at sites of lateral cell contact in cultured Madin-Darby canine kidney (MDCK) cells (Hammerton et al., 1991).

A large number of integral membrane proteins have been shown to interact with the membrane skeleton. These include voltage-dependent sodium channels (Srinivasan et al., 1988), epithelial sodium channels (Rotin et al., 1994), anion exchangers (Morgans and Kopito, 1988), the Na^+-Ca^{2+} exchanger (Li et al., 1993), IP_3 receptors (Bourguignon and Jin, 1995), the ryanodine receptor (Bourguignon, 1995), CD45 (Iida et al., 1994), CD44 (Lokeshwar and Bourguignon, 1992), and members of the L1 family of neural cell adhesion molecules (Davis et al., 1993; Davis and Bennett, 1994). Some of these interacting membrane proteins may serve as positional cues that direct polarized assembly of the membrane skeleton. Others may respond to a polarized membrane skeleton, as in the case of the sodium pump. The elucidation of mechanisms that govern these complex protein interactions is likely to shed light on fundamental principles of plasma membrane organization.

Here we describe the progress made by using the fruit fly as an experimental system in which to study the membrane skeleton. First, we summarize the identification and characterization of α spectrin mutations in non-erythroid cells. These genetic studies have established that there is an essential role for the membrane skeleton during normal development. Another novel feature of *Drosophila* as a model system is the ability to study the process of polarized membrane skeleton assembly in *Drosophila* S2 tissue culture cells. We describe recent studies showing that individual interactions between the membrane and membrane skeleton have distinctly different consequences for plasma membrane organization. While some of the basic properties of the membrane skeleton will be summarized here, the reader is refered to several excellent reviews that provide a more comprehensive description of the membrane skeleton in diverse cell types (Ahn and Kunkel, 1993; Bennett, 1990; Bennett and Gilligan, 1993; Campbell, 1995; Lux and Palek, 1995).

Overview of the Membrane Skeleton

The major components of the membrane skeleton and their interactions with one another (shown schematically in Fig. 1) are conserved between distant species. The spectrin molecule is the primary structural component of the membrane skeleton. Spectrin is a rod-shaped tetramer composed of α and β subunits. Each subunit consists largely of a series of 106 amino acid spectrin repeats (shown as ellipses). There are also nonrepetitive sequences in each subunit that are responsible for some of the unique protein interactions of the native molecule (*rectangles*). These include an SH3 domain near the middle of the α subunit and an EF hand Ca^{2+} binding domain at the carboxy terminus. An actin binding domain is found at the amino terminus of the β spectrin subunit. In the native molecule the calcium binding domain of the α subunit lies adjacent to the amino terminal domain of the β subunit. Based on the effects of calcium on the related protein α actinin, which has a similar arrangement of calcium binding and actin binding sites (Blanchard et al., 1989), the carboxy terminal domain of α spectrin is thought to confer calcium sensitivity on the actin binding activity of β spectrin (Wasenius et al., 1989; Dubreuil et al., 1991). Thus spectrin is a bivalent actin crosslinking molecule whose activity may be regulated by the calcium concentration in the cell.

The actin crosslinking activity of spectrin allows it to form a protein network beneath the plasma membrane. In the human erythrocyte, the network is based on junctional complexes in which 6 spectrin tetramers converge on short actin filaments to form a long-range geodesic structure. The network has been directly observed by electron microscopy in the erythrocyte (Byers and Branton, 1985; Liu et al.,

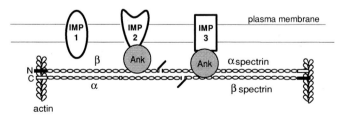

Figure 1. Conserved features of the membrane skeleton in vertebrates and *Drosophila*. Spectrin and actin are components of a protein network at the cytoplasmic face of the plasma membrane. The network is attached to integral membrane proteins (*IMPs*) by ankyrin and by direct interactions with the β subunit of spectrin (*IMP-1*). See text for details.

1987). The similarities between spectrin molecules found in erythrocytes and other "non-erythroid" cells, and their conserved protein interactions indicate that spectrin and actin are capable of forming a similar protein network in other cell types.

There are multiple attachment points between the spectrin membrane skeleton and the lipid bilayer. The best characterized link occurs through ankyrin, which is also a peripheral membrane protein. The ankyrin molecule can be divided into distinct spectrin-binding and membrane-binding domains (Bennett, 1992). The spectrin-binding domain interacts with spectrin repeat 15 of the β subunit, near the center of the spectrin tetramer (Kennedy et al., 1991). The membrane binding domain of ankyrin is composed of 24 imperfect repeats of a canonical 33 amino acid ankyrin repeat motif (Lux et al., 1990; Lambert et al., 1990). Several membrane proteins, including the anion exchanger (Davis and Bennett, 1990), neurofascin (Davis and Bennett, 1994), and voltage-dependent sodium channels (Srinivasan et al., 1992), are known to interact with ankyrin via this domain.

There are also ankyrin-independent membrane binding sites that link the membrane skeleton to the plasma membrane. For example, there are binding sites on the spectrin molecule that directly associate with integral membrane proteins (Steiner and Bennett, 1988; Lombardo et al., 1994; *IMP-1* in Fig. 1). Another class of interactions occurs indirectly through membrane skeleton-associated proteins other than ankyrin. For example, protein 4.1 in the human erythrocyte associates with the spectrin-actin junction and forms a complex with the p55 protein and the integral membrane protein glycophorin C (Marfatia et al., 1994).

The Membrane Skeleton in *Drosophila*

Drosophila provides a unique experimental system in which powerful genetic tools can be used to study protein function. Spectrin was initially discovered in *Drosophila* by Dan Kiehart and colleagues in the course of characterizing actin-binding proteins from *Drosophila* tissue culture cells (Dubreuil et al., 1987). A reverse genetic strategy was subsequently used to identify mutations in spectrin. The initial characterization of these mutants has provided important new insights into the function of the membrane skeleton.

Spectrin was purified from *Drosophila* S3 tissue culture cells using standard biochemical methods (Dubreuil et al., 1987). The appearance of the molecule by electron microscopy, its subunit composition, and its actin binding activity are virtually identical to vertebrate spectrins. An antibody was produced against the purified protein and shown to cross-react with vertebrate spectrins. The antibody was also used to isolate spectrin clones from *Drosophila* cDNA expression libraries (Byers et al., 1987). The sequence of complete *Drosophila* α and β spectrin cDNAs (Dubreuil et al., 1989; Byers et al., 1992) revealed 63 and 49% identity to their vertebrate counterparts, respectively. There are two isoforms of spectrin in *Drosophila* that utilize the same α subunit in combination with one of two distinct β spectrin gene products to form αβ and αβ$_H$ spectrin tetramers. The β$_H$ spectrin isoform was named because of its unusually high molecular weight (~430k) relative to other β spectrins that have been studied. The appearance of αβ$_H$ spectrin by electron microscopy and partial sequence analysis of the β$_H$ subunit indicate that it is a conserved member of the spectrin protein family (Dubreuil et al., 1990). Interestingly a

β_H homolog was recently identified in *C. elegans* (Austin et al., 1995), suggesting that β_H spectrin is an ancient, conserved member of the spectrin superfamily.

A single ankyrin gene has been identified in *Drosophila.* Ankyrin was initially identified by polymerase chain reaction using oligonucleotide primers based on the sequence of mammalian ankyrins (Dubreuil and Yu, 1994). The PCR product was used as a probe to isolate cDNA clones. The complete cDNA sequence of *Drosophila* ankyrin revealed that its domain structure and amino acid sequence are conserved relative to vertebrate ankyrins. An antibody produced against a recombinant fragment of *Drosophila* ankyrin identified a single 170k product of the *Drosophila* ankyrin gene. Co-immunoprecipitations were used to demonstrate an interaction between *Drosophila* spectrin and ankyrin. Antibody staining revealed that *Drosophila* ankyrin and spectrin are both associated with the plasma membrane of *Drosophila* cells (described below). Thus, the evidence so far is consistent with a conserved role for ankyrin in attaching spectrin and actin to the plasma membrane (Fig. 1).

Genetic Studies of the Membrane Skeleton in *Drosophila*

Thus far, genetic studies of the *Drosophila* membrane skeleton have been limited to the α spectrin gene. The reverse genetic approach entails the identification of a protein of interest followed by cloning of DNA sequences representing that protein (Rubin, 1988). Hybridization of the DNA probe to *Drosophila* polytene chromosomes allows one to identify the chromosomal address of the gene of interest. Once the locus has been identified, one can use a battery of standard genetic approaches to recover the desired mutants. Using this approach, the *Drosophila* α spectrin gene was found to reside on the third chromosome at position 62B (Byers et al., 1987). The chromosomal region had previously been subjected to a saturation mutagenesis in the course of characterizing neighboring genes on the third chromosome (Sliter et al., 1989). Two strategies were used to identify the mutant complementation group that encodes α spectrin, among the neighboring unrelated complementation groups (Lee et al., 1993). First, the pattern of α spectrin staining was examined in Western blots of total proteins from flies carrying candidate mutations. Second, a rescue transgene, built from α spectrin cDNA, was stably incorporated into germ-line DNA of transformed flies. The transgene was tested for its ability to rescue homozygous mutations in the α spectrin region that otherwise cause death early in development. These strategies identified a complementation group that includes several different lethal α spectrin alleles. Some of these alleles encode mutant protein products with altered electrophoretic mobility in Western blots.

All of the α spectrin mutant alleles were found to be lethal during early *Drosophila* larval development in homozygotes. Following fertilization of the egg, normal embryonic development takes place within the egg shell during the first day. Subsequent larval development is subdivided into three stages called instars, which are the periods between molts that allow the larva to increase in size. The first instar larva hatches on day 2 after fertilization. Second and third instar development takes place over several additional days after which the organism pupates. In the pupa, adult structures grow to replace larval tissues. The new adult emerges at the end of a total of 10 days of development in the laboratory. The α spectrin mutants

die during the first larval instar indicating that there must be an essential requirement for spectrin function in the larva. Interestingly, the elaborate processes of embryonic development that take place prior to larval hatching do not appear to require the presence of a functional α spectrin gene. Interpretation of this result is
complicated somewhat by the finding that there is a maternal contribution of α
spectrin that may provide partial function during early development (Pesacreta et
al., 1989). However, by the time lethality is observed α spectrin is no longer detectable in the mutant larvae (Dubreuil and Yu, 1994; Lee et al., 1993).

The molecular lesion in two of the lethal α spectrin alleles was found to be due
to frameshift mutations that introduce premature stop codons into the α spectrin
coding sequence (Lee et al., 1993). One of these alleles, RG41, has a frameshift
near the beginning of the coding sequence and consequently produces no significant protein product. Therefore RG41 is considered to be a null α spectrin mutation. Another allele, RG35, has a frameshift near the very end of the coding sequence, within the calcium binding EF hand domain. This mutant produces a nearly
intact protein product that only lacks the carboxy-terminal domain. Interestingly,
the phenotypes of both of these mutants are virtually identical: death during early
larval development. The molecular lesion in the RG35 mutant may affect the activity of the neighboring actin binding domain found in the β spectrin subunit and may
therefore compromise actin crosslinking activity. Similarly, the complete loss of the
α subunit in the RG41 mutant is expected to affect the actin crosslinking activity of
the native molecule by preventing spectrin tetramer formation. A point mutation in
the α subunit that specifically blocks spectrin tetramer formation also constitutes a
knockout of α spectrin function (Deng et al., 1995). Thus, the essential function of
the spectrin molecule appears to involve its actin-crosslinking activity.

The cause of death in α spectrin mutant larvae is not known. However, the
goal of our experimental strategy is to examine the cellular consequences of spectrin deficiency. The inability to pinpoint the exact cause of death is not a significant
obstacle. One of the most conspicuous structures present in the first instar *Drosophila* larva is the digestive tract. The larval gut is a simple epithelial tube that can
be divided into subdomains based on the distinctive appearance of cells in each region. The characteristics of the anterior midgut have proven useful for examining
the cellular consequences of α spectrin deficiency. The cuprophilic cells, which accumulate dietary copper, alternate along the anterior midgut with interstitial cells
that are distinct in appearance. The cuprophilic cells are spheroidal with a peculiar
apical invagination that is continuous with the gut lumen. The apical invagination is
covered with actin-rich microvilli, much like the enterocytes found in vertebrates.
The apical domain of these cells stains brightly with rhodamine phalloidin in a characteristic pattern that reflects the regular spacing of cuprophilic cells (Fig. 2 *A*).

Staining of the midgut with an anti–α spectrin antibody reveals the characteristic pattern of plasma membrane staining (Fig. 2 *B*) found in many other *Drosophila*
cell types (Pesacreta et al., 1989). As expected, the α spectrin staining pattern is absent in the RG41 α spectrin mutant (Fig. 2 *D*). Staining of the midgut of RG41 mutants with rhodamine phalloidin reveals a striking perturbation of cell organization
relative to wild type (Fig. 2 *C*). The apical domain of cuprophilic cells, which forms
a flattened disk shaped structure in wild type larvae, is grossly misshapen in the mutants. Moreover, the long range organization of alternating cuprophilic and interstitial cells is conspicuously disrupted. Electron microscopy of the cuprophilic cells re-

Actin   α Spectrin

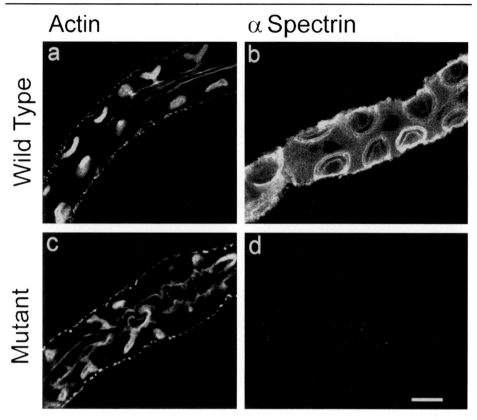

Figure 2. A mutation in *Drosophila* α spectrin results in cell shape and interaction defects in the larval digestive tract. The cuprophilic region of the anterior midgut from first instar wild type larvae (*A* and *B*) and RG41 α spectrin mutant larvae (*C* and *D*) were stained with rhodamine phalloidin (*A* and *C*) to detect filamentous actin, or with rabbit anti–α spectrin antibody followed by fluorescent secondary antibody (*B* and *D*). Reproduced from *The Journal of Cell Biology*, 1993, vol. 123, pp. 1797–1809 by copyright permission of the Rockefeller University Press. Bar = 10 µM.

vealed that the normally close apposition of plasma membranes along the lateral margins of cells is also disrupted, leaving large gaps between cells that are not detected in the wild-type gut (Lee et al., 1993). Thus, α spectrin mutations cause defects in cell shape, disruption of tissue organization, and a defect in cell–cell adhesion. These results are consistent with an important structural role for spectrin at the plasma membrane.

An Organization Role for the Membrane Skeleton

One of the ways spectrin may normally function at the plasma membrane is by providing a protein scaffold that mechanically stabilizes the lipid bilayer. However, the phenotypes of the α spectrin mutants may also be due to more subtle defects in plasma membrane organization. To further investigate the ability of the membrane skeleton to impose order on the plasma membrane, it will be important to develop a better understanding of how order is imposed on the membrane skeleton. For example, we would like to know how the assembly of the membrane skeleton is targeted to particular membrane domains.

Little is known yet about the mechanisms that govern membrane skeleton assembly in cells. One obstacle has been the fact that in most cells that have been examined the membrane skeleton is already associated with the plasma membrane. Thus it has not been possible to identify the cues that promote the initial assembly of the membrane skeleton or the cues that sort and target multiple membrane skeleton isoforms within cells.

Membrane Skeleton Assembly in *Drosophila* S2 Tissue Culture Cells

Drosophila S2 tissue culture cells are small spherical cells that grow in suspension culture in defined medium supplemented with fetal calf serum. The S2 line was originally recovered from primary cultures of dissociated *Drosophila* embryos (Schneider, 1972). Little is known about the developmental origin or destination of these cells in the embryo. Most studies have simply used S2 cells as a source of material for biochemical studies or for expression of transfected cDNAs.

Initial immunolocalization studies revealed that ankyrin was not detectably associated with the plasma membrane of S2 cells, despite evidence from Western blots showing that spectrin and ankyrin are abundantly expressed (Dubreuil and Yu, 1994). We hypothesized that the absence of ankyrin at the plasma membrane of S2 cells was due to the lack of an appropriate membrane receptor to recruit ankyrin from the cytoplasm. If so, then the expression of such a receptor as a transgene in S2 cells was expected to recruit ankyrin to the membrane. At the time these experiments were carried out, a new class of ankyrin binding integral membrane proteins had been identified in rat brain extracts (Davis et al., 1993). Subsequent studies revealed that there is a family of vertebrate neural cell adhesion molecules related to human L1 that share a conserved interaction with ankyrin through their cytoplasmic domains (Davis and Bennett, 1994).

A *Drosophila* homolog of human L1 was first discovered by Goodman and colleagues in a search for molecules that mediate cell recognition and adhesion in the nervous system (Bieber et al., 1989). The *Drosophila* protein neuroglian resembles L1 family members in its domain organization (Fig. 3) and exhibits 29% overall sequence identity to human L1 (Hortsch, 1996). The homophilic extracellular domain of L1 and neuroglian are composed of 6 distal immunoglobulin-like domains and 5 proximal fibronectin type III repeats (Fig. 3). Some vertebrate family members have a proline and threonine-rich domain in place of the first fibronectin repeat but are otherwise identical in domain organization to L1 and neuroglian (Davis et al., 1993). The greatest sequence conservation between diverse family members is found in the cytoplasmic domain of the molecule.

The homophilic adhesive activity of L1 family members is also conserved in *Drosophila* neuroglian. While S2 cells normally grow as a single cell suspension, cells expressing a cDNA-based neuroglian transgene form large aggregates. There are two isoforms of neuroglian that share the conserved cytoplasmic sequence found in other L1 family members, but they differ in the length of the cytoplasmic

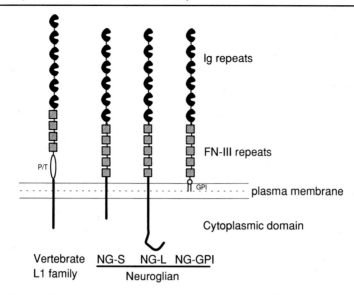

Figure 3. *Drosophila* neuroglian is a conserved member of the L1 family of vertebrate neural cell adhesion molecules. Vertebrate L1 molecules are conserved in domain structure and amino acid sequence relative to *Drosophila* neuroglian, although some family members such as neurofascin lack the membrane-proximal fibronectin type III domain and instead have a proline-threonine rich domain (*P/T*). Two natural isoforms of neuroglian differ in the mRNA splicing of the cytoplasmic domain coding sequence. A synthetic isoform was engineered so that a glycosyl phosphatidylinositol membrane linkage replaces the normal transmembrane and cytoplasmic sequences of neuroglian.

domain (Hortsch et al., 1990). A third form of neuroglian was engineered by Hortsch and colleagues (1995) by replacing the transmembrane and cytoplasmic sequences of neuroglian with a signal for glycosyl phosphatidyl inositol linkage to the lipid bilayer (Fig. 3). The lipid-linked form of neuroglian exhibits robust adhesive activity despite its lack of transmembrane and cytoplasmic sequences.

We compared the distribution of ankyrin in control S2 cells to its distribution in neuroglian-expressing S2 cells (Dubreuil et al., 1996). This experiment directly tested the effect of an ankyrin binding membrane protein on the cellular distribution of the membrane skeleton. In one experiment, the distributions of ankyrin and neuroglian were examined in a population of cells during the initial stage of aggregation in response to neuroglian expression (Fig. 4). Some cells in the culture failed to express neuroglian altogether (Fig. 4 *B, double arrowhead*). These cells exhibited faint ankyrin staining throughout the cytoplasm as well as a punctate perinuclear distribution (Fig. 4 *A, double arrowhead*). In contrast, neuroglian expressing cells that formed cell-cell contacts (Fig. 4 *B, arrow*) exhibited a striking codistribution of ankyrin at sites of contact (Fig. 4 *A, arrow*). Identical results were obtained with either of the two natural isoforms of neuroglian. Thus the distribution of ankyrin in S2 cells responds to the presence of an ankyrin binding integral membrane protein.

Further experiments revealed that the recruitment of ankyrin to sites of neuroglian-mediated cell contact was due to a direct interaction between ankyrin and the cytoplasmic domain of neuroglian (Dubreuil et al., 1996). First, ankyrin was not recruited to sites of cell contact upon expression of the GPI-linked form of neuroglian that lacks the conserved cytoplasmic domain (Fig. 3). Thus, recruitment of ankyrin to the plasma membrane was not a secondary consequence of cell aggregation. Second, yeast two-hybrid analysis was used to demonstrate that recombinant fragments of ankyrin and neuroglian physically interact with one another. These results demonstrated that the biochemical interaction between L1 family members and ankyrin first detected in vertebrate systems is conserved in *Drosophila.*

The above experiments revealed that one class of membrane-cytoskeleton interaction could be studied in isolation in S2 cells. We next used immunolocalization to examine the downstream effects of ankyrin recruitment on other interacting components of the plasma membrane and membrane skeleton. The distribution of β spectrin in S2 cells before and after expression of neuroglian was similar to the distribution of ankyrin (Dubreuil et al., 1996). Prior to formation of cell-cell contacts, β spectrin was detected in a diffuse distribution in the cytoplasm with no detectable staining of the plasma membrane. However, once neuroglian-expressing cells formed aggregates, β spectrin was concentrated together with ankyrin at sites of cell contact. The distribution of the sodium pump also conformed to the distribution of ankyrin and spectrin in neuroglian-expressing S2 cells (unpublished observation). Thus neuroglian appears to act through ankyrin to recruit assembly of the spectrin membrane skeleton at sites of cell–cell contact. Targeting of the membrane skeleton to these sites leads to a further recruitment of the sodium pump to the same membrane domain. These results are consistent with studies in vertebrate systems showing that ankyrin, spectrin, and the sodium pump redistribute to sites of cell–cell contact in response to E-cadherin–mediated cell adhesion (McNeill et al., 1990). Therefore, cell adhesion appears to be an important mechanism for the segregation of the membrane skeleton and other interacting membrane proteins within a specialized cell surface domain.

Neuroglian behaves as a signal-transducing molecule that selectively recruits membrane skeleton assembly to sites of cell–cell contact. Previous studies demonstrated that E-cadherin becomes concentrated at sites of cell–cell contact in the course of cell aggregation (e.g., McNeill et al., 1990). While the biochemical link between E-cadherin and membrane skeleton assembly is not yet known, the segregated distribution of the adhesion molecule is consistent with the polar assembly of the membrane skeleton. In contrast, double immunofluorescent staining revealed that, while ankyrin was selectively recruited to sites of cell–cell contact, neuroglian was detectable over the entire surface of aggregated cells (Fig. 4, *A* and *B*, *arrow*). Ankyrin was not detected at the surface of single cells that expressed high levels of neuroglian (Fig. 4 *B*, *arrowhead*). The presence of neuroglian at the surface of non-aggregated cells, or at noncontact regions of aggregated cells, was insufficient to recruit ankyrin to these sites. Instead it appears that neuroglian is somehow activated by cell adhesion to recruit membrane skeleton assembly.

Implications for Membrane Skeleton Assembly and Function

The unique properties of *Drosophila* S2 cells have made it possible to study individual interactions between the plasma membrane and the membrane skeleton, essentially in isolation. Our results establish that neuroglian can recruit polarized assembly of the membrane skeleton via ankyrin, as indicated by an arrow pointing toward ankyrin in Fig. 5. In contrast, the sodium pump represents a class of membrane protein that responds to the membrane skeleton, indicated by an arrow pointing toward the sodium pump. Direct interactions between spectrin and the plasma membrane also occur, but their consequences are not yet known, as indicated by double arrows.

The proposed organizational role of the membrane skeleton is based in part on its multivalent nature. Each spectrin tetramer can potentially interact with two ankyrin molecules (Fig. 1). Thus, conceivably, one ankyrin molecule may target the membrane skeleton while the second ankyrin may recruit additional membrane binding proteins to form a specialized membrane domain. Longer range interactions could also contribute by allowing one ankyrin molecule, associated with one spectrin molecule, to recruit additional spectrins (and ankyrins) through spectrin–actin network formation. Short range interactions may also be important. Recent mapping studies revealed that distinct sites on the ankyrin molecule are responsible for interactions with the voltage-dependent sodium channel and neurofascin (Michaely and Bennett, 1995). Therefore, it is possible for a single ankyrin molecule to behave as a coordinating center that simultaneously links the membrane skeleton to its positional cues and responding membrane proteins.

We previously demonstrated that the polarized distribution of the sodium pump in epithelial cells is unaffected by a mutation in α spectrin (Lee et al., 1993). Subsequent studies revealed that the loss of α spectrin in these mutants had little if any effect on the accumulation of the other spectrin subunits, ankyrin or neuroglian (Dubreuil, 1996). If ankyrin has an organizational role at the plasma membrane, by itself or through short range interactions with β spectrin, then that role may have escaped detection in our previous genetic studies of α spectrin. An important future

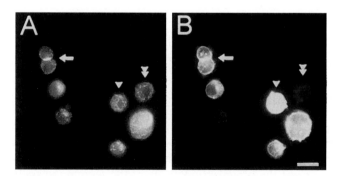

Figure 4. Neuroglian functions as a potent inducer of polarized membrane skeleton assembly at sites of cell–cell contact. A population of transfected S2 cells was induced to express neuroglian before indirect immunofluorescent staining with antibodies against *Drosophila* neuroglian (*A*) and *Drosophila* ankyrin (*B*). Ankyrin was not detectable at the plasma membrane of single cells (*A, arrowheads*) but was recruited to sites of cell contact in aggregating S2 cells (*arrow*). In contrast, neuroglian was detectable at adherent and nonadherent surfaces of cells (*B*). Double arrowhead indicates a cell that did not detectably express neuroglian. Bar = 10 μM.

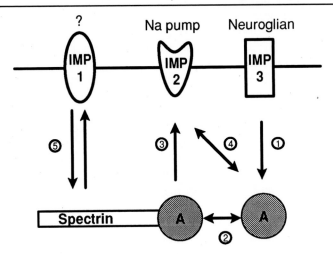

Figure 5. 11A model for multiple interactions between the plasma membrane and membrane skeleton. Three classes of integral membrane proteins (*IMPs*) are shown. There are membrane binding sites that directly interact with the β subunit of spectrin (*IMP-1*). Other membrane proteins interact with the membrane skeleton through ankyrin (*IMPs 2* and *3*). Neuroglian (*IMP-3*) is an example of a membrane binding site that provides positional information for membrane skeleton assembly (*1, downward arrow*). In contrast, the sodium pump (*IMP-2*) appears to receive positional information from the membrane skeleton (*3, upward arrow*). The multivalent nature of the membrane skeleton (*2*) provides a possible mechanism to link these two different ankyrin binding membrane proteins to one another. Ankyrin itself is multivalent (*4*), providing a further mechanism for coupling IMPs. The consequences of the direct interaction between spectrin and the plasma membrane (*5*) are unknown but may potentially transmit positional information in either direction between the membrane and membrane skeleton.

goal will be to examine the effects of ankyrin and β spectrin mutations on cell surface polarity.

It will also be important in future studies to categorize the membrane proteins that interact with spectrin and ankyrin according to their role in membrane skeleton assembly. For example, we would like to know if sodium channels, anion exchangers, and other membrane proteins have an active role in recruiting membrane skeleton assembly (as with neuroglian), or if instead they respond to the state of membrane skeleton assembly (as with the sodium pump). *Drosophila* S2 cells will provide a useful experimental system in which to ask these questions. Ultimately it will be possible to test the organizational role of the membrane skeleton by using the powerful tools of *Drosophila* genetics.

Acknowledgments

We thank Jason Frankel for comments on the manuscript.

Supported by NIH GM49301 and a grant from the American Heart Association Chicago Affiliate.

References

Ahn, A.H., and L.M. Kunkel. 1993. The structural and functional diversity of dystrophin. *Nat. Genet.* 3:283–291.

Austin, J., C. McKeown, and A. Patel. 1995. SMA-1 encodes a spectrin required for morphogenesis of the *C. elegans* embryo. *Mol. Biol. Cell.* 6:270a.

Bennett, V. 1990. Spectrin based membrane skeleton: a multipotential adaptor between membrane and cytoplasm. *Physiol. Rev.* 70:1029–1065.

Bennett, V. 1992. Ankyrins. *J. Biol. Chem.* 267:8703–8706.

Bennett, V., and D.M. Gilligan. 1993. The spectrin-based membrane skeleton and micronscale organization of the plasma membrane. *Annu. Rev. Cell Biol.* 9:27–66.

Bieber, A.J., P.M. Snow, M. Hortsch, N.H. Patel, J.R. Jacobs, Z.R. Traquina, J. Schilling, and C.S. Goodman. 1989. Drosophila neuroglian: a member of the immunoglobulin superfamily with extensive homology to the vertebrate neural adhesion molecule L1. *Cell.* 59:447–460.

Blanchard, A., V. Ohanian, and D. Critchley. 1989. The structure and function of alpha-actinin. *J. Musc. Res. Cell Motil.* 10:280–289.

Bourguignon, L.Y.W., and H. Jin. 1995. Identification of the ankyrin-binding domain of the mouse T-lymphoma cell inositol 1,4,5-triphosphate (IP3) receptor and its role in the regulation of IP3-mediated internal Ca^{2+} release. *J. Biol. Chem.* 270:7257–7260.

Bourguignon, L.Y.W. 1995. Ryanodine receptor-ankyrin interaction regulates internal Ca^{2+} release in mouse T-lymphoma cells. *J. Biol. Chem.* 270:17917–17922.

Byers, T.J., and D. Branton. 1985. Visualization of the protein associations in the erythrocyte membrane skeleton. *Proc. Natl. Acad. Sci. USA.* 82:6153–6157.

Byers, T.J., R.R. Dubreuil, D. Branton, D.P. Kiehart, and L.S.B. Goldstein. 1987. Drosophila spectrin II. Conserved features of the alpha subunit are revealed by analysis of cDNA clones and fusion proteins. *J. Cell Biol.* 105:2103–2110.

Byers, T.J., E. Brandin, E. Winograd, R. Lue, and D. Branton. 1992. The complete sequence of Drosophila beta spectrin reveals supra-motifs comprising eight 106-residue segments. *Proc. Natl. Acad. Sci. USA.* 89:6187–6191.

Campbell, K.P. 1995. Three muscular dystrophies: loss of cytoskeleton-extracellular matrix linkage. *Cell.* 80:675–679.

Craig, S.W., and J.V. Pardo. 1983. Gamma actin, spectrin, and intermediate filament proteins colocalize with vinculin at costameres, myofibril-to-sarcolemma attachment sites. *Cell Motil.* 3:449–462.

Davis, L., and V. Bennett. 1990. Mapping the binding sites of human erythrocyte ankyrin for the anion exchanger and spectrin. *J. Biol. Chem.* 265:10589–10596.

Davis, J.Q., T. McLaughlin, and V. Bennett. 1993. Ankyrin-binding proteins related to nervous system cell adhesion molecules: candidates to provide transmembrane and intercellular connections in adult brain. *J. Cell Biol.* 121:121–133.

Davis, J.Q., and V. Bennett. 1994. Ankyrin binding activity shared by the neurofascin/L1/NrCAM family of nervous system cell adhesion molecules. *J. Biol. Chem.* 269:27163–27166.

Deng, H., J.K. Lee, L.S.B. Goldstein, and D. Branton. 1995. Drosophila development requires spectrin network formation. *J. Cell Biol.* 128:71–79.

Drenckhahn, D., K. Schluter, D.P. Allen, and V. Bennett. 1985. Colocalization of band 3 with ankyrin and spectrin at the basal membrane of intercalated cells in the rat kidney. *Science.* 230:1287–1289.

Drubin, D.G., and W. J. Nelson. 1996. Origins of cell polarity. *Cell.* 84:335–344.

Dubreuil, R., T.J. Byers, D. Branton, L.S.B. Goldstein, and D.P. Kiehart. 1987. Drosophila spectrin I. Characterization of the purified protein. *J. Cell Biol.* 105:2095–2102.

Dubreuil, R.R., T.J. Byers, A.L. Sillman, D. Bar-Zvi, L.S.B. Goldstein, and D. Branton. 1989. The complete sequence of Drosophila alpha spectrin: conservation of structural domains between alpha spectrins and alpha actinin. *J. Cell Biol.* 109:2197–2206.

Dubreuil, R.R., T.J. Byers, C.T. Stewart, and D.P. Kiehart. 1990. A beta spectrin isoform from Drosophila (betaH) is similar in size to vertebrate dystrophin. *J. Cell Biol.* 111:1849–1858.

Dubreuil, R.R., E. Brandin, J.H.S. Reisberg, L.S.B. Goldstein, and D. Branton. 1991. Structure, calmodulin-binding, and calcium-binding properties of recombinant alpha spectrin polypeptides. *J. Biol. Chem.* 266:7189–7193.

Dubreuil, R.R. 1991. Structure and evolution of the actin crosslinking proteins. *BioEssays.* 13:219–226.

Dubreuil, R.R., and J. Yu. 1994. Ankyrin and beta spectrin accumulate independently of alpha spectrin in Drosophila. *Proc. Natl. Acad. Sci. USA.* 91:10285–10289.

Dubreuil, R.R., G.R. MacVicar, S. Dissanayake, C. Liu, D. Homer, and M. Hortsch. 1996. Neuroglian-mediated adhesion induces assembly of the membrane skeleton at cell contact sites. *J. Cell Biol.* 133:647–655.

Dubreuil, R.R. 1996. Molecular and genetic dissection of the membrane skeleton in Drosophila. *In* Current Topics in Membranes, Vol. 43. W.J. Nelson, editor. Academic Press, San Diego. 147–167.

Ervasti, J.M., K. Ohlendieck, S.D. Kahl, M.G. Gaver, and K.P. Campbell. 1990. Deficiency of a glycoprotein component of the dystrophin complex in dystrophic muscle. *Nature.* 345:315–319.

Hammerton, R.W., K.A. Krzeminski, R.W. Mays, T.A. Ryan, D.A. Wollner, and W.J. Nelson. 1991. Mechanism for regulating cell surface distribution of Na^+, K^+-ATPase in polarized epithelial cells. *Science.* 254:847–850.

Hartwig, J.H., and D.J. Kwiatkowski. 1991. Actin-binding proteins. *Curr. Opin. Cell Biol.* 3:87–97.

Hortsch, M., A.J. Bieber, N.H. Patel, and C.S. Goodman. 1990. Differential splicing generates a nervous system-specific form of Drosophila neuroglian. *Neuron.* 4:697–709.

Hortsch, M., Y.-M.E. Wang, Y. Marikar, and A.J. Bieber. 1995. The cytoplasmic domain of the Drosophila cell adhesion molecule nueroglian is not essential for its homophilic adhesive properties. *J. Biol. Chem.* 270:18809–18817.

Hortsch, M. 1996. The L1 family of neural cell adhesion molecules. Old proteins performing new tricks. *Neuron.* 17:587–593.

Iida, N., V.B. Lokeshwar, and L.Y.W. Bourguignon. 1994. Mapping the fodrin binding domain of CD45, a leukocyte membrane-associated tyrosine phosphatase. *J. Biol. Chem.* 269:28576–28583.

Kennedy, S.P., S.L. Warren, B.G. Forget, and J.S. Morrow. 1991. Ankyrin binds to the 15th repetitive unit of erythroid and nonerythroid beta spectrin. *J. Cell Biol.* 114:267–277.

Koenig, M., A.O. Monaco, and L.M. Kunkel. 1988. The complete sequence of dystrophin predicts a rod-shaped cytoskeletal protein. *Cell.* 53:219–228.

Lambert, S., H. Yu, J.T. Prchal, J. Lawler, P. Ruff, D. Speicher, M.C. Cheung, Y.W. Kan, and J. Palek. 1990. cDNA sequence for human erythrocyte ankyrin. *Proc. Natl. Acad. Sci. USA.* 87:1730–1734.

Lazarides, E., and W.J. Nelson. 1983. Erythrocyte and brain forms of spectrin in cerebellum: distinct membrane-cytoskeleton domains in neurons. *Science.* 220:1295–1297.

Lee, J., R. Coyne, R.R. Dubreuil, L.S.B. Goldstein, and D. Branton. 1993. Cell shape and interaction defects in alpha-spectrin mutants of *Drosophila melanogaster. J. Cell Biol.* 123: 1797–1809.

Li, Z., E.P. Burke, J.S. Frank, V. Bennett, and K.D. Phillipson. 1993. The cardiac Na^+-Ca^+ exchanger binds to the cytoskeletal protein ankyrin. *J. Biol. Chem.* 268:11489–11491.

Liu, S.-C., L.H. Derick, and J. Palek. 1987. Visualization of the hexagonal lattice in the erythrocyte membrane skeleton. *J. Cell Biol.* 104:527–536.

Lokeshwar, V.B., and L.Y.W. Bourguignon. 1992. The lymphoma transmembrane glycoprotein GP85 (CD44) is a novel guanine nucleotide-binding protein which regulates GP85 (CD44)-ankyrin interaction. *J. Biol. Chem.* 267:22073–22078.

Lombardo, C.R., S.A. Weed, S.P. Kennedy, B.G. Forget, and J.S. Morrow. 1994. BetaII-spectrin (fodrin) and beta IE2-spectrin (muscle) contain NH_2- and COOH-terminal membrane association domains (MAD-1 and MAD2). *J. Biol. Chem.* 269:29212–29219.

Lux, S.E., K.M. John, and V. Bennett. 1990. Analysis of cDNA for human erythrocyte ankyrin indicates a repeated structure with homology to tissue-differentiation and cell-cycle control proteins. *Nature.* 344:36–42.

Lux, S.E., and J. Palek. 1995. Disorders of the Red Cell Membrane. *In* Blood: Principles and Practice of Hematology. R.I. Handin, S.E. Lux, and T.P. Stossel, editors. J.B. Lippincott Co., Philadelphia. 1701–1818.

Marfatia, S.M., R.A. Lue, D. Branton, and A.H. Chishti. 1994. In vitro binding studies suggest a membrane-associated complex between erythroid p55, protein 4.1 and glycophorin C. *J. Biol. Chem.* 269:8631–8634.

Matsudaira, P. 1991. Modular organization of actin crosslinking proteins. *TIBS.* 16:87–92.

McNeill, H., M. Ozawa, R. Kemler, and W.J. Nelson. 1990. Novel function of the cell adhesion molecule uvomorulin as an inducer of cell surface polarity. *Cell.* 62:309–316.

Michaely, P., and V. Bennett. 1995. Mechanism for binding site diversity on ankyrin. *J. Biol. Chem.* 270:31298–31302.

Morgans, C.W., and R.R. Kopito. 1993. Association of the brain anion exchanger, AE3, with the repeat domain of ankyrin. *J. Cell Sci.* 105:1137–1142.

Nelson, W.J., and P.J. Veshnock. 1986. Dynamics of membrane skeleton (fodrin) organization during development of polarity in Madin-Darby canine kidney epithelial cells. *J. Cell Biol.* 103:1751–1765.

Nelson, W.J., and P.J. Veshnock. 1987. Ankyrin binding to (Na^+ & K^+) ATPase and implications for the organization of membrane domains in polarized cells. *Nature.* 328:533–536.

Pesacreta, T.C., T.J. Byers, R.R. Dubreuil, D.P. Keihart, and D. Branton. 1989. Drosophila spectrin: the membrane skeleton during embryogenesis. *J. Cell Biol.* 108:1697–1709.

Rotin, D., D. Bar-Sagi, H. O'Brodovich, J. Merilainen, V.P. Lehto, C.M. Canessa, B.C. Rossier, and G.P. Downey. 1994. An SH3 binding region in the epithelial Na$^+$ channel (alpharENaC) mediates its localization at the apical membrane. *EMBO J.* 13:4440–4450.

Rubin, G.M. 1988. Drosophila melanogaster as an experimental organism. *Science.* 240:1453–1459.

Schneider, I. 1972. Drosophila cell and tissue culture. *In* The Genetics and Biology of Drosophila, Vol. 2A. M. Ashburner, T.R.F. Wright, editors. Academic Press, New York. 266–315.

Sliter, T.J., V.C. Henrich, R.L. Tucker, and L.I. Gilbert. 1989. The genetics of the Dras3-Roughened-ecdysoneless chromosomal region (62B3-4 to 62D3-4) in Drosophila melanogaster: analysis of recessive lethal mutations. *Genetics.* 123:327–336.

Srinivasan, Y., L. Elmer, J. Davis, V. Bennett, and K. Angelides. 1988. Ankyrin and spectrin associate with voltage-dependent sodium channels in brain. *Nature.* 333:177–180.

Srinivasan, Y., M. Lewallen, and K.J. Angelides. 1992. Mapping the binding site on ankyrin for the voltage-dependent sodium channel from brain. *J. Biol. Chem.* 267:7483–7489.

Steiner, J.P., and V. Bennett. 1988. Ankyrin-independent membrane protein-binding sites for brain and erythrocyte spectrin. *J. Biol. Chem.* 263:14417–14425.

Wasenius, V.-M., M. Saraste, P. Salven, M. Eramaa, L. Holm, and V.-P. Lehto. 1989. Primary structure of the brain alpha-spectrin. *J. Cell Biol.* 108:79–93.

Molecular Architecture of the Specialized Axonal Membrane at the Node of Ranvier

Vann Bennett, Stephen Lambert, Jonathan Q. Davis, and Xu Zhang

Howard Hughes Medical Institute and Departments of Cell Biology and Biochemistry, Duke University Medical Center, Durham, North Carolina 27710

Introduction

This chapter will focus on progress of our laboratory in resolving molecular components involved in membrane-cytoskeletal interactions at the axonal membrane of nodes of Ranvier in the vertebrate nervous system. Nodes of Ranvier are the sites on myelinated axons, first discovered by Ranvier (1874), where insulating layers of myelin are interrupted and are the sites of ion conductance required for propagation of action potentials. Myelinated axons and their nodes of Ranvier are an important evolutionary advance of vertebrates that permits rapid, saltatory propagation of action potentials without an increase in axonal diameter. This physiological achievement is based on adaptations at a molecular level resulting in an elegant cooperation between glial cells and nerve axons. Nodes of Ranvier are comprised of two adjacent specializations of the axonal plasma membrane characterized by high local concentrations of voltage-regulated ion channels: a paranodal region underlying the paranodal processes of glial cells which contains fast 4-AP–sensitive K^+ channels, and a nodal axon segment between paranodes which is enriched in voltage-dependent sodium channels (Waxman and Ritchie, 1993). A characteristic feature of the cytoplasmic surface of the nodal axon is a dense plaque of material resolved in transmission electron micrographs (Robertson, 1959; Berthold and Rydmark, 1995). The paranodal domain of axons is tightly opposed to paranodal loops of either Schwann cells in the peripheral nervous system or oligodendrocytes in the central nervous system. The nodal axon segment of adult axons is in close contact with specialized microvilli of either Schwann cells in the peripheral nervous system (Berthold and Rydmark, 1995) or astrocytes in the central nervous system (Black et al., 1995).

Nodes of Ranvier are of considerable clinical interest due to their involvement in pathological conditions including peripheral neuropathies (Griffin et al., 1996; Sima et al., 1993), axonal ischemic injury (Waxman et al., 1992), and trauma (Maxwell et al., 1991). Nodes also are the sites of regeneration of damaged peripheral nerve axons (Fawcett and Keynes, 1990). Myelinated axons and nodes of Ranvier also exemplify basic issues for cell biologists: formation of polarized cell domains, assembly of integral proteins into lateral membrane domains, and formation of morphological structures that require cooperation between distinct types of cells.

A clue as to how the node may be established and/or maintained comes from observations that the voltage-dependent sodium channel co-purifies and binds with high affinity to the peripheral membrane protein ankyrin (Srinivasan et al., 1988). Ankyrin is localized at the nodal axonal plasma membrane and is a component of the electron-dense undercoating (Kordeli et al., 1990). The levels of ankyrin at the node of Ranvier are in the range of 1,000–2,000 molecules per square micron based on rough estimates of fluorescence intensity, and thus approximately stoichiometric with the voltage-dependent sodium channel. Ankyrin was initially characterized in human erythrocytes where it forms a high affinity link between the cytoplasmic domain of the anion exchanger and the spectrin/actin network (reviewed in Bennett and Gilligan, 1993). Ankyrin has a general role in other tissues as an adaptor between a variety of integral membrane proteins and the spectrin skeleton. Ankyrin-binding proteins relevant to axons include the voltage-dependent sodium channel (Srinivasan et al., 1988), Na/K ATPase (Ariyasu et al., 1985; Nelson and Veshnock, 1987), and the L1/neurofascin/NrCAM family of cell adhesion molecules (Davis et al., 1993; Davis and Bennett, 1994; Dubreuil et al., 1996). Ankyrin and ankyrin-binding membrane proteins thus potentially are capable of mediating key interactions involved in localization of ion channels and coordinated interactions between axons and glial cells.

Overview of Ankyrin Genes, Functional Domains, and Spliceoforms

The ankyrin gene family of mammals currently includes three members: ankyrin$_R$ (ANK1) (Lux et al., 1990; Lambert et al., 1990), first discovered in erythrocytes, but also expressed in brain (Lambert and Bennett, 1993a, b) and muscle (Birkenmeier et al., 1993); ankyrin$_B$ (ANK2) (Otto et al., 1991), the major form of ankyrin in brain; and a recently discovered ankyrin$_G$ (ANK3) (Kordeli et al., 1995; Peters et al., 1995; Devorajan et al., 1996), which is found in most tissues and includes alternatively spliced forms targeted to nodes of Ranvier (see below). Diseases attributed to ankyrins in humans include cases of hereditary spherocytosis resulting from decreased expression and/or expression of mutated forms of ankyrin$_R$ (Lux and Palek, 1995). The nb/nb mutation in mice results in a nearly complete deficiency of ankyrin$_R$ with a phenotype of severe anemia and degeneration of a subset of Purkinje cell neurons accompanied by signs of cerebellar dysfunction (Peters et al., 1991). Ankyrin also is expressed in *Drosophila* (Dubreuil and Yu, 1994) and in the nematode *C. elegans* (Otsuka et al., 1995). Mutations in the ankyrin gene in *C. elegans* results in the unc44 phenotype resulting from abnormal axonal guidance during development (Otsuka et al., 1995).

Ankyrins are modular proteins comprised of three domains conserved among the different family members as well as specialized domains found in alternatively spliced isoforms (see Fig. 1). The conserved domains are an NH$_2$-terminal 89–95 kD membrane-binding domain, a 62-kD spectrin-binding domain, and a 12-kD "death domain." The death domain is followed by a regulatory domain subject to alternative splicing in the case of ankyrin$_R$ that modulates both binding of the anion exchanger and spectrin (Hall and Bennett, 1987; Davis et al., 1992). Death domains were first reported in proteins such as Fas and the tumor necrosis factor receptor which participate in apoptosis pathways (Cleveland and Ihle, 1995). These domains can associate with related death domains in other proteins. The protein interactions

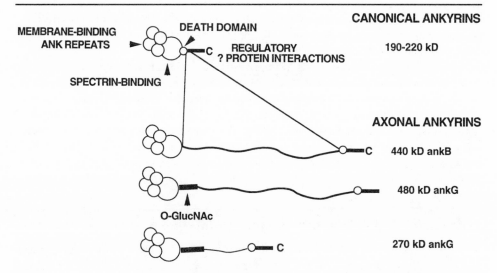

Figure 1. Schematic model for domain organization of some of the ankyrins expressed in brain. The predicted inserted tail domain in axonal ankyrins is not drawn to scale but could be as long as 0.5 microns if fully extended. Not shown in this figure are the alternatively spliced forms of ankyrin$_G$ and ankyrin$_B$ missing portions of the membrane-binding domain.

of the ankyrin death domain could involve self-association and/or interactions with other proteins and are not yet resolved.

Ankyrin genes are subject to tissue-dependent alternative splicing resulting in deletion as well as insertion of functional domains. Ankyrin$_B$ and ankyrin$_G$ include isoforms missing the membrane-binding domain (Kunimoto et al., 1991; Kordeli et al., 1995; Peters et al., 1995; Deverajan et al., 1996). The functions of these truncated ankyrins are not known, but are likely to involve an intracellular as opposed to a plasma membrane role. Ankyrin$_G$ transcripts include at least three other small forms that remain to be defined in terms of sequence.

The ankyrin$_B$ gene also encodes an alternate transcript with an insertion of 6 kb placed between the spectrin-binding domain and death domain. The large transcript of ankyrin$_B$ encodes a 440-kD polypeptide with a 220-kD sequence inserted between the membrane/spectrin binding domains and COOH-terminal domain (Kunimoto et al., 1991; Chan et al., 1993). Much of the inserted sequence has the configuration of an extended random coil based on physical properties of expressed polypeptides (Chan et al., 1993). This sequence is likely to be highly phosphorylated and has 220 predicted sites for phosphorylation by protein kinases (casein kinase 2, protein kinase C, and proline-directed protein kinase). Many of the predicted phosphorylation sites for protein kinase C are located in a series of 15 12-residue serine-rich repeats. The properties of the inserted sequence suggest a model for 440-kD ankyrin$_B$ where the globular membrane-associated head domain is separated from the death domain by an extended filamentous tail domain encoded by the inserted sequence. The length of the tail domain could be up to 0.5 microns if fully extended, which is a distance that in principal could be resolved in the light microscope. Functions of the inserted sequence are not yet known, but one obvious idea is that it connects molecules interacting with the death domain and membrane-binding domains. Given the potential distance spanned by the inserted sequence, these molecules could be in distinct membrane domains or even different cellular compartments.

440-kD ankyrin$_B$ is localized in unmyelinated and premyelinated axons (Chan et al., 1993; Kunimoto et al., 1991; Kunimoto, 1995), presumably due to some form of axonal targeting. 440-kD ankyrin$_B$ is abundant before birth in rats, but is rapidly lost during myelination, and is restricted to unmyelinated axons in adult animals (Chan et al., 1993; Kunimoto et al., 1991). The mechanism for down-regulation of 440-kD ankyrin$_B$ may involve signaling between glial cells and neurons, since hypomyelinated mutant mice exhibit an increase in 440-kD ankyrin$_B$ (Chan et al., 1993).

Ankyrin Is Multivalent with Respect to Integral Membrane Proteins

Ankyrins interact through the membrane-binding domain with multiple membrane proteins with apparently unrelated primary sequence. Currently known ankyrin-binding proteins include ion channels, calcium release channels, and cell adhesion molecules (reviewed in Bennett and Gilligan, 1993) (Fig. 2). The diversity of potential binding partners for ankyrin raises the question of whether these proteins all bind to a common site, as the case for calmodulin and its binding proteins, or whether ankyrin contains more than one and perhaps multiple distinct binding sites. The membrane-binding domain of ankyrin, which is responsible for most to the binding interactions, is comprised of 24 consecutive 33-residue repeats termed ank repeats. Ank repeats are present in many types of proteins including transcription factors, molecules such as Notch that determine cell fate, and inhibitors of cell

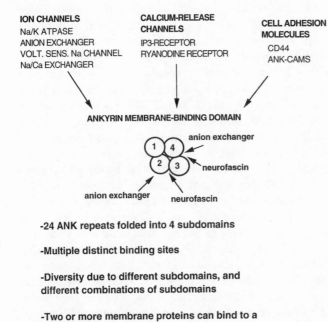

Figure 2. Ankyrin binds to diverse membrane proteins via distinct sites formed from ank repeats. Data is summarized from Michaely and Bennett, 1993; 1995*a*, *b*.

cycle-dependent protein kinases (Michaely and Bennett, 1992; Bork, 1993). Ank repeats in these proteins have been implicated in protein-protein interactions. However, no additional common feature has yet been attributed to proteins that bind to Ank repeats.

Analysis of native ankyrin and recombinant polypeptides by protease sensitivity and circular dicroism spectroscopy has demonstrated that the 24 ANK repeats are organized into four 6-repeat folding domains (Michaely and Bennett, 1993). The role of these 6-repeat domains in protein recognition has been addressed by detailed studies of association of ankyrin with the anion exchanger cytoplasmic domain and neurofascin (Michaely and Bennett, 1995a, b). Important conclusions from these studies is that ankyrin has multiple binding sites and that ankyrin can accommodate more than one membrane protein at the same time (Michaely and Bennett, 1995a, b). Diversity in binding sites on ankyrin results from several factors: utilization of different subdomains, different combinations of subdomains, and distinct sites within the same subdomain pair (Fig. 2). The anion exchanger associates with two sites that are cooperatively coupled, one comprised of subdomains three plus four, and the other by subdomain two. The corresponding ankyrin-binding sites on the anion exchanger also are likely to be distinct, raising the possibility that under certain conditions ankyrin can form linear arrays of anion exchanger in the plane of the plasma membrane. Ankyrin also has two sites for neurofascin, although these do not exhibit cooperativity. One site is located on subdomains three plus four, and the other is on subdomains two plus three. Ankyrin can simultaneously bind the anion exchanger and neurofascin based on lack of competition between these proteins. Ability of ankyrin to form hetero-complexes with cell adhesion molecules and an ion channel potentially has important implications for the node of Ranvier, and will be discussed below.

Ankyrin-binding Membrane Proteins at the Node of Ranvier

The voltage-dependent sodium channel was first discovered to have ankyrin-binding activity due to co-purification of ankyrin and spectrin during isolation of the sodium channel (Srinivasan et al., 1988). Pure ankyrin was subsequently found to associate with the purified sodium channel reconstituted into liposomes with a K_d of 20–50 nM. The voltage-dependent sodium channel colocalizes with ankyrin at the ultrastructural level in postsynaptic membrane infoldings of the neuromuscular junction (Flucher and Daniels, 1989), and both proteins have been independently localized at nodes of Ranvier and axon initial segments (Kordeli et al., 1990; Davis et al., 1996; Lambert, S., J.Q. Davis, and V. Bennett, manuscript submitted for publication). The voltage-dependent sodium channel has three membrane-spanning subunits (Catterall, 1995), including the alpha subunit with channel activity and a beta-2 subunit with sequence homology to the cell adhesion molecule F11/contactin (Isom et al., 1995). It is not known which subunit binds to ankyrin or which combinations of isoforms of the sodium channel subunits are localized at the node of Ranvier.

The major class of ankyrin-binding proteins in brain is a group of cell adhesion molecules in the Ig/FnIII superfamily which includes members localized at the node of Ranvier (Davis et al., 1993; Davis and Bennett, 1994; Davis et al., 1996). Ankyrin-binding cell adhesion molecules are neurofascin, L1, NrCAM, and Ng-CAM in vertebrates and neuroglian in *Drosophila* (Sonderegger and Rathgen, 1992; Rathgen and Jessel, 1991; Grumet, 1991; Hortsch and Goodman, 1991) and have in common a conserved cytoplasmic domain now known to contain the ankyrin-binding site (Davis and Bennett, 1994). These proteins are believed to be involved in neurite outgrowth, axonal fasciculation and targeting, cell migration, and synaptogenesis during embryonic and postnatal development. Other activities reported for members of the superfamily include participation in signal transduction pathways across membranes (Williams et al., 1994; Ignelzi et al., 1994). Associations of the extracellular domains of these proteins occur as a result of self-association, as well as interactions with other members of the Ig/FN111 superfamily and with extracellular molecules such as members of the tenascin family (Sonderegger and Rathgen, 1992; Morales et al., 1993; Felsenfeld et al., 1994).

The genes encoding neurofascin, L1, NrCAM, and NgCAM are subject to considerable diversity due to alternative exon usage. A complete understanding of this family will require definition of polypeptides in terms of actual exons. This level of understanding is some years away, but has begun with neurofascin. Neurofascin has five potential splice sites (Davis et al., 1993; Volkmer et al., 1992; Davis et al., 1996). Two of the major forms of neurofascin, termed 186-kD and 155-kD neurofascin, respectively, have been defined in terms of exon usage by characterization of full-length cDNAs, and exhibit difference at each of the five splice sites (Davis et al., 1996). 186-kD neurofascin lacks the third FNIII domain but has a mucin-like domain inserted between the FNIII domains and the plasma membrane. 155-kD neurofascin, in contrast, has a complete set of four FNIII domains but lacks the mucin-like sequence.

Antibodies to 186-kD neurofascin strongly stain axonal initial segments in the CNS and nodes of Ranvier in both central and peripheral neurons in a pattern similar to the nodal form of ankyrin (see below). Antibodies to 155-kD neurofascin stain neuronal plasma membranes in a more uniform fashion, do not stain mature nodes of Ranvier in central or peripheral nerves, but stain other structures which are consistent with bundles of unmyelinated axons. Localization of 186-kD neurofascin at nodes of Ranvier and axon initial segments suggests the possibility of a specific role for the mucin-like domain in lateral and/or transcellular interactions involving voltage-sensitive Na channels in these excitable membrane domains.

186-kD neurofascin is the first nervous system cell adhesion molecules to be assigned to the node of Ranvier and axon initial segments and to be defined in terms of actual exons. L1, another member of the family, is expressed in unmyelinated axons and disappears with the onset of myelination (Martini, 1994). Neurofascin is not the only cell adhesion molecule at these sites, however. NrCAM also is present at nodes of Ranvier as well as unmyelinated axons (Davis et al., 1996). The precise exons present in the nodal and unmyelinated forms of NrCAM have not been defined.

Specialized Isoforms of Ankyrin$_G$ at the Node of Ranvier and Axon Initial Segments

The discovery that the voltage-dependent sodium channel associated with ankyrin, and that a distinct isoform/gene of ankyrin was localized at nodes of Ranvier (Kordeli et al., 1990; Kordeli and Bennett, 1991) prompted a search for the ankyrin iso-form with a distribution at sites known to have high local concentrations of the channel such as the node of Ranvier and axon initial segments. The gene encoding nodal ankyrin was identified as ankyrin$_G$ (G for general expression) based on char-acterization of a third ankyrin cDNA distinct from the known ankyrins (Kordeli et al., 1995). Ankyrin$_G$ actually includes multiple transcripts expressed in most tissues as well as the isoform present at nodes of Ranvier and was independently discov-ered in a search for kidney ankyrins (Peters et al., 1995; Deverajan et al., 1996). Ankyrin isoforms targeted to the node of Ranvier have distinguishing features of a predicted extended tail domain as well as a serine/threonine–rich stretch of se-quence contiguous to the spectrin-binding domain. The predicted polypeptides have MW of 270 and 480 kD, with the difference in size resulting from a deletion of 190 kD of tail domain sequence. Antibodies raised against the serine-rich domain, and spectrin-binding domain which react with both 270- and 480-kD polypeptides label nodes of Ranvier in peripheral and central axons as well as axon initial seg-ments (Kordeli et al., 1995). In addition, antibodies specific for the 480-kD polypeptide also label these structures (Zhang and Bennett, 1996). The 480-kD form of ankyrin$_G$ thus is a component of nodes of Ranvier and axon initial seg-ments, and the 270-kD polypeptide may also be present. The 270-kD polypeptide is expressed later in development than the 480-kD form and is likely to have a distinct function (S. Lambert, unpublished data).

The serine/threonine–rich domain is a unique feature of nodal ankyrins and has recently been demonstrated to be O-glycosylated with GlcNAc monosaccha-rides (Zhang and Bennett, 1996). Other axonal proteins modified by O-linked GlcNAc include neurofilament subunits (Dong et al., 1993). Consequences of O-glycosylation are not known but could represent a mechanism to cap potential phosphorylation sites (Haltiwanger et al., 1992) or to mediate protein-protein inter-actions. It is of interest in this regard that phosphorylation of neurofilaments is de-creased at nodes of Ranvier (Mata et al., 1992). A protein-GlucNAC–specific anti-body preferentially stains the nodal axon segment, suggesting that glycosylation is a distinguishing feature of this cell domain (Zhang and Bennett, 1996). One hypothe-sis to explain these observations is that a sugar transferase is localized at the node of Ranvier and participates in modifications that distinguish the node from other areas along the axon. Alternatively, the O-linked GlucNAc modification of ankyrin and possibly other proteins could occur in the cell body and be involved in their tar-geting to nodes.

Ankyrin$_G$ isoforms are components of the amorphous submembrane coat ob-served in transmission electron micrographs (Robertson, 1959) and are the first ex-ample of a cytoplasmic protein selectively targeted to nodes of Ranvier and axon initial segments. These proteins with their specialized domains could eventually provide useful clues to pathways for assembly of other components at the node. Spectrin has previously been reported to be concentrated at nodes of Ranvier (Koenig and Repasky, 1985), but spectrin also is located between nodes by immu-nofluorescence and immunogold electron microscopy (Trapp et al., 1989; Lambert, Davis, and Bennett, manuscript submitted for publication). 440-kD ankyrin$_B$ and 480/270 kD are examples of a relatively small number of proteins known to be se-lectively targeted to axons. Other axonal proteins include GAP43 (Goslin et al., 1988) and the microtubule-associated protein tau (Binder et al., 1985). It will be of interest to determine whether axonal ankyrins are transported by fast and/or slow transport mechanisms. It also will be important to determine which features of the primary structure are responsible for directing ankyrin to axons and to nodes of Ranvier.

Nodal forms of the cell adhesion molecules neurofascin and NrCAM are candidates to form a molecular complex with the specialized nodal isoform of ankyrin$_G$. The proposed ankyrin-neurofascin/NrCAM complex would be predicted to be configured as an extended rod-shaped structure extending from the extracellular space through the plasma membrane and up to 200 nm into the axoplasm (Fig. 3). The rationale for this structure is based on direct visualization of neurofascin as a 40–60 nm rod by electron microscopy (Davis et al., 1993), and a conceptual model for ankyrin$_G$ as a ball and chain based on homology between the COOH-terminal portion of 480-kD ankyrin$_G$ and the random-coil sequence present in 440-kD ankyrin$_B$ (Chan et al., 1993; Kordeli et al., 1995). NrCAM is likely to have the same general shape as neurofascin, since both proteins contain the same number of copies of independently folded Ig and FNIII domains. Ichimura and Ellisman have reported visualization of transmembrane filaments at the node of Ranvier that are strikingly similar in dimensions to those predicted for a neurofascin/NrCAM–480-kD ankyrin$_G$ complex (Ichimura and Ellisman, 1991). The observed filaments extend 40–80 nm from Schwann cell microvillar processes in the extracellular space to the axonal membrane and continue into the axoplasm and appear to contact cytoskeletal structures. Ankyrin has been demonstrated to bind to microtubules (Bennett and Davis, 1981) and intermediate filaments (Georgatos and Marchesi, 1985), and, in principle, could be responsible for the cytoskeletal contacts resolved by electron microscopy. It will be important to directly evaluate by immunogold labeling the relationship of the transcellular filaments of Ichimura and Ellisman to ankyrin and ankyrin-binding adhesion molecules.

Coordinate Recruitment of Ankyrin$_G$ and Ankyrin-binding Proteins during Morphogenesis of the Node of Ranvier

Developmental studies suggest that ankyrin, neurofascin, and the voltage-dependent sodium channel associate into microdomains of axons as early events in morphogenesis of the node of Ranvier (Lambert et al., 1995). 480-kD ankyrin$_G$ and 440-kD

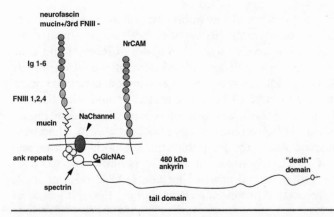

Figure 3. Model for organization of ankyrin and ankyrin-binding proteins at the node of Ranvier.

ankyrin$_B$ are coexpressed in premyelinated axons at early stages of axonal growth during embryonic life. During myelination of the sciatic nerve in the postnatal period, 440-kD ankyrin$_B$ disappears and 480/270-ankyrin$_G$ redistributes from a continuous distribution along premyelinated axons to localized patches adjacent to the ends of myelin-associated glycoprotein (MAG)-staining processes of myelinating Schwann cells. All foci of ankyrin$_G$ also contain neurofascin, NrCAM, and the voltage-dependent sodium channel. However, neurofascin and NrCAM clusters precede ankyrin$_G$ at foci adjacent to the MAG-processes and are candidates to either recruit or stabilize ankyrin$_G$ at these sites.

Further observations at later stages of sciatic nerve development suggest pairs of ankyrin/neurofascin/NrCAM/sodium channel clusters associated with adjacent Schwann cells flanking the presumed nodal site fuse to form the mature node of Ranvier. Studies of nodal development in the hyomyelinating mutant mouse *trembler* reveal that formation of compact myelin is not required for the formation of foci of ankyrin/neurofascin/sodium channels, but may be necessary for the formation of mature nodes of Ranvier. Ankyrin-based clusters of sodium channels could have partial function in saltatory conduction of the action potential and could explain the relatively mild phenotype of Trembler and the related disease of Marie-Charcot-Tooth type 1A in humans. Interestingly, similar clusters of sodium channels have been reported in regenerating peripheral nerve (Dugandzija-Novakovic et al., 1995), which may represent a reversion to an earlier developmental stage of the axon as a response to injury.

Colocalization of ankyrin with ankyrin-binding proteins in early developmental intermediates in assembly of nodes of Ranvier are gratifying results that support in vitro evidence for association between ankyrin and its target proteins. The possibility that ankyrin could be stabilizing heterocomplexes between the voltage-dependent sodium channel and neurofascin or NrCAM is suggested by the colocalization data and the in vitro evidence that the ankyrin membrane-binding domain has several independent binding sites and has the potential to form hetero-complexes between neurofascin and at least one ion channel (Michaely and Bennett, 1995*a*, *b*). A potential benefit of coupling the sodium channel to a cell adhesion molecule would be to target the sodium channel at sites defined by extracellular cues. The molecular nature of the external signal(s) is currently unknown but could be contributed in the peripheral nervous system by extracellular matrix molecules such as tenascin$_R$ and/or specialized glial cell processes that extend into the nodal cleft (Black and Waxman, 1988). In addition, negative signals favoring disassembly of ankyrin-membrane interactions could be provided by the MAG-staining glial cell processes.

Future Directions

These studies have provided a set of interacting proteins involved in membrane-cytoskeletal connections that are targeted to nodes of Ranvier and are defined in terms of actual exon usage in the case of ankyrin$_G$ and 186-kD neurofascin. Utilization of these molecules as probes for early stages of nodal development is just beginning and has already resolved early intermediate in the assembly pathway. Other studies of mechanisms of axonal transport and targeting of these molecules to nodes are possible and may provide clues as to the molecular basis for formation and maintenance of these specialized membrane domains. One major challenge for future work will be to extend predictions of ankyrin activity in promoting lateral and transcellular protein interactions based on in vitro biochemical analysis to actual biology of neurons. Experimental systems have been developed for formation of nodes of Ranvier utilizing co-cultures of neurons and Schwann cells (Bunge and Wood, 1987). In principal, this type of tissue culture system in conjunction with expression of antisense or dominant-negative constructs could be used to evaluate contributions of individual proteins and interactions. Other strategies could involve gene disruption by homologous recombination, or tissue-specific expression of dominant-negative constricts in mice.

A second, more fundamental challenge will be to fully elucidate the signalling pathways initiated by contact between axons and glial cells, which may be mediated at least in part by members of the family of ankyrin-binding cell adhesion molecules. These molecules or their neighbors could be responsible for axonal stimulation of the myelination program by Schwann cells, and Schwann cell regulation of local organization of neurofilaments and possibly other proteins (deWaegh et al., 1992). It is pertinent in this regard that binding of ankyrin to neurofascin has recently been found to be inhibited by tyrosine phosphorylation (Garver, T., S. Turia, Q. Ren, and V. Bennett, manuscript submitted for publication). Missing information includes the ligand system that activates phosphorylation of neurofascin, and possible regulation at the level of dephosphorylation of neurofascin. Potential clinical benefits of these studies could result from insight into mechanisms of nerve regeneration and repair following injury.

Acknowledgments

This research was supported in part by a grant from the National Institutes of Health. Brenda Sampson is gratefully acknowledged for help in preparing the manuscript.

References

Ariyasu, R.G., J.A. Nichol, and M.H. Ellisman. 1985. Localization of sodium/potassium adenosine triphosphatase in multiple cell types of the murine nervous system with antibodies raised against the enzyme from kidney. *J. Neurosci.* 5:2581–2596.

Bennett, V., and J. Davis. 1981. Erythrocyte ankyrin: immunoreactive analogues are associated with mitotic structures in cultured cells and with microtubules in brain. *Proc. Natl. Acad. Sci. USA.* 78:7550–7554.

Bennett, V., and D.M. Gilligan. 1993. The spectrin-based membrane skeleton and micronscale organization of the plasma membrane. *Annu. Rev. Cell Biol.* 9:27–66.

Birkenmeier, C.S., R.A. White, L.L. Peters, E.J. Hall, S.E. Lux, and J.E. Barker. 1993. Complex patterns of sequence variation and multiple 5' and 3' ends are found among transcripts of the erythroid ankyrin gene. *J. Biol. Chem.* 268:9533–9540.

Berthold, C.H., and M. Rydmark. 1995. Morphology of normal peripheral axons. *In* The Axon. G.S. Waxman, J.D. Kocsis, and P.K. Stys, editors. Oxford University Press, New York/ Oxford. 13–48.

Binder, L.I., A. Frankfurter, and L.I. Rebhun. 1985. The distribution of tau in the mammalian central nervous system. *J. Cell Biol.* 101:1371–1378.

Black, J.A., H. Sontheimer, Y. Oh, and S.G. Waxman. 1995. The oligodendrocyte, the perinodal astrocyte, and the central node of Ranvier. *In* The Axon. G.S. Waxman, J.D. Kocsis, and P.K. Stys, editors. Oxford University Press, New York/Oxford. 116–143.

Black, J.A., and S. Waxman. 1988. The perinodal astrocyte. *Glia.* 1:1169–1183.

Bork, P. 1993. Hundreds of ankyrin-like repeats in functionally diverse proteins: mobile modules that cross phyla horizontally? *Proteins.* 17:363–374.

Bunge, R.P., and P.M. Wood. 1987. Tissue culture studies of interactions between axons and myelinating cells of the central and peripheral nervous system. *Prog. Brain Res.* 71:43–152.

Catterall, W.A. 1995. Structure and function of voltage-gated ion channels. *Annu. Rev. Biochem.* 64:493–531.

Chan, W., E. Kordeli, and V. Bennett. 1993. 440-kD ankyrin$_B$: structure of the major developmentally regulated domain and selective localization in unmyelinated axons. *J. Cell Biol.* 123:1463–1473.

Cleveland, J.L., and J.N. Ihle. 1995. Contenders in Fas/TNF death signaling. *Cell.* 81:479–482.

Davis, L.H., J.Q. Davis, and V. Bennett. 1992. Ankyrin regulation: an alternatively spliced segment of the regulatory domain functions as an intramolecular modulator. *J. Biol. Chem.* 267:18966–18972.

Davis, J., and V. Bennett. 1994. Ankyrin-binding activity shared by the neurofascin/L1/NrCam family of nervous system cell adhesion molecules. *J. Biol. Chem.* 269:27163–27166.

Davis, J., S. Lambert, and V. Bennett. 1996. Molecular composition of the node of Ranvier: identification of ankyrin-binding cell adhesion molecules neurofascin (mucin+/third FNIII domain-) and NrCAm at nodal axon segments. *J. Cell Biol.* In press.

Davis, J.Q., T. McLaughlin, and V. Bennett. 1993. Ankyrin-binding proteins related to nervous system cell adhesion molecules: candidates to provide transmembrane and intercellular connections in adult brain. *J. Cell Biol.* 232:121–133.

Deverajan, P., P. Stabach, A.S. Mann, T. Ardito, M. Kashgarian, and J. Morrow. 1996. Identification of a small cytoplasmic ankyrin (AnkG119) in the kidney and muscle that binds beta spectrin and associates with the Golgi apparatus. *J. Cell Biol.* 133:819–830.

deWaegh, S., V.M.-Y. Lee, and S.T. Brady. 1992. Local modulation of neurofilament phosphorylation, axonal caliber, and slow axonal transport by myelinating Schwann cells. *Cell.* 68:451–463.

Dong, D., Z.-S. Xu, M. Chevrier, R. Cotter, D. Cleveland, and G.W. Hart. 1993. Glycosylation of mammalian neurofilaments. *J. Biol. Chem.* 268:16679–16687.

Dubreuil, R., and J. Yu. 1994. Ankyrin and beta-spectrin accumulate independently of alphaspectrin in *Drosophila. Proc. Natl. Acad. Sci. USA.* 91:10285–10289.

Dubreuil, R.R., R. MacVicar, S. Dissanayake, C. Liu, D. Homer, and M. Hortsch. 1996. Neuroglian-mediated cell adhesion induces assembly of the membrane skeleton at cell contact sites. *J. Cell Biol.* 133:647–655.

Dugandzija-Novakovic, S., A.G. Koszowski, S.R. Levinson, and P. Shrager. 1995. Clustering of Na$^+$ channels and node of Ranvier formation in remyelinating axons. *J. Neurosci.* 15:492–503.

Fawcett, J.W., and R.J. Keynes. 1990. Peripheral nerve regeneration. *Annu. Rev. Neurosci.* 13:43–60.

Felsenfeld, D., M. Hynes, K. Skoler, A. Furley, and T.M. Jessel. 1994. TAG-1 can mediate homophilic binding but neurite outgrowth on TAG-1 requires an L1-like molecule and beta1 integrins. *Neuron.* 12:675–690.

Flucher, B., and M. Daniels. 1989. Membrane proteins in the neuromuscular junction. Distribution of Na channels and ankyrin is complementary to acetylcholine receptors and the 43-kD protein. *Neuron.* 3:163–175.

Georgatos, S., and V. Marchesi. 1985. The binding of vimentin to human erythrocyte membranes: a model system for the study of intermediate filament-membrane interactions. *J. Cell Biol.* 100:1955–1961.

Goslin, K., J. Schreyer, P. Skene, and G. Banker. 1988. Development of neuronal polarity: GAP43 distinguishes axonal from dendritic growth cones. *Nature.* 336:672–674.

Griffin, J.W., C.Y. Li, C. Macko, T.W. Ho, S. Hsieh, P. Xue, F.A. Wang, D.R. Cornblath, G.M. McKhann, and A.K. Asbury. 1996. Early nodal changes in the acute motor axonal neuropathy pattern of the Guillain-Barre syndrome. *J. Neurocytol.* 25:33–51.

Grumet, M. 1991. Cell adhesion molecules and their subgroups in the nervous system. *Curr. Opin. Neurobiol.* 1:370–376.

Hall, T.G., and V. Bennett. 1987. Regulatory domains of erythrocyte ankyrin. *J. Biol. Chem.* 262:10537–10545.

Haltiwanger, R.S., M. Blomberg, and G.W. Hart. 1992. Glycosylation of nuclear and cytoplasmic fractions. *J. Biol. Chem.* 267:9005–9013.

Hortsch, M., and C. Goodman. 1991. Cell and substrate adhesion molecules. *Annu. Rev. Cell Biol.* 7:505–557.

Ichimura, T., and M.H. Ellisman. 1991. Three-dimensional fine structure of cytoskeletal-membrane interactions at nodes of Ranvier. *J. Neurocytol.* 20:667–681.

Ignelzi, M., D. Miller, P. Soriano, and P.F. Maness. 1994. Impaired neurite outgrowth of src-minus cerebellar neurons on the cell adhesion molecule L1. *Neuron.* 12:873–884.

Isom, L.L., D.S. Ragsdale, K. DeJongh, R.E. Westenboek, B.F. Reber, T. Scheuer, and W.A. Catterall. 1995. Structure and function of the beta-2 subunit of brain sodium channels, a transmembrane glycoprotein with a CAM motif. *Cell.* 83:433–442.

Jordon, C., B. Puschel, R. Koob, and D. Drenckhahn. 1995. Identification of a binding motif for ankyrin on the α-subunit of Na,K-ATPase. *J. Biol. Chem.* 270:29971–29975.

Koenig, E., and E. Repasky. 1985. A regional analysis of alpha-spectrin in the isolated Mauthner neuron and in isolated axons of the goldfish and rabbit. *J. Neurosci.* 5:705–714.

Kordeli, E., and V. Bennett. 1991. Distinct ankyrin isoforms at neuron cell bodies and nodes of Ranvier resolved using erythrocyte ankyrin-deficient mice. *J. Cell Biol.* 114:1243–1259.

Kordeli, E., J. Davis, B. Trapp, and V. Bennett. 1990. An isoform of ankyrin is localized at nodes of Ranvier in myelinated axons of central and peripheral nerves. *J. Cell Biol.* 110: 1341–1352.

Kordeli, E., S. Lambert, and V. Bennett. 1995. Ankyrin$_G$: a new ankyrin gene with neural-specific isoforms localized at the axonal initial segment and node of Ranvier. *J. Biol. Chem.* 270:2352–2359.

Kunimoto, M. 1995. A neuron-specific isoform of brain ankyrin, 440 kD ankyrin$_B$, is targeted to axons of rat cerebellar neurons. *J. Cell Biol.* 131:1821–1830.

Kunimoto, M., E. Otto, and V. Bennett. 1991. A new 440-kDa isoform is the major ankyrin in neonatal rat brain. *J. Cell Biol.* 115:1319–1331.

Lambert, S., and V. Bennett. 1993a. From anemia to cerebellar dysfunction. A review of the ankyrin gene family. *Eur. J. Biochem.* 211:1–6.

Lambert, S., and V. Bennett. 1993b. Post-mitotic expression of ankyrin$_R$ and beta$_R$-spectrin in discrete neuronal populations of the rat brain. *J. Neurosci.* 13:3725–3735.

Lambert, S., J.Q. Davis, and V. Bennett. 1996. manuscript in preparation.

Lambert, S., J. Davis, P. Michaely, and V. Bennett. 1995. Ankyrin clustering in the coordinate recruitment of ion channels and adhesion molecules during morphogenesis of the node of Ranvier. *Mol. Biol. Cell.* 6:98a.

Lambert, S., H. Yu, J. Prchal, J. Lawler, P. Ruff, D. Speicher, M. Cheung, Y. Kan, and J. Palek. 1990. cDNA sequence for human erythrocyte ankyrin. *Proc. Natl. Acad. Sci. USA.* 87:1730–1734.

Lux, S.E., and J. Palek. 1995. Disorders of the red cell membrane. *In* Blood: Principles and Practice of Hematology. R.I. Handin, S.E. Lux, and T.P. Stossel, editors. J.B. Lippincott Co., Philadelphia. 1701–1808.

Lux, S.E., K.M. John, and V. Bennett. 1990. Analysis of cDNA for human erythrocyte ankyrin indicates a repeated structure with homology to tissue-differentiation and cell-cycle control proteins. *Nature.* 344:36–42.

Martini, R. 1994. Expression and functional roles of neuronal cell adhesion molecules and extracellular matrix components during development and regeneration of peripheral nerves. *J. Neurocytol.* 23:1–28.

Mata, M., N. Kupina, and D.J. Fink. 1992. Phosphorylation-dependent neurofilament epitopes are reduced as the node of Ranvier. *J. Neurocytol.* 21:199–210.

Maxwell, W.L., A. Irvine, J. Graham, T. Adams, R. Gennarelli, R. Tipperman, and M. Sturatis. 1991. Focal axonal injury: the early axonal response to stretch. *J. Neurocytol.* 20:157–164.

Michaely, P., and V. Bennett. 1992. ANK repeats: a ubiquitous motif involved in macromolecular recognition. *Trends Cell Biol.* 2:127–129.

Michaely, P., and V. Bennett. 1993. The membrane-binding domain of ankyrin contains four independently-folded subdomains each comprised of six ankyrin repeats. *J. Biol. Chem.* 268:22703–22709.

Michaely, P., and V. Bennett. 1995a. The ANK repeats of erythrocyte ankyrin form two distinct but cooperative binding sites for the erythrocyte anion exchanger. *J. Biol. Chem.* 270:22050–22057.

Michaely, P., and V. Bennett. 1995b. Mechanism for binding site diversity on ankyrin: comparison of binding sites on ankyrin for neurofascin and the Cl^-/HCO_3^- anion exchanger. *J. Biol. Chem.* 270:31298–31302.

Morales, G., M. Hubert, T. Brummendorf, U. Treubert, A. Tarnok, U. Schwarz, and F. Rathgen. 1993. Induction of axonal growth by heterophilic interactions between the cell surface recognition proteins F11 and Nr-CAM/Bravo. *Neuron.* 11:1113–1122.

Nelson, W.J., and P.J. Veshnock. 1987. Ankyrin binding to the (Na^+/K^+) ATPase and implications for the organization of membrane domains in polarized cells. *Nature.* 328:533–535.

Otsuka, A., R. Franco, B. Yang, K. Shim, L. Tang, Y. Zhang, P. Boontrakulpoontawee, A. Jeyaprakash, E. Hedgecock, V. Wheaton, and A. Sobery. 1995. An ankyrin-related gene (*unc-44*) is necessary for proper axonal guidance in *Caenorhabditis elegans. J. Cell Biol.* 129:1081–1092.

Otto, E., M. Kunimoto, T. McLaughlin, and V. Bennett. 1991. Isolation and characterization of cDNAs encoding human brain ankyrins reveal a family of alternatively-spliced genes. *J. Cell Biol.* 114:241–253.

Peters, L.L., C.S. Birkenmeirer, R.T. Bronson, R.A. White, S.E. Lux, E. Otto, V. Bennett, A. Higgins, and J.E. Barker. 1991. Purkinje cell degeneration associated with erythroid ankyrin deficiency in NB/NB mice. *J. Cell Biol.* 114:1233–1241.

Peters, L.L., K.M. John, F.M. Lu, E.M. Eicher, A. Higgins, M. Yialamas, L.C. Turtzo, A. Otsuka, and S.E. Lux. 1995. Ank3 (epithelial ankyrin), a widely distributed new member of the ankyrin gene family and the major ankyrin in kidney, is expressed in alternatively spliced forms, including forms that lack the repeat domain. *J. Cell Biol.* 130:313–321.

Ranvier, L. 1874. De quelque faits relatifs a l'histologie et a la physiologique des muscl stries. *Arch. Physiol. Norm. Path.* 2(Ser.1):5–15.

Rathjen, F.G., and T.M. Jessel. 1991. Glycoproteins that regulate the growth and guidance of vertebrate axons: domains and dynamics of the immunoglobulin/ fibronectin type III subfamily. *Semin. Neurosci.* 3:297–307.

Robertson, J.D. 1959. Preliminary observations on the ultrastructure of nodes of Ranvier. Z. Zellforsch. *Mikrosk. Anat.* 50:553–560.

Sima, A.A., A. Prasher, V. Nathaniel, M.R. Werb, and D.A. Greene. 1993. Overt diabetic neuropathy: repair of axo-glial dysjunction and axonal atrophy by aldose reductase inhibition and its correlation to improvement in nerve conduction velocity. *Diabetic Med.* 10:115–121.

Sonderegger, P., and F.G. Rathgen. 1992. Regulation of axonal growth in the vertebrate nervous system by interactions between glycoproteins belonging to two subgroups of the immunoglobulin superfamily. *J. Cell Biol.* 119:1387–1394.

Srinivasan, Y., L. Elmer, J. Davis, V. Bennett, and K. Angelides. 1988. Ankyrin and spectrin associate with voltage-dependent sodium channels in brain. *Nature.* 333:177–180.

Trapp, B.D., S.B. Andrews, A. Wong, M. O'Connell, and J.W. Griffin. 1989. Co-localization of the myelin-associated glycoprotein and the microfilament components, F-actin and spectrin, in Schwann cells of myelinated nerve fibers. *J. Neurocytol.* 18:47–60.

Waxman, S.G., J.A. Black, P.K. Stys, and B.R. Ransom. 1992. Ultrastructural concomitants of anoxic injury and early post-anoxic recovery in rat optic nerve. *Brain Res.* 574:105–119.

Waxman, S.G., and J.M. Ritchie. 1993. Molecular dissection of the myelinated axon. *Ann. Neurol.* 33:121–136.

Williams, E.J., J. Furness, F.S. Walsh, and P. Doherty. 1994. Activation of the FGF receptor underlies neurite outgrowth stimulated by L1, N-CAM and N-cadherin. *Neuron.* 13:583–594.

Volkmer, H., B. Hassel, J.M. Wolff, R. Frank, and F.G. Rathjen. 1992. Structure of the axonal surface recognition molecule neurofascin and its relationship to a neural subgroup of the immunoglobulin superfamily. *J. Cell Biol.* 118:149–161.

Zhang, X., and V. Bennett. 1996. Identification of O-linked N-acetylglucosamine modification of ankyrin$_G$ isoforms targeted to nodes of Ranvier. *J. Biol. Chem.* 271:31391–31398.

Chapter 4

Specialized Membrane Domains and
Their Cytoskeletal Connections

Specificity of Desmosomal Plaque Protein Interactions with Intermediate Filaments: Keeping Adhesive Junctions Segregated

Kathleen J. Green,*‡§ Elayne A. Bornslaeger,*§ Andrew P. Kowalczyk,‡§
Helena L. Palka,*§ and Suzanne M. Norvell*§

*Departments of *Pathology, ‡Dermatology, and §R.H. Lurie Cancer Center,
Northwestern University Medical School, Chicago, Illinois 60611*

Introduction

Members of the intermediate filament (IF) multigene family are major constituents of the karyoskeleton and cytoskeleton in vertebrate and some invertebrate organisms. In contrast to microfilaments and microtubules, IF can be composed of many different types of polypeptide building blocks that fall into 5 classes exhibiting tissue- and cell type–specific patterns of expression. It is thought that this diversity reflects a functional specificity closely linked to the requirements of each tissue. (For a review of intermediate filaments see Fuchs and Weber, 1994; Heins and Aebi, 1994; Klymkowsky, 1995; Steinert and Roop, 1988.)

IFs are anchored both at the cell nucleus and the inner side of the plasma membrane and thus link these cellular elements indirectly through an anastomosing network that extends throughout the cytoplasm (Cowin and Burke, 1996; Goldman et al., 1985). This cytoarchitectural feature plays a critical role in imparting mechanical strength to tissues. Mutations in keratin IF polypeptides found in specific layers of the epidermis lead to cell fragility and cell cytolysis, giving rise to epidermal blistering (Fuchs, 1994; McLean and Lane, 1995; Steinert et al., 1994). The anchorage points for IF networks at the plasma membrane in epithelial tissues are intercellular junctions known as desmosomes (Collins and Garrod, 1994; Cowin and Burke, 1996; Cowin et al., 1985; Garrod, 1993; Green and Jones, 1996; Schmidt et al., 1994; Schwarz et al., 1990; Staehelin, 1974). Desmosomes are not restricted to epithelial cell membranes; they also associate with desmin-containing IF in cardiac muscle, and with vimentin-containing IF in meningeal cells as well as dendritic reticulum cells of the lymph node and spleen (Schmidt et al., 1994; Schwarz et al., 1990). By coupling stress-resisting IF networks to sites of tight adhesion, these membrane adhesive spot welds are thought to impart additional strength to the tissue.

Desmosome Structure: The Desmosomal Cadherins and Plakoglobin

Ultrastructurally, desmosomes appear as mirror image electron dense tripartate plaques that sandwich a specialized membrane core. The adjacent plasma membranes are separated by a 30-nm extracellular space bisected by a "central dense stratum" (Farquhar and Palade, 1963; Schwarz et al., 1990; Staehelin, 1974; Steinberg et al., 1987). The structural complexity of the desmosome is matched by its molecular complexity. Not only do individual desmosomes have numerous components, their composition varies from tissue to tissue. The transmembrane desmosomal components were demonstrated several years ago to be related to the cadherin family of calcium-dependent cell–cell adhesion molecules, and now form their own class called the desmosomal cadherins (Fig. 1) (Buxton and Magee, 1992; Cowin and Mechanic, 1994; Koch and Franke, 1994). The cadherin-like nature of these molecules is consistent with their presumed role in cell–cell adhesion. Additional evidence bolstering the idea that desmosomal cadherins function in adhesion comes from the existence of a class of autoimmune diseases called pemphigus, in which antibodies directed against the desmoglein subclass of desmosomal cadherins lead to loss of cell–cell adhesion and the development of oral and epidermal blisters (Amagai, 1995; Stanley, 1993; 1995). Somewhat surprisingly, however, experimental evidence suggests that desmosomal cadherins may not function independently as adhesion molecules, unlike their classic cadherin counterparts. This conclusion is based on the fact that, unlike E-cadherin (Nagafuchi et al., 1987), desmoglein and desmocollin alone are unable to confer adhesive properties on L cell fibroblasts as assessed by aggregation in cell suspension (Amagai et al., 1994; Chidgey et al., 1996; Kowalczyk et al., 1996). This may be due to the complexity of cadherin forms both within a single desmosome, as well as within tissues, and possibly the requirement for additional noncadherin molecules in the adhesive complex. For example, each desmosome is thought to contain members of at least two cadherin subclasses, the desmogleins and desmocollins. So far at least three desmoglein genes and three desmocollin genes have been identified, and these are expressed in a tissue- and differentiation-specific pattern (Buxton et al., 1993; 1994). Furthermore, each desmocollin gene gives rise to two alternatively spliced products, increasing the possible complexity of the desmosomal adhesive complex (Parker et al., 1991). Finally, noncadherin molecules such as the GPI-linked E48 antigen may contribute to the adhesive properties of desmosomes (Brakenhoff et al., 1995).

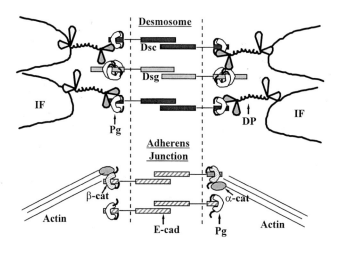

Figure 1. Comparison of adhesive junction structure. A diagram of a desmosome (*top*) is compared with an adherens junction (*bottom*). Only the major constitutive components for each junction referred to in the present study are shown. In the desmosome these include: desmocollin (*Dsc*), desmoglein (*Dsg*), plakoglobin (*Pg*), desmoplakin (*DP*), and intermediate filaments (*IF*). In the adherens junction these include: E-cadherin (*E-cad*), β-catenin (*β-cat*), α-catenin (*α-cat*), and plakoglobin (*Pg*).

Although the mechanism by which desmosomal cadherins mediate adhesion is an important unanswered question, this paper will focus on an equally intriguing question. That is, how does the desmosomal plaque mediate specific attachment to intermediate filaments and not actin-containing microfilaments? This issue is highlighted by the fact that one of the major plaque components of the desmosome, the armadillo repeat protein known as plakoglobin (Cowin, 1994; Cowin et al., 1986), has been shown to interact not only with the cytoplasmic tail of the desmosomal cadherins (Korman et al., 1989; Kowalczyk et al., 1994; Mathur et al., 1994; Roh and Stanley, 1995*b*; Troyanovsky et al., 1994*a,b*) but also that of the classic cadherins (Aberle et al., 1994; Hulsken et al., 1994; Jou et al., 1995; Knudsen and Wheelock, 1992; Peifer et al., 1992). Classic cadherins form the core of the microfilament-associated adherens junction which are assembled in close proximity to desmosomes in epithelial cells (Fig. 1) (Birchmeier, 1993; Green et al., 1987; Tsukita et al., 1990). Thus the question arises as to how specificity of cytoskeletal attachment is actually established at a molecular level.

Desmoplakin: A Putative Cytoskeletal Linker

To examine how cytoskeletal anchorage is achieved during intercellular junction assembly, we have been investigating protein–protein interactions that form the link responsible for IF attachment to the desmosome. Desmoplakin (DP) is the most abundant of the desmosomal plaque molecules, and a constitutive component of desmosomes in all tissues (Green and Stappenbeck, 1994; Schwarz et al., 1990). Two alternatively spliced forms thought to originate from a single gene have been identified, DPI and II (Virata et al., 1992). DPI (332 kD as determined from the cloned cDNA) is predicted to be a long rod-shaped molecule with a central α-helical coiled-coiled rod domain similar to that found in other fibrous and cytoskeletal molecules such as myosin (Green et al., 1990; 1992; O'Keefe et al., 1989). This central domain is flanked by two globular end domains shown by domain mapping studies to have separate functions. The COOH terminus contains sequences required for association with IF networks of the vimentin and keratin type, as well as a short regulatory stretch of 68 amino acids at the very end of the COOH terminus (Stappenbeck et al., 1993; Stappenbeck and Green, 1992). These final 68 residues are required for binding to keratin, but not vimentin, IF and also contain a consensus site for phosphorylation by cAMP-dependent protein kinase that regulates the interaction of this domain with keratin filament networks (Stappenbeck et al., 1994). More recent studies have demonstrated that DP's interaction with IF polypeptides is direct (Kouklis et al., 1994; Meng et al., 1995). The NH_2 terminus of DP, on the other hand, is required for association of this molecule with the desmosomal plaque, although the direct binding partners for this domain have not been identified (Stappenbeck et al., 1993).

Although the earlier domain mapping studies strongly supported the idea that DP could directly link IF with the desmosomal plaque, they fell short of demonstrating that DP is required for IF anchorage. In addition, the mechanism by which DP associates specifically with the desmosomal plaque and is excluded from adherens junctions is not understood. Finally, the possibility that DP itself plays a direct role in preventing microfilaments from being recruited to desmosomal plaques has not been tested. These latter two questions are important considering that plakoglobin, a prime candidate for providing a link between desmoplakin and the IF cytoskeleton, is present in both adherens junctions and desmosomes (Cowin et al., 1986). The experiments described below focus on the function and binding partners of individual desmosomal molecules and their constituent domains using two major approaches: (*1*) expression of dominant negative mutant proteins in epithelial cells that assemble intercellular junctions and (*2*) reconstitution of junctional complexes in fibroblasts that do not normally express junctional components. Together these studies are beginning to shed light not only on the function of individual components in desmosome assembly, but also how the choice of binding partners may govern the specificity of junction assembly leading to attachment of the correct filament system to the correct plasma membrane domains.

Results and Discussion

Desmoplakin's Role in Linking Intermediate Filaments to the Desmosomal Plaque

To directly test whether DP is required for the specific association of IF with the desmosomal plaque, a polypeptide derived from the DP NH_2-terminal domain was stably expressed in A-431 epithelial cells, which assemble both desmosomes and adherens junctions (Bornslaeger et al., 1996). Our rationale was that expression of such a polypeptide lacking the COOH-terminal IF binding domain of DP should compete for desmosomal binding partners with endogenous DP, and thus lead to its displacement from cell–cell borders. If indeed full length DP is required for IF anchorage at the plaque, then the consequence of perturbing the endogenous molecule should be an uncoupling of the IF network from the desmosome.

A number of stable cell lines expressing the DP NH_2-terminal polypeptide (DP-NTP) were generated and analyzed biochemically and morphologically. As shown in Fig. 2, DP-NTP was generated from NH_2-terminal DP cDNAs containing a stop codon at amino acid residue 585 (Genbank #M77830), giving rise in each case to a 70-kD polypeptide which was either untagged (Fig. 3, *A–D*) or FLAG epitope-tagged at its NH_2 terminus (Figs. 3, *E* and *F*, and 4). DP-NTP accumulated at cell–cell borders suggesting that this polypeptide contained sufficient information to associate with endogenous desmosomal components. This interpretation was consistent with the colocalization of DP-NTP with endogenous desmoglein, desmocollin, and plakoglobin (see Fig. 5, Bornslaeger et al., 1996). Immunoblotting with an antibody that recognized both DP-NTP and endogenous DP (NW161) revealed that the ectopically expressed DP-NTP was present at approximately threefold higher levels than endogenous full length DP. In addition, DP-NTP generally appeared to localize more continuously along cell borders than is typical for DP in control cells (Figs. 3 and 4).

As predicted, the subcellular distribution of endogenous DP was severely perturbed, with the degree of perturbation varying somewhat within a population of cells. In most cases, the regularly spaced punctate pattern at cell–cell borders typical of control cells (Fig. 3 *A*) was altered, often replaced with a smaller number of large aggregates containing endogenous DP and DP-NTP, still associated with keratin-containing filament bundles (Fig. 3, *C* and *D*). In the most extreme examples, endogenous DP appeared to be largely absent from cell borders, although FLAG-

Generation of N-terminal DP Disrupter Polypeptide

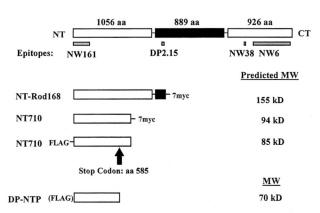

Figure 2. Schematic diagram of desmoplakin (DP) protein structure and DP NH$_2$-terminal polypeptides. Directly below a diagram of DPI are the DP antibodies used in this study and Bornslaeger et al. (1996), along with a designation of the regions containing the epitopes to which they are directed. Below this are the constructs utilized to generate DP-NTP expressing A-431 cell lines. As shown here, several different constructs were employed in the study, but in each case a stop codon at residue 585 led to the generation of a 70-kD protein designated DP-NTP. This stop codon apparently arose spontaneously during the bacterial replication of a DP plasmid precursor. For constructs with a 7myc tag at the COOH terminus, read-through beyond the stop codon led to the generation of small amounts of full length product with the predicted MWs shown at the right; these products were detected only using an antibody 9E10 (available from ATCC, Rockville, MD), previously shown to detect the 7myc tag with extremely high sensitivity (Stappenbeck et al., 1993) and not with the NW161 antibody. (For further details see Bornslaeger et al., 1996).

tagged or untagged DP-NTP remained abundantly distributed along the same borders (Fig. 3, *E* and *F*). Importantly, at the light microscope level, keratin IF bundles appeared to be anchored at cell borders only where endogenous DP was localized.

Ultrastructural analysis revealed the presence of numerous junction-like structures in cell lines expressing DP-NTP (see Fig. 7, Bornslaeger et al., 1996). Many of these structures appeared to lack an inner plaque and were completely devoid of attached IF, confirming the immunofluorescence results described above. However, some junctional structures were still associated with sparse, loosely packed IF suggesting the possibility that small amounts of endogenous DP might still be present, something that would not be unexpected given that we had employed a dominant negative approach. Immunogold microscopy using antibody DP2.15 directed against the DP rod domain did in fact confirm the presence of small amounts of endogenous DP in some of these junctions (see Fig. 7, Bornslaeger et al., 1996). Together, the immunofluorescence and electron microscopical results suggested that a range of junctional structures apparently co-exist in these lines, from oversized desmosomes containing large amounts of endogenous DP and associated with IF bundles to structures containing DP-NTP but little endogenous DP and largely lacking associated IF.

Together these data strongly support the idea that DP is required for anchorage of IF bundles to the desmosomal plaque. Nevertheless, the contribution of other IF linking molecules also reported to be present in desmosomes cannot be ruled out. In fact, recent results suggesting that DP's interactions with keratins in vitro is enhanced by IFAP300 are consistent with the possibility that a complex of proteins in the desmosomal plaque is involved in filament anchorage

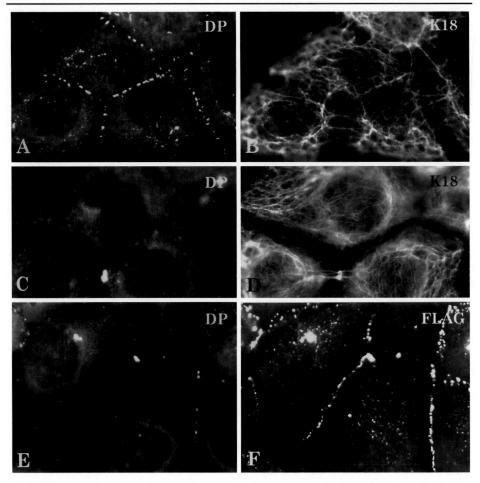

Figure 3. Indirect double label immunofluorescence analysis of endogenous DP, keratin IF, and DP-NTP in control and DP-NTP-expressing A-431 cell lines. The antigens recognized are designated in the upper right hand corner of each panel and are: endogenous desmoplakin (*DP*) recognized with rabbit polyclonal NW6, keratin 18 (*K18*) recognized by mouse monoclonal KSB17.2 (Sigma Chemical Co., St. Louis, MO), and the FLAG epitope tag recognized by the monoclonal M2 antibody (Kodak). *A* and *B* represent control A-431 cell lines expressing only the puromycin resistance plasmid used for selection in these experiments. *C* and *D* represent lines expressing the NTRod168 construct, and *E* and *F* are FLAG.NT710 lines. Note extensive interaction of keratin IF with desmosomes at cell borders in the control cells and the loss of anchored keratin bundles at cell–cell borders in panels *C* and *D*, except for the single site where endogenous DP remains. Panels *E* and *F* demonstrate that although endogenous DP is largely perturbed at cell borders, abundant junctional staining for DP-NTP recognized by the FLAG epitope tag antibody is still seen. (Magnifications: *A* and *B* = 7,000×; *C* and *D* = 5,700×; *E* and *F* = 5,500×) (Panels *A–D* are adapted from Bornslaeger et al., 1996).

(Skalli et al., 1995). Thus we should consider the possibility that in DP-NTP–expressing lines, disruption of endogenous DP might also displace other potential IF-binding molecules such as IFAP300, thus leading to a complete loss of filament attachment.

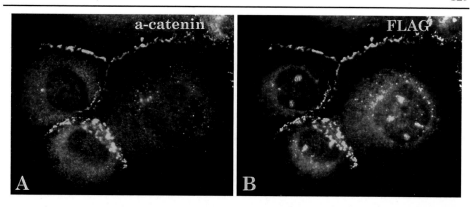

Figure 4. Indirect double label immunofluorescence analysis of DP-NTP and adherens junction components in DP-NTP expressing cells. An FLAG.NT710 line reacted with a polyclonal antibody directed against α-catenin (*A*) and monoclonal antibody M2 against the FLAG epitope (*B*). Note the extensive colocalization of these components. Endogenous desmoglein, desmocollin and plakoglobin also colocalize (not shown) with adherens junction components in these cells, unlike the situation in control A-431 cells. (Magnification: 4,000×).

Possible Role of Desmoplakin in Providing Specificity of Filament Attachment and Recruitment of Desmosomal versus Adherens Junction Components

Although IF were disconnected from junctional plaques in cell lines expressing DP-NTP, the peripheral band of actin normally present in these cells still remained, and if anything, appeared to be more robust (see Fig. 8, Bornslaeger et al., 1996). To assess the effect of DP-NTP expression and loss of IF anchorage on adherens junction components, we examined the distribution of α-catenin, β-catenin, and E-cadherin in these cell lines. In contrast to control cells, where desmosomal and adherens junction markers are largely distinct, in DP-NTP expressing lines, DP-NTP and endogenous desmosomal components colocalized extensively with α-catenin, E-cadherin, and to a lesser extent, β-catenin (Fig. 4 and Bornslaeger et al., 1996). These observations suggest that the normal segregation of junctional components that occurs during adherens junction and desmosome assembly is compromised in DP-NTP expressing cells. Immunogold electron microscopy confirmed that α-catenin and DP-NTP are intermixed extensively in the junctional structures present in these cells (see Fig. 10, Bornslaeger et al., 1996).

The mechanism by which this intermixing occurs is still a mystery. It seems likely that the expression of DP-NTP leads to a perturbation of protein interactions that normally prevent recruitment of classic cadherin–plakoglobin complexes into desmosomes. Whether this perturbation is due to a loss of sequences from the DP NH$_2$ terminus or rod that are required for preventing certain protein–protein interactions, or whether it is due to the loss of IF attachment normally facilitated by the missing COOH terminus is not known. What candidate binding partners in the plaque might contribute to junctional mixing in DP-NTP expressing cells? Although a number of possible scenarios can be envisioned, the most likely hinge on the role of the common junctional component plakoglobin. The involvement of plakoglobin would also be most easily explained if it is capable of forming a complex with DP. Previous work by Troyanovsky et al. (1994*b*) have demonstrated the importance of plakoglobin along with DP in linking IF to the cell surface. However, the mechanism by which plakoglobin facilitates this link has not been elucidated. In order to do so, we must determine whether plakoglobin binds directly to DP, and if so, whether this interaction first requires an association with a desmosomal cadherin.

L Cell Fibroblasts Used as an In Vivo Reconstitution System for Desmosomal Assembly Complexes

Stoichiometry of plakoglobin–cadherin complexes. To examine whether DP-NTP can associate directly with the desmosomal cadherin–plakoglobin complex and to address how adherens junction and desmosomal components are normally segregated, we have developed a "reconstitution" system in which multiple junctional components are ectopically expressed in L cell fibroblasts, which do not normally assemble junctions. In this system, plakoglobin associates and coimmunoprecipitates with both desmogleins and desmocollins. Formation of this complex leads to the metabolic stabilization of plakoglobin, and its recruitment to the plasma membrane where it colocalizes in a diffuse distribution with the desmosomal cadherins (Kowalczyk et al., 1994). We took advantage of the myc epitope tag at the COOH terminus of each protein to directly compare the amount of plakoglobin that coimmunoprecipitated with desmocollin or desmoglein, and determined that the stoichiometry of the plakoglobin:desmoglein complex is ∼6:1 (Kowalczyk et al., 1996), a ratio similar to that determined by metabolic labeling of cells transiently overexpressing these components (Witcher et al., 1996). This is dramatically different from the 1:1 stoichiometry exhibited by the plakoglobin:desmocollin complex (Kowalczyk et al., 1996) or by plakoglobin:classic cadherin complexes (Ozawa and Kemler, 1992).

The mechanism by which this stoichiometry is achieved at a molecular level is not understood, nor is its functional significance. However, several observations could potentially be explained by a scenario in which the desmoglein tail competes with desmocollin for binding to plakoglobin. First, the desmoglein tail exhibits a behavior distinct from the desmocollin tail when these domains are stably expressed in chimeric form with a gap junction connexin protein in A-431 cells. In these experiments, the desmocollin tail recruited plakoglobin, DP, and IF to junctional plaques at the plasma membrane. However, the desmoglein tail in this system disrupted endogenous desmosomes, resulting in diffuse DP staining and loss of IF anchorage (Troyanovsky et al., 1993). In other experiments carried out in our own laboratory, we have generated A-431 cell lines stably expressing different levels of full length desmoglein 1 (these cells normally express desmoglein 2 and some desmoglein 3 but not desmoglein 1). It appears that moderate amounts of desmoglein 1 can incorporate into endogenous junctions, whereas higher amounts lead to disruption of these junctions (Norvell and Green, manuscript in preparation). One possible explanation for these results is that the desmoglein tail might sequester plakoglobin, resulting in a decrease in the pool available for binding to desmocollin. Arguing against such a scenario, at least in the case of the desmoglein–connexin chimera described above, is that a cytosolic pool of plakoglobin still exists in A-431 cells expressing the desmoglein chimera (Troyanovsky et al., 1993). Although it is possible that this pool is in a form unavailable for binding to desmocollin, a satisfactory explanation for the ability of desmoglein to disrupt desmosome assembly has yet to be experimentally determined. In addition, the possible significance for normal junction assembly of the dramatic difference between the plakoglobin:desmoglein stoichiometry and the stoichiometry of other plakoglobin:cadherin complexes is not yet known.

Interaction of the DP NH$_2$ terminus with desmosomal cadherin–plakoglobin complexes. As stated above, one important unanswered question is whether DP can form a complex with plakoglobin, as this molecule is a good candidate for mediating interactions between the DP NH$_2$ terminus and the desmosomal cadherins. To address this question, DP-NTP was co-expressed in L cells with plakoglobin and either desmoglein or desmocollin. A dramatic alteration in the normally diffuse distribution of desmoglein and plakoglobin (Fig. 5 *A*) to a punctate distribution containing all three components was observed (Fig. 5 *B*). Importantly, an antibody directed against the DP NH$_2$ terminus coimmunoprecipitated plakoglobin in a complex with DP-NTP, suggesting that plakoglobin provides a link to the cadherin-based desmosomal core. Plakoglobin deletions in which the NH$_2$ or COOH terminus was removed also associated with DP, indicating that the central armadillo repeats of plakoglobin contain the sequences required for complex formation (Kowalczyk et al., manuscript in preparation). Together these results suggest not only that desmocomal cadherins and plakoglobin form a complex with DP-NTP, but that DP-NTP is also capable of clustering these transmembrane desmosomal cadherin complexes.

The observation that DP-(NTP)–induced clustering occurs with either desmocollin or desmoglein is interesting given that the chimeric connexin–desmosomal cadherin tail proteins described above differed in their ability to recruit desmosomal plaque components in A-431 cells (Troyanovsky et al., 1993; 1994*a, b*). In L cells desmocollin and desmoglein behaved similarly, suggesting that either of these desmosomal cadherins may be able to recruit DP and set up the linkage with the IF cytoskeleton.

The formation of desmosomal protein clusters in L cells may represent a stage of junction assembly that occurs normally in epithelial cells. Clustering appears not to depend on cadherin-based cell–cell adhesion, which does not occur in L cells expressing desmoglein or desmocollin alone (Kowalczyk et al., 1996). Thus, rather than leading to the formation of typical junctions at cell–cell contact points, clustering leads to the assembly of punctate structures on cell membranes and within the cell (presumably associated with vesicles containing desmosomal cadherins). Ultrastructural analysis currently underway should reveal whether these clusters appear similar to the half-desmosomal structures recently reported by Demlehner et al. that occur in the absence of cell–cell contact in epithelial cells (Demlehner et al., 1995).

The similarity of these clusters to nascent junctional plaques was more easily assessed by employing a chimeric molecule with the E-cadherin extracellular domain and a desmosomal cadherin cytoplasmic domain. The E-cadherin extracellular domain has been demonstrated previously to confer adhesive properties on L cells, and functions even when attached to the desmoglein 3 cytoplasmic tail (Roh and Stanley, 1995*a*). When we expressed an E-cadherin/desmoglein 1 chimera along with plakoglobin and DP-NTP, punctate structures reminiscent of epithelial cell desmosomes were observed at some cell–cell interfaces. Thus, the fibroblast recon-

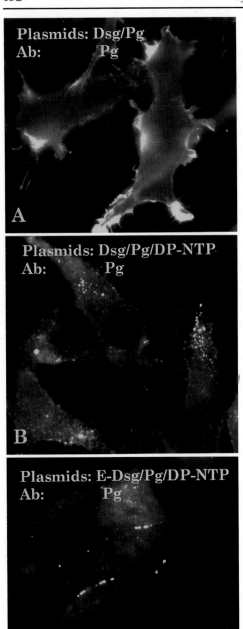

Figure 5. Immunofluorescence analysis of L cell lines stably expressing different combinations of intercellular junction components. The plasmid combinations are designated in the upper left of each panel. In each case, plakoglobin is being recognized using a monoclonal antibody 11E4. Note the presence of punctate clusters in cells expressing DP-NTP in conjunction with plakoglobin (*Pg*) and desmoglein (*Dsg*) (*B*) or with plakoglobin and a chimeric cadherin constructed of an E-cadherin extracellular domain and a desmoglein cytoplasmic domain (*E-Dsg*) (*C*). In addition to plakoglobin, DP-NTP and the cadherins are also present in these structures (not shown). However, plakoglobin is diffuse on the cell surface in cells only expressing desmoglein and plakoglobin (*A*). (Magnifications: *A* = 8,800×; *B* = 8,400×; *C* = 9,700×).

stitution system promises to continue to provide a tool for defining the molecules necessary and sufficient for generating a junction de novo.

It remains to be definitively demonstrated that DP absolutely requires plakoglobin for binding to desmosomal cadherins or whether it can bind directly to the desmosomal cadherin tails. Along these lines, Troyanovsky et al. reported the re-

cruitment of DP to desmocollin connexin chimeras even in the absence of the plakoglobin binding site of desmocollin, suggesting that a juxtamembrane sequence in
the desmocollin tail is sufficient to recruit DP (Troyanovsky et al., 1994*b*). Experiments are ongoing to test whether plakoglobin is required for DP binding to desmocollin or desmoglein using the L cell reconstitution system.

Plakoglobin Binding Partners Govern Specificity of Junction Assembly and Interaction with the Cytoskeleton

Plakoglobin is capable of binding directly to a number of proteins in the junctional
plaque and the cytoplasm. This molecule has been demonstrated to bind directly to
both α-catenin, a linker to the actin cytoskeleton found in adherens junctions
(Rimm et al., 1995), and classic cadherins (Aberle et al., 1994; Jou et al., 1995; Sacco
et al., 1995). As described above, plakoglobin also binds directly to desmoglein and
desmocollin, and perhaps desmoplakin. Intriguingly, plakoglobin, like β-catenin,
binds to proteins not found in the junctional complex, most notably the APC tumor
supressor gene product, and may be involved in cell signaling and growth control
(reviewed in Klymkowsky and Parr, 1995). Thus, the question arises as to how plakoglobin and other members of the armadillo family are targeted to junctions versus the cytoplasm, and once targeted, how specificity of junction assembly and cytoskeletal interaction is achieved. One possibility is that plakoglobin contains
multiple overlapping binding sites for direct binding partners in different junctions.

Supporting this theory is the fact that the binding site for α-catenin on plakoglobin (Aberle et al., 1996; Sacco et al., 1995) overlaps extensively with a region
shown to be critical for binding to desmoglein (Chitaev et al., 1996; Wahl et al.,
1996; Witcher et al., 1996). Consistent with these findings, α-catenin is unable to associate with plakoglobin when it is complexed with a desmosomal cadherin. Since
α-catenin is required either by itself (Rimm et al., 1995) or through its interaction
with α-actinin (Knudsen et al., 1995) for association with actin, its failure to associate with plakoglobin:desmosomal cadherin complexes would be predicted to prevent
anchorage to the microfilament cytoskeleton at these plasma membrane domains
(Kowalcyzk et al., manuscript in preparation). In addition, in L cells co-expressing a
chimeric cadherin consisting of the desmoglein extracellular domain and an E-cadherin tail along with plakoglobin and DP-NTP, the chimera remained largely diffuse on the cell surface. This observation suggests that the DP NH_2 terminus associates with and preferentially clusters desmosomal cadherin–plakoglobin complexes
but not classic cadherin–plakoglobin complexes. Thus, an additional mechanism by
which junctional specificity might be achieved could be that plakoglobin bound to
E-cadherin may be unable to efficiently bind DP and/or vice versa.

Summary and Model of Desmosome Assembly and Specificity

The experiments reviewed here provide the framework for modeling how the desmosomal plaque is assembled and interacts with IF but not microfilaments. In this
model plakoglobin plays a central role at the intersection of possible protein–protein
interactions, with multiple overlapping binding sites for constituents of both junctions. The data so far suggest that identity of the cadherin tail to which plakoglobin
binds plays an important role in specifying whether a microfilament- (i.e., α-catenin) or IF-linking molecule (i.e., DP) is recruited to the complex. It may also be the

case that plakoglobin–α-catenin or plakoglobin–DP interactions inhibit plakoglobin binding to desmosomal or classic cadherin tails, respectively. In this regard, the potential direct partnership of DP and plakoglobin and mapping of sequences required for such an interaction warrant further investigation. Dissection of plakoglobin:DP binding regions may reveal additional overlapping sites that explain why DP is unable to cluster complexes containing the E-cadherin tail. For instance, overlap of DP and classic cadherin binding sites in the central region of the armadillo repeats of plakoglobin could explain this result.

The dynamics that actually occur in a cell to determine which protein–protein interactions are favored at a particular time and location are not currently understood. A possible contributing factor could be the phosphorylation state of junctional components. A number of observations suggest that protein phosphorylation can regulate intercellular junction assembly, maintenance and function (Balsamo et al., 1996; Citi, 1992; Kinch et al., 1995; Pasdar et al., 1995; Sheu et al., 1989; Staddon et al., 1995; Volberg et al., 1992), and it is known that plakoglobin, the desmosomal cadherins (with the exception of the shorter spliced form of desmocollin), and desmoplakin are phosphoproteins (Cowin, 1994; Mueller and Franke, 1983; Parrish et al., 1990; Stappenbeck et al., 1994). Furthermore, it has been demonstrated that desmosome assembly is inhibited by okadaic acid, suggesting that phosphatase activity is required for normal desmosome formation (Pasdar et al., 1995), and that the interaction of DP's COOH terminus with keratin IF networks is inhibited by phosphorylation of a serine 23 residues from the end of the molecule.

The idea that specific protein interactions within the cadherin–catenin complex might be regulated by phosphorylation is supported by mutational analysis of the E-cadherin catenin binding domain in which substitution of a serine cluster abolished phosphorylation and affected complex formation with catenins and led to a loss of cadherin-mediated adhesion (Stappert and Kemler, 1994). Tyrosine phosphorylation of β-catenin in ras-transformed breast epithelia also led to its decreased association with E-cadherin and an increase in its detergent solubility (Kinch et al., 1995). The idea that kinases and phosphatases may play a role in regulating the assembly and function of intercellular junctions has recently been bolstered by the identification of receptor protein tyrosine phosphatases that associate with cadherins (Balsamo et al., 1996; Brady-Kalnay et al., 1995; Fuchs et al., 1996). Although the evidence that phosphorylation can regulate protein–protein interactions in cadherin-based complexes is accumulating, whether phosphorylation can actually govern the specificity of such interactions must await future investigation.

Acknowledgments

The authors thank all those who graciously contributed reagents for this study, including M. Wheelock, K. Johnson, J. Ortin, P. McCrea, and M. Takeichi. Thanks go also to other members of the Green lab, in particular Mitch Denning, for useful discussions and critical analysis of the manuscript.

This work was supported by grants to K. Green from the NIH (AR41836 and AR43380), the March of Dimes Birth Defects Foundation, and the Council For Tobacco Research, USA, Inc. K. Green is an American Cancer Society Faculty Research Awardee. A. Kowalczyk has been supported by a fellowship from the Dermatology Foundation. H. Palka is supported in part by NIH Training Grant T32 CA09560-10, and S. Norvell is supported in part by NIH Training Grant T32 ES07284-01.

References

Aberle, H., S. Butz, J. Stappert, H. Weissig, R. Kemler, and H. Hoschuetzky. 1994. Assembly of the cadherin-catenin complex in vitro with recombinant proteins. *J. Cell Sci.* 107:3655–3663.

Aberle, H., H. Schwartz, H. Hoschuetzky, and R. Kemler. 1996. Single amino acid substitutions in proteins of the armadillo gene family abolish their binding to α-catenin. *J. Biol. Chem.* 271:1520–1526.

Amagai, M. 1995. Adhesion meolcules. I. Keratinocyte-keratinocyte interactions; cadherins and pemphigus. *J. Invest. Derm.* 104:146–152.

Amagai, M., S. Karpati, V. Klaus-Kovtun, M.C. Udey, and J.R. Stanley. 1994. The extracellular domain of pemphigus vulgaris antigen (desmoglein 3) mediates weak homophilic adhesion. *J. Invest. Derm.* 102:402–408.

Balsamo, J., T. Leung, H. Ernst, M.K.B. Zanin, S. Hoffman, and J. Lilien. 1996. Regulated binding of a PTP1B-like-phosphatase to N-cadherin: control of cadherin-mediated adhesion by dephosphorylation of β-catenin. *J. Cell Biol.* 134:801–813.

Birchmeier, W. 1993. Undercoat-constitutive proteins of cell–cell adherens junctions. *Jpn. J. Cancer Res.* 84:1332–1333.

Bornslaeger, E.B., C.M. Corcoran, T.S. Stappenbeck, and K.J. Green. 1996. Breaking the connection: displacement of the desmosomal plaque protein desmoplakin from cell–cell interfaces disrupts anchorage of intermediate filament bundles and alters intercellular junction assembly. *J. Cell Biol.* 134:985–1002.

Brady-Kalnay, S.M., D.L. Rimm, and N.K. Tonks. 1995. Receptor protein tyrosine phosphatase PTPμ associates with cadherins and catenins in vivo. *J. Cell Biol.* 130:977–986.

Brakenhoff, R.H., M. Gerretsen, E.M.C. Knippels, M.V. Dijk, H.V. Essen, D.O. Weghuis, R.J. Sinke, G.B. Snow, and G.A.M.S.V. Dongen. 1995. The human E48 antigen, highly homologous to the murine Ly-6 antigen ThB, is a GPI-anchored molecule apparently involved in keratinocyte cell–cell adhesion. *J. Cell Biol.* 129:1677–1689.

Buxton, R.S., P. Cowin, W.W. Franke, D.R. Garrod, K.J. Green, I.A. King, P.J. Koch, A.I. Magee, D.A. Rees, J.R. Stanley, and M.S. Steinberg. 1993. Nomenclature of the desmosomal cadherins. *J. Cell Biol.* 121:481–483.

Buxton, R.S., and A.I. Magee. 1992. Structure and interactions of desmosomal and other cadherins. *Sem. Cell Biol.* 3:157–167.

Buxton, R.S., A.I. Magee, I.A. King, and J. Arnemann. 1994. Desmosomal genes. *In* Molecular Biology of Desmosomes and Hemidesmosomes. J.E. Collins and D.R. Garrod, editors, R.G. Landes Co., Austin, TX. 1–17.

Chidgey, M.A.J., J.P. Clarke, and D.R. Garrod. 1996. Expression of full-length desmosomal glycoproteins (desmocollins) is not sufficient to confer strong adhesion on transfected L929 cells. *J. Invest. Derm.* 106:689–695.

Chitaev, N.A., R.E. Leube, R.B. Troyanovsky, L.G. Eshkind, W.W. Franke, and S.M. Troyanovsky. 1996. The binding of plakoglobin to desmosomal cadherins: patterns of binding sites and topogenic potential. *J. Cell Biol.* 133:359–369.

Citi, S. 1992. Protein kinase inhibitors prevent junction dissociation induced by low extracellular calcium in MDCK epithelial cells. *J. Cell Biol.* 117:169–178.

Collins, J.E., and D.R. Garrod. 1994. Molecular Biology of Desmosomes and Hemidesmosomes. R.G. Landes Co., Austin, TX. pp. 131.

Cowin, P. 1994. Plakoglobin. *In* Molecular Biology of Desmosomes and Hemidesmosomes. J.E. Collins and D.R. Garrod, editors. R.G. Landes Co., Austin, TX. 53–68.

Cowin, P., and B. Burke. 1996. Cytoskeleton-membrane interactions. *Curr. Opin. Cell Biol.* 8:56–65.

Cowin, P., W.W. Franke, C. Grund, H.-P. Kapprell, and J. Kartenbeck. 1985. The desmosome-intermediate filament complex. *In* The Cell in Contact. G.M. Edelman and J.-P. Thiery, editors. John Wiley and Sons, New York. 427–460.

Cowin, P., H.-P. Kapprell, W.W. Franke, J. Tamkun, and R.O. Hynes. 1986. Plakoglobin: a protein common to different kinds of intercellular adhering junctions. *Cell.* 46:1063–1073.

Cowin, P., and S. Mechanic. 1994. Desmosomal cadherins and their cytoplasmic interactions. *In* Molecular Mechanisms of Epithelial Cell Junctions: From Development to Disease. S. Citi, editor. R.G. Landes Co., Austin, TX. 141–155.

Demlehner, M.P., S. Schafer, C. Grund, and W.W. Franke. 1995. Continual assembly of half-desmosomal structures in the absence of cell contacts and their frustrated endocytosis: a coordinated Sisyphus cycle. *J. Cell Biol.* 131:745–760.

Farquhar, M.G., and G.E. Palade. 1963. Junctional complexes in various epithelia. *J. Cell Biol.* 17:375–412.

Fuchs, E. 1994. Intermediate filaments and disease: mutations that cripple cell strength. *J. Cell Biol.* 125:511–516.

Fuchs, E., and K. Weber. 1994. Intermediate filaments: structure, dynamics, function and disease. *Annu. Rev. Biochem.* 63:345–382.

Fuchs, M., T. Muller, M.M. Lerch, and A. Ullrich. 1996. Association of human protein tyrosine phosphatase kappa with members of the armadillo family. *J. Biol. Chem.* 271:16712–16719.

Garrod, D.R. 1993. Desmosomes and hemidesmosomes. *Curr. Opin. Cell Biol.* 5:30–40.

Goldman, R.D., A. Goldman, K. Green, J. Jones, N. Lieska, and H.-Y. Yang. 1985. Intermediate filaments: possible functions as cytoskeletal connecting links between the nucleus and the cell surface. *Ann. NY Acad. Sci.* 455:1–17.

Green, K.J., B. Geiger, J.C.R. Jones, J.C. Talian, and R.D. Goldman. 1987. The relationship between intermediate filaments and microfilaments prior to and during the formation of desmosomes and adherens-type junctions in mouse epidermal keratinocytes. *J. Cell Biol.* 104:1389–1402.

Green, K.J., and J.C.R. Jones. 1996. Desmosomes and hemidesmosomes: structure and function of molecular components. *FASEB J.* 10:871–881.

Green, K.J., D.A.D. Parry, P.M. Steinert, M.L.A. Virata, R.M. Wagner, B.D. Angst, and L.A. Nilles. 1990. Structure of the human desmoplakins: implications for function in the desmosomal plaque. *J. Biol. Chem.* 265:2603–2612.

Green, K.J., and T.S. Stappenbeck. 1994. The desmosomal plaque: role in attachment of intermediate filaments to the cell surface. *In* Molecular Mechanisms of Epithelial Cell Junctions: From Development to Disease. S. Citi, editor. R.G. Landes Co., Austin, TX. 157–171.

Green, K.J., M.L.A. Virata, G.W. Elgart, J.R. Stanley, and D.A.D. Parry. 1992. Comparative structural analysis of desmoplakin, bullous pemphigoid antigen and plectin: members of a new gene family involved in organization of intermediate filaments. *Int. J. Biol. Macromol.* 14:145–153.

Heins, S., and U. Aebi. 1994. Making heads and tails of intermediate filament assembly, dynamics and networks. *Curr. Opin. Cell Biol.* 6:25–33.

Hulsken, J., W. Birchmeier, and J. Behrens. 1994. E-cadherin and APC compete for the interaction with β-catenin and the cytoskeleton. *J. Cell Biol.* 127:2061–2069.

Jou, T.S., D.B. Stewart, J. Stappert, W.J. Nelson, and J.A. Marrs. 1995. Genetic and biochemical dissection of protein linkages in the cadherin-catenin complex. *Proc. Natl. Acad. Sci. USA.* 92:5067–5071.

Kinch, M.S., G.J. Clark, C.J. Der, and K. Burridge. 1995. Tyrosine phosphorylation regulates the adhesions of ras-transformed breast epithelia. *J. Cell Biol.* 130:461–471.

Klymkowsky, M.W. 1995. Intermediate filaments: new proteins, some answers, more questions. *Curr. Opin. Cell Biol.* 7:46–54.

Klymkowsky, M.W., and B. Parr. 1995. The body language of cells: the intimate connection between cell adhesion and behavior. *Cell.* 83:5–8.

Knudsen, K.A., A.P. Soler, K.R. Johnson, and M.J. Wheelock. 1995. Interaction of α-actinin with the cadherin/catenin cell–cell adhesion complex via α-catenin. *J. Cell Biol.* 130:67–77.

Knudsen, K.A., and M.J. Wheelock. 1992. Plakoglobin, or an 83-kD homologue distinct from beta-catenin interacts with E-cadherin and N-cadherin. *J. Cell Biol.* 118:671–679.

Koch, P.J., and W.W. Franke. 1994. Desmosomal cadherins: another growing multigene family of adhesion molecules. *Curr. Opin. Cell. Biol.* 6:682–687.

Korman, N.J., R.W. Eyre, V. Klaus-Kovtun, and J.R. Stanley. 1989. Demonstration of an adhering-junction molecule (plakoglobin) in the autoantigens of pemphigus foliaceus and pemphigus vulgaris. *N. Engl. J. Med.* 321:631–635.

Kouklis, P.D., E. Hutton, and E. Fuchs. 1994. Making a connection: direct binding between keratin intermediate filaments and desmosomal proteins. *J. Cell Biol.* 127:1049–1060.

Kowalczyk, A.P., J.E. Borgwardt, and K.J. Green. 1996. Analysis of desmosomal cadherin-adhesive function and stoichiometry of desmosomal cadherin-plakoglobin complexes. *J. Invest. Derm.* 107:293–300.

Kowalczyk, A.P., H.L. Palka, H.H. Luu, L.A. Nilles, J.E. Anderson, M.J. Wheelock, and K.J. Green. 1994. Posttranslational regulation of plakoglobin expression: influence of the desmosomal cadherins on plakoglobin metabolic stability. *J. Biol. Chem.* 269:31214–31223.

Mathur, M., L. Goodwin, and P. Cowin. 1994. Interactions of the cytoplasmic domain of the desmosomal cadherin Dsg1 with plakoglobin. *J. Biol. Chem.* 269:14075–14080.

McLean, W.H.I., and E.B. Lane. 1995. Intermediate filaments in disease. *Curr. Opin. Cell Biol.* 7:118–125.

Meng, J.J., K.J. Green, and W. Ip. 1995. The molecular basis for the interaction between intermediate filament (IF) proteins and desmoplakin. *Mol. Biol. Cell.* 6:376a.

Mueller, H., and W.W. Franke. 1983. Biochemical and immunological characterization of desmoplakins I and II, the major polypeptides of the desmosomal plaque. *J. Mol. Biol.* 163:647–671.

Nagafuchi, A., Y. Shirayoshi, K. Okasaki, K. Yamada, and M. Takeichi. 1987. Transformation of cell adhesion properties by exogenously introduced E-cadherin cDNA. *Nature.* 329:341–343.

O'Keefe, E.J., H.P. Erickson, and V. Bennett. 1989. Desmoplakin I and desmoplakin II. Purification and characterization. *J. Biol. Chem.* 264:8310–8318.

Ozawa, M., and R. Kemler. 1992. Molecular organization of the uvomorulin-catenin complex. *J. Cell Biol.* 116:989–996.

Parker, A.E., G.N. Wheeler, J. Arnemann, S. Pidsley, P. Ataliotis, C.L. Thomas, D.A. Rees, A.I. Magee, and R.S. Buxton. 1991. Desmosomal glycoproteins II and III: cadherin-like junctional molecules generated by alternative splicing. *J. Biol. Chem.* 266:10438–10445.

Parrish, E.P., J.E. Marston, D.L. Mattey, H.R. Measures, R. Venning, and D.R. Garrod. 1990. Size heterogeneity, phosphorylation and transmembrane organisation of desmosomal glycoproteins 2 and 3 (desmocollins) in MDCK cells. *J. Cell Sci.* 96:239–248.

Pasdar, M., Z. Li, and H. Chan. 1995. Desmosome assembly and disassembly are regulated by reversible protein phosphorylation in cultured epithelial cells. *Cell Motil. Cytoskel.* 30:108–121.

Peifer, M., P.D. McCrea, K.J. Green, E. Wieschaus, and B.M. Gumbiner. 1992. The vertebrate adhesive junction proteins β-catenin and plakoglobin and the Drosophila segment polarity gene armadillo form a multigene family with similar properties. *J. Cell Biol.* 118:681–691.

Rimm, D.L., E.R. Koslov, P. Kebriaei, C.D. Cianci, and J.S. Morrow. 1995. α1(E)-catenin is an actin-binding and -bundling protein mediating the attachment of F-actin to the membrane adhesion complex. *Proc. Natl. Acad. Sci. USA.* 92:8813–8817.

Roh, J.-Y., and J.R. Stanley. 1995a. Intracellular domain of desmoglein 3 (pemphigus vulgaris antigen) confers adhesive function on the extracellular domain of E-cadherin without binding catenins. *J. Cell Biol.* 128:939–947.

Roh, J.-Y., and J.R. Stanley. 1995b. Plakoglobin binding by human Dsg3 (pemphigus vulgaris antigen) in keratinocytes requires the cadherin-like intracytoplasmic segment. *J. Invest. Derm.* 104:720–724.

Sacco, P.A., T.M. McGranahan, M.J. Wheelock, and K.R. Johnson. 1995. Indentification of plakoglobin domains required for association with N-cadherin and alpha-catenin. *J. Biol. Chem.* 270:20201–20206.

Schmidt, A., H.W. Heid, S. Schafer, U.A. Nuber, R. Zimbelmann, and W.W. Franke. 1994. Desmosomes and cytoskeletal architecture in epithelial differentiation: cell type-specific plaque components and intermediate filament anchorage. *Eur. J. Cell Biol.* 65:229–245.

Schwarz, M.A., K. Owaribe, J. Kartenbeck, and W.W. Franke. 1990. Desmosomes and hemidesmosomes: constitutive molecular components. *Annu. Rev. Cell Biol.* 6:461–491.

Sheu, H.-M., Y. Kitajima, and H. Yaoita. 1989. Involvement of protein kinase C in translocation of desmoplakins from cytosol to plasma membrane during desmosome formation in human squamous cell carcinoma cells grown in low to normal calcium concentration. *Exp. Cell Res.* 185:176–190.

Skalli, O., P.M. Steinert, and R.D. Goldman. 1995. Mechanisms involved in anchoring keratin intermediate filaments (IF) to desmosomes. *Mol. Biol. Cell.* 6:192a.

Staddon, J.M., K. Herrenknecht, C. Smales, and L.L. Rubin. 1995. Evidence that tyrosine phosphorylation may increase tight junction permeability. *J. Cell Sci.* 108:609–619.

Staehelin, L.A. 1974. Structure and function of intercellular junctions. *Int. Rev. Cytol.* 39:191–283.

Stanley, J.R. 1993. Cell adhesion molecules as targets of autoantibodies in pemphigus and pemphigoid, bullous diseases due to defective epidermal cell adhesion. *Adv. Immunol.* 51:291–325.

Stanley, J.R. 1995. Autoantibodies against adhesion molecules and structures in blistering skin diseases. *J. Exp. Med.* 181:1–4.

Stappenbeck, T.S., E.A. Bornslaeger, C.M. Corcoran, H.H. Luu, M.L.A. Virata, and K.J. Green. 1993. Functional analysis of desmoplakin domains: specification of the interaction with keratin versus vimentin intermediate filament networks. *J. Cell Biol.* 123:691–705.

Stappenbeck, T.S., and K.J. Green. 1992. The desmoplakin carboxyl terminus coaligns with and specifically disrupts intermediate filament networks when expressed in cultured cells. *J. Cell Biol.* 116:1197–1209.

Stappenbeck, T.S., J.A. Lamb, C.M. Corcoran, and K.J. Green. 1994. Phosphorylation of the desmoplakin COOH terminus negatively regulates its interaction with keratin intermediate filament networks. *J. Biol. Chem.* 269:29351–29354.

Stappert, J., and R. Kemler. 1994. A short core region of E-cadherin is essential for catenin binding and is highly phosphorylated. *Cell Adh. Comm.* 2:319–327.

Steinberg, M.S., H. Shida, G.J. Giudice, M. Shida, N.H. Patel, and O.W. Blaschuk. 1987. On the molecular organization, diversity and functions of desmosomal proteins. *Ciba Found. Symp.* 125:3–25.

Steinert, P.M., A.C.T. North, and D.A.D. Parry. 1994. Structural features of keratin intermediate filaments. *J. Invest. Derm. Suppl.* 103:19s–24s.

Steinert, P.M., and D.R. Roop. 1988. Molecular and cellular biology of intermediate filaments. *Annu. Rev. Biochem.* 57:593–625.

Troyanovsky, S.M., L.G. Eshkind, R.B. Troyanovsky, R.E. Leube, and W.W. Franke. 1993. Contributions of cytoplasmic domains of desmosomal cadherins to desmosome assembly and intermediate filament anchorage. *Cell.* 72:561–574.

Troyanovsky, S.M., R.B. Troyanovsky, L.G. Eshkind, V.A. Krutovskikh, R.E. Leube, and W.W. Franke. 1994a. Identification of the plakoglobin-binding domain in desmoglein and its role in plaque assembly and intermediate filament anchorage. *J. Cell Biol.* 127:151–160.

Troyanovsky, S.M., R.B. Troyanovsky, L.G. Eshkind, R.E. Leube, and W.W. Franke. 1994b. Identification of amino acid sequence motifs in desmocollin, a desmosomal glycoprotein, that are required for plakoglobin binding and plaque formation. *Proc. Natl. Acad. Sci. USA.* 91:10790–10794.

Tsukita, S., S. Tsukita, and A. Nagafuchi. 1990. The undercoat of adherens junctions: a key specialized structure in organogenesis and carcinogenesis. *Cell Struct. Funct.* 15:7–12.

Virata, M.L.A., R.M. Wagner, D.A.D. Parry, and K.J. Green. 1992. Molecular structure of the human desmoplakin I and II amino terminus. *Proc. Natl. Acad. Sci. USA.* 89:544–548.

Volberg, T., Y. Zick, R. Dror, I. Sabanay, C. Gilon, A. Levitzki, and B. Geiger. 1992. The effect of tyrosine-specific protein phosphorylation on the assembly of adherens-type junctions. *EMBO J.* 11:1733–1742.

Wahl, J.K., P.A. Sacco, T.M. McGranahan-Sadler, L.M. Sauppe, M.J. Wheelock, and K.R. Johnson. 1996. Plakoglobin domains that define its association with the desmosomal cadherins and the classical cadherins: identification of unique and shared domains. *J. Cell Sci.* 109:1143–1154.

Witcher, L.L., R. Collins, S. Puttagunta, S.E. Mechanic, M. Munson, B. Gumbiner, and P. Cowin. 1996. Desmosomal cadherin binding domains of plakoglobin. *J. Biol. Chem.* 271:10904–10909.

Intermediate Filament Linker Proteins

Elaine Fuchs, Yanmin Yang, James Dowling, Panos Kouklis,
Elizabeth Smith, Lifei Guo, and Qian-Chun Yu

*Howard Hughes Medical Institute, Department of Molecular Genetics and Cell
Biology, The University of Chicago, Chicago, Illinois 60637*

BPAG1e, BPAG1n, plectin, and desmoplakin constitute a small group of large
coiled-coil proteins that have the capacity to associate with intermediate filaments
(IFs) (Foisner et al., 1991; Green et al., 1992; Stappenbeck and Green, 1992; Kouklis et al., 1994; Yang et al., 1996). The 230-kD epidermal form of BPAG1
(BPAG1e) was first identified by antisera from patients with the autoimmune disorder, Bullous Pemphigoid (BP; Sugi et al., 1989; Tamai et al., 1993; Stanley, 1993).
BPAG1e is specifically expressed in stratified squamous epithelia, where it localizes to the intracellular portion of the hemidesmosomal plaque (Stanley, 1993).
BPAG1e has been implicated in linking the keratin IF network to hemidesmosomes, which are integrin-mediated cell substratum adherens junctions that attach
the basal layer of the stratified tissue to the underlying basement membrane (Stanley, 1993). Plectin decorates IF networks and is expressed ubiquitously, implicated
as a global stabilizer and integrator of the cytoskeleton (Foisner et al., 1991). Desmoplakin is expressed predominantly in muscle and stratified epithelia, where it localizes to desmosomes and appears to link IFs to the desmosomal plaque (Stappenbeck and Green, 1992; Kouklis et al., 1994). Desmosomes are cadherin-mediated
cell-cell junctions that interconnect muscle or epithelial cells within their respective
tissues (Schwarz et al., 1990; Garrod et al., 1993). Recently, we identified a new
form of BPAG1, BPAG1n, which is exclusively expressed in neurons (Yang et al.,
1996; see also Brown et al., 1995*b*). This form has the remarkable capacity to link
actin microfilaments to neurofilaments (Yang et al., 1996). This small, but novel
group of IF linker proteins is especially interesting in light of genetic studies that
identify defects in the BPAG1 gene as the cause of *dystonia musculorum*, a well-
known neurological disorder in mice (Guo et al., 1995; Brown et al., 1995a), and defects in the plectin gene as the cause of Epidermolysis Bullosa Simplex with Muscular Dystrophy in humans (Smith et al., 1996; McLean et al., 1996). In this report, we
discuss recent research that has illuminated the importance of this group of proteins
in cell biology.

Attachments between the IF Linker Proteins and IF Networks

The IF linker proteins have a large central coiled-coil domain that is flanked by
globular NH_2- and COOH-terminal domains (Green et al., 1992). Since antibodies
to these proteins localize to or near IF networks, it was reasonable to propose that
these proteins might associate with IFs. Deletion mutagenesis studies confirmed
this association, and delineated the globular COOH-terminal domain as the site for

this interaction (Stappenbeck and Green, 1992; Wiche et al., 1993; Kouklis et al., 1994; Yang et al., 1996). Important in this regard is a significant level of sequence similarity in this domain, particularly in a group of tandem repeats of a 19-residue motif (Green et al., 1992). The COOH-terminal segment of plectin contains six of these repeats, while that of desmoplakin has three and BPAG1e and BPAG1n have two. Based on mutagenesis studies, these repeats seem to play some role in IF associations (Wiche et al., 1993).

The best evidence that IF linker proteins directly associate with IFs stems from biochemical studies on a bacterially expressed and fplc chromatography–purified desmoplakin tail segment (Kouklis et al., 1994). When the desmoplakin tail domain was biotinylated and used in overlay assays, it selectively bound to the head domains of type II epidermal keratins (Kouklis et al., 1994). This method did not identify an association between the desmoplakin tail segment and either the simple epithelial keratins or vimentin, despite the fact that the tail segment localizes to these IF networks in transfected cells in vitro (Stappenbeck and Green, 1992; Kouklis et al., 1994). Taken together, it seems likely that the interactions between IF linker proteins and different IF networks may be tailored to suit the specific structural needs of each cytoskeleton. Moreover, recent evidence suggests that cells may be able to regulate the forming and breaking of these interactions by a mechanism involving phosphorylation of the tail segment of the linker protein (Stappenbeck et al., 1994). Thus, the associations appear to be dynamic as well as specific.

Associations between IF Linker Proteins and Other Cellular Components

Given that the tail segment of IF linker proteins is involved in association to the IF network, it is perhaps not too surprising that recent studies have begun to implicate the head domain of IF linker proteins in associations with other components of cells. An epitope-tagged, amino-terminal head segment of desmoplakin has been shown to associate specifically with the desmosomal plaque in transfected epithelial cells, and when expressed at sufficiently high levels, causes a disruption of the plaque (Bornslaeger et al., 1996; E. Smith and E. Fuchs, unpublished results). Which desmosomal proteins associate with the desmoplakin head segment remains to be elucidated.

The only other head segment studied in detail is that of BPAG1n, which contains 214 amino acid residues that share a high degree of sequence similarity to the actin binding domain of β-spectrin and α-actinin (Yang et al., 1996). To assess whether this domain has the capacity to associate with the actin cytoskeleton, we used DNA transfection to express an epitope-tagged amino-terminal segment of BPAG1n in SW13 cells, a cell line that has no cytoplasmic IF network (Sarria et al., 1990). Antibodies against the BPAG1n head domain and against β-actin colocalized along the actin stress fibers and at the cortical actin cytoskeleton underlying the plasma membrane of these cells (Yang et al., 1996). The specificity of BPAG1n's amino-terminal domain for actin filaments was shown using biochemical studies, which gave a K_d of BPAG1n's binding to actin of 2.1×10^{-7} M (Yang et al., 1996). This is within the range of that observed for β-spectrin and other proteins with actin binding domains (Lo et al., 1994; Jongstra-Bilen et al., 1992).

Interestingly, plectin's head domain, but not that of BPAG1e or desmoplakin, shares sequence homology with that of the actin binding domain in BPAG1n (Yang et al., 1996). Antibodies against plectin localize not only to intermediate filaments, but also to actin stress fibers and focal contacts in cultured cells (Foisner et al., 1991; Seifert et al., 1992). That plectin may act as a crosslinker between actin and intermediate filament networks has recently been proposed by Foisner et al. (1995), who showed by immunoelectron microscopy that antibodies against plectin label crossbridge-like structures that connect IFs with actin microfilaments in cytoskeletal extracts from cultured glioma cells expressing endogenous plectin. Our recent studies on BPAG1n, coupled with the sequence similarities in the head domains of BPAG1n and plectin, suggest that plectin also associates directly with the actin cytoskeleton.

Molecular Demonstrations that IF Linker Proteins Are Able to Perform as Integrators of IF Cytoskeletons and Other Cellular Components

In view of a bona fide actin binding domain at one end of BPAG1n's coiled-coil rod segment and a neurofilament binding domain at the other end, we recently examined the ability of these associations to take place simultaneously (Yang et al., 1996). In this case, we used gene transfection of SW13 cells to show that a recombinant 280-kD, full-length BPAG1n protein colocalized with both a recombinant neurofilament cytoskeleton and the actin cytoskeleton as judged by triple immunofluorescence. In the absence of BPAG1n, the neurofilaments did not colocalize along the actin network, suggesting that BPAG1n is specifically able to influence IF architecture through its dual association with actin microfilaments.

Direct evidence demonstrating a true linking function is still lacking for desmoplakin, BPAG1e, and plectin. Given the ability of desmoplakin and BPAG1e to localize to desmosomes and hemidesmosomes, respectively, it seems likely that desmoplakin will have the capacity to simultaneously bind to IFs and at least one member of the desmosomal plaque, while BPAG1e will be able to directly link IFs to hemidesmosomes. Given the sequence similarities between plectin and BPAG1n, it seems likely that plectin, like BPAG1n, will function at least in part in connecting the actin and IF cytoskeletons in cells. Plectin has also been shown to bind to lamin B, microtubule-associated proteins MAP1 and MAP2, α-spectrin, and fodrin (Foisner et al., 1991). It remains to be determined whether these interactions occur in vivo, and if so, how many of these associations can occur simultaneously and involve separate binding domains.

Mutations in IF Linker Proteins Cause Severe Genetic Disorders

Several recent studies have led to the realization that IF linker proteins serve vital functions in cells and tissues. The first genetic study leading to this conclusion came from Guo et al. (1995), who used gene targeting to ablate the BPAG1 gene in mice. The targeting vector was engineered such that both BPAG1e and BPAG1n were ablated. Since these two proteins are mutually exclusive in their expression patterns (Yang et al., 1996), the functions of both proteins could be elucidated by analysis of the knockout mice (Guo et al., 1995).

Basal epidermal cells lacking BPAG1e have hemidesmosomes with seemingly normal structure, but with severed connections to the keratin IF cytoskeleton (Guo et al., 1995). It was both interesting and surprising that the localization of other known hemidesmosomal proteins seemed unaffected by removal of BPAG1e, suggesting that IFs, but not other hemidesmosomal components, require BPAG1e for their anchorage to the plaque.

The severing of keratin filaments from hemidesmosomes did not seem to compromise the ability of the outer surface of the plaque to attach to anchoring filaments (Guo et al., 1995). Thus, the number and structure of anchoring filaments associated with each plaque was indistinguishable from normal. This behavior seemed to differ from actin-mediated cell substratum attachment, where treatment of cells with cytochalasin D to disrupt the actin cytoskeleton prevented cell substratum attachment (for review, see Luna and Hitt, 1992). Moreover, since the removal of BPAG1e did not perturb cell-substratum adhesion, it is unlikely that BPAG1 antibodies are directly responsible for initiating the epidermal-dermal blistering seen in bullous pemphigoid patients.

The selective severing of connections between keratin IFs and hemidesmosomes in the BPAG1e null skin did lead to the rupturing of basal epidermal cells upon mechanical stress (Guo et al., 1995). Thus, skin blistering was seen in these knockout mice, but it was blistering due to intraepidermal cell cytolysis, typical of the human disorder, epidermolysis bullosa simplex (EBS), rather than epidermal-dermal separation, typical of the autoimmune disease, bullous pemphigoid (Fitzpatrick et al., 1993; Guo et al., 1995). The EBS phenotype in these mice was milder than that which is created when the keratin network is perturbed, either through gene ablation or through dominant negative mutations (Vassar et al., 1991; Coulombe et al., 1991; Bonifas et al., 1991; Rugg et al., 1994; Chan et al., 1994). This is readily explained by the fact that the IF network is only partially perturbed when the IF network is severed from hemidesmosomes. In contrast, when the IF network is completely perturbed through disruption of filament assembly, cells become fragile and prone to rupturing upon physical stress.

Ablating the BPAG1 gene gave rise to neurologic degeneration identical to that seen in the *dystonia musculorum* (*dt/dt*) mouse (Guo et al., 1995), a well-known autosomal recessive defect (Duchen and Strich, 1964; Duchen et al., 1976; Messer and Strominger, 1980). This observation preceded the discovery of BPAG1n (Brown et al., 1995*a,b*; Yang et al., 1996), and therefore initially came as a complete surprise. However, one of two spontaneously derived *dt/dt* mouse strains was found to lack BPAG1e and mating the *dt* mice with the BPAG1 knockout mice provided further evidence that we had inadvertently discovered the basis for this well-known genetic defect in mice. The cloning of BPAG1n and its localization to the nervous system (Brown et al., 1995*a,b*; Yang et al., 1996) has subsequently led to an understanding of the neurologic disorder in these animals.

In the *dt/dt* mice (Janota, 1972; Sotelo and Guenet, 1988) and in the BPAG1 knockout mice (Guo et al., 1995; Dowling et al., 1997), axons within the peripheral nervous system of 4-week-old mice degenerate. Severely affected are the large primary sensory axons, while more modest degeneration is seen in second order sensory neurons and in a few motor neurons. Studies on mouse chimeras composed of *dt/dt* and wild-type cells indicate that the defect is intrinsic (although not necessarily exclusively) to sensory neurons (Campbell and Peterson, 1992).

In view of our studies illuminating BPAG1n as an actin-IF linker protein, we reexamined the pathology of our BPAG1 knockout animals, and discovered that the NF network within the sensory axons is markedly altered when BPAG1n is missing (Yang et al., 1996; Dowling et al., 1997). This finding strongly implies that BPAG1n does indeed perform an actin-NF linkage in vivo and provides a graphic illustration of the vital features of this novel type of cytoskeletal connector protein. This offers a partial explanation as to why the axonal cytoskeleton of the primary sensory neurons is the primary target of degeneration in these mice, as first suggested by Campbell and Peterson (1992) upon analysis of chimeric nervous systems of mice generated from a mixture of *dt/dt* and wild-type ES cells. The BPAG1n null mice tell us that without these connections, neurofilaments become less abundant and the remaining ones are disorganized, leading to axon fragility and degeneration. In this regard, it may be BPAG1n's role is to keep NFs in their place in the axon, a likely prerequisite for normal axonal transport and function.

What does the BPAG1 knockout mouse and analysis of BPAG1n tell us about human genetic disease? Recent studies on a group of rare patients with a combined EBS and neuromuscular syndrome has begun to deliver clues (Smith et al., 1996; McLean et al., 1996). Most cases of EBS are autosomal dominant and have a pathology which is largely restricted to the skin (Fine et al., 1991). A rare autosomal recessive form of EBS manifests itself not only as a skin blistering disorder, but also as muscular dystrophy (Smith et al., 1996 and references therein). This phenotype is similar to that seen in BPAG1 knockout mice (Guo et al., 1995). Immuno-fluorescence studies revealed that skin sections of these patients did label with antibodies against BPAG1e, but they did not label with antibodies against HD1, a hemidesmosomal protein that colocalizes with BPAG1e (Owaribe et al., 1990; Smith et al., 1996; McLean et al., 1996). Intriguingly, HD1 not only appears to be plectin, but in addition, patients with EBS and muscular dystrophy were found to have premature termination mutations in both alleles of the human plectin gene (Smith et al., 1996; McLean et al., 1996).

These most exciting recent findings underscore the importance of mouse knockout technology in bringing us closer to an understanding between cell biology and human genetic disease. They also emphasize the importance of IF linker proteins in cytoskeletal integrity and cellular function. These linker proteins are clearly critical for the survival of epidermal cells and neurons, which devote most of their protein synthesizing capacity to producing an extensive cytoskeletal network. They now also appear to be vital for the integrity of muscle cells, where the IF network has often been overshadowed by the actin and myosin cytoskeletons. As we continue to learn more about the functions of IFs and their associated connections, we are brought to the realization that the IF cytoskeleton is significantly more important to life than we had ever previously imagined.

References

Bonifas, J.M., A.L. Rothman, and E.H. Epstein. 1991. Epidermolysis bullosa simplex: evidence in two families for keratin gene abnormalities. *Science.* 254:1202–1205.

Bornslaeger, E.A., C.M. Corcoran, T.S. Stappenbeck, and K.J. Green. 1996. Breaking the connection: displacement of the desmosomal plaque protein desmoplakin from cell-cell in-

146 _Cytoskeletal Regulation of Membrane Function_

terfaces disrupts anchorage of intermediate filament bundles and alters intercellular junction assembly. _J. Cell Biol._ 134:985–1001.

Brown, A., G. Bernier, M. Mathieu, J. Rossant, and R. Kothary. 1995a. The mouse dystonia musculorum gene is a neural isoform of bullous pemphigoid antigen 1. _Nature Genet._ 10:301–306.

Brown, A., G. Dalpe, M. Mathieu, and R. Kothary. 1995b. Cloning and characterization of the neural isoforms of human dystonin. _Genomics._ 29:777–780.

Campbell, R.M., and A.C. Peterson. 1992. An intrinsic neuronal defect operates in dystonia musculorum: a study of dt/dt<==>+/+ chimeras. _Neuron._ 9:693–703.

Chan, Y.-M., Q.-C. Yu, A. Christiano, J. Uitto, R.S. Kucherlapati, J. LeBlanc-Straceski, and E. Fuchs. 1994. Mutations in the non-helical linker segment L1-2 of keratin 5 in patients with Weber-Cockayne Epidermolysis Bullosa Simplex. _J. Cell Sci._ 107:765–774.

Coulombe, P.A., M.E. Hutton, A. Letai, A. Hebert, A.S. Paller, and E. Fuchs. 1991. Point mutations in human keratin 14 genes of epidermolysis bullosa simplex patients: genetic and functional analyses. _Cell._ 66:1301–1311.

Dowling, J., Y. Yang, R. Wollmann, L.F. Reichardt, and E. Fuchs. 1997. Developmental expression of BPAG1-n: insights into the spastic ataxia and gross neurologic degeneration in Dystonia musculorum mice. _Dev. Biol._ In press.

Duchen, L.W. 1976. Dystonia Musculorum: an inherited disease of the nervous system in the mouse. _Adv. Neurol._ 14:353–365.

Duchen, L.W., and S.J. Strich. 1964. Clinical and pathological studies of an hereditary neuropathy in mice (dystonia musculorum). _Brain._ 87:367–378.

Fine, J.D., E.A. Bauer, R.A. Briggaman, D.M. Carter, R.A.J. Eady, N.B. Esterly, K.A. Holbrook, S. Hurwitz, L. Johnson, A. Lin, R. Pearson, and V.P. Sybert. 1991. Revised clinical and laboratory criteria for subtypes of inherited epidermolysis bullosa. _J. Am. Acad. Dermatol._ 24:119–135.

Fitzpatrick, T.B., A.Z. Eisen, K. Wolff, I.M. Freedberg, and K.F. Austen. 1993. Dermatology in General Medicine, Fourth ed., Vol. I and II. McGraw-Hill, Inc., New York.

Foisner, R., P. Traub, and G. Wiche. 1991. Protein kinase A- and protein kinase C-regulated interaction of plectin with lamin B and vimentin. _Proc. Natl. Acad. Sci. USA._ 88:3812–3816.

Foisner, R., W. Bohn, K. Mannweiler, and G. Wiche. 1995. Distribution and ultrastructure of plectin arrays in subclones of rat glioma C_6 cells differing in intermediate filament protein (vimentin) expression. _J. Struct. Biol._ 115:304–317.

Garrod, D.R. 1993. Desmosomes and hemidesmosomes. _Curr. Opin. Cell Biol._ 5:30–40.

Green, K.J., M.L.A. Virata, G.W. Elgart, J.R. Stanley, and D.A.D. Parry. 1992. Comparative structural analysis of desmoplakin, bullous pemphigoid antigen and plectin: members of a new gene family involved in organization of intermediate filaments. _Int. J. Biol. Macromol._ 14:145–153.

Guo, L., L. Degenstein, J. Dowling, Q.-C. Yu, R. Wollmann, B. Perman, and E. Fuchs. 1995. Gene targeting of BPAG1: abnormalities in mechanical strength and cell migration in stratified squamous epithelia and severe neurologic degeneration. _Cell._ 81:233–243.

Janota, I. 1972. Ultrastructural studies of an hereditary sensory neuropathy in mice (Dystonia Musculorum). _Brain._ 95:529–536.

Jongstra-Bilen, J., P.A. Janmey, J.H. Hartwig, S. Galca, and J. Jongstra. 1992. The lymphocyte-specific protein LSP1 binds to F-actin and to the cytoskeleton through its COOH-terminal basic domain. *J. Cell Biol.* 118:1443–1453.

Kouklis, P., E. Hutton, and E. Fuchs. 1994. Making the connection: keratin intermediate filaments and desmosome proteins. *J. Cell Biol.* 127:1049–1060.

Lo, S.H., P.A. Janmey, J.H. Hartwig, and L.B. Chen. 1994. Interactions of tensin with actin and identification of its three distinct actin-binding domains. *J. Cell Biol.* 125:1067–1075.

Luna, E.J., and A.L. Hitt. 1992. Cytoskeleton-plasma membrane interactions. *Science.* 258:955–963.

McLean, W.H.I., L. Pulkkinen, F.J.D. Smith, E.L. Rugg, E.B. Lane, F. Bullrich, R.E. Burgeson, S. Amano, D.L. Hudson, K. Owaribe, et al. 1996. Loss of plectin causes epidermolysis bulllosa with muscular dystrophy: cDNA cloning and genomic organization. *Genes Dev.* 10:1724–1735.

Messer, A., and N.L. Strominger. 1980. An allele of the mouse mutant dystonia musculorum exhibits lesions in red nucleus and striatum. *Neuroscience.* 5:543–549.

Owaribe, K., J. Kartenbeck, S. Stumpp, T.M. Magin, T. Krieg, L.A. Diaz, and W.W. Franke. 1990. The hemidesmosomal plaque I. Characterization of a major constituent protein as a differentiation marker for certain forms of epithelia. *Differentiation.* 45:207–220.

Rugg, E.L., W.H.I. McLean, E.B. Lane, R. Pitera, J.R. McMillan, P.J.C. Dopping-Hepenstal, H.A. Navsaria, I.M. Leigh, and R.A.J. Eady. 1994. A functional "knockout" of human keratin 14. *Genes Dev.* 8:2563–2573.

Sarria, A.J., S.K. Nordeen, and R.M. Evans. 1990. Regulated expression of vimentin cDNA in cells in the presence and absence of a preexisting vimentin filament network. *J. Cell Biol.* 111:553–565.

Schwarz, M.A., K. Owaribe, J. Kartenbeck, and W.W. Franke. 1990. Desmosomes and hemidesmosomes: constitutive molecular components. *Annu. Rev. Cell Biol.* 6:461–491.

Seifert, G.J., D. Lawson, and G. Wiche. 1992. Immunolocalization of the intermediate filament-associated protein plectin at focal contacts and actin stress fibers. *Eur. J. Cell Biol.* 59:138–147.

Smith, F.J.D., R.A.J. Eady, I.M. Leigh, J.R. McMillan, E.L. Rugg, D.P. Kelsell, S.P. Bryant, N.K. Spurr, J.F. Geddes, G. Kirtschig, et al. 1996. Plectin deficiency results in muscular dystrophy with epidermolysis bullosa. *Nature Genet.* 13:450–457.

Sotelo, C., and J.L. Guenet. 1988. Pathologic changes in the CNS of dystonia musculorum mutant mouse: an animal model for human spinocerebellar ataxia. *Neuroscience.* 27:403–424.

Stanley, J.R. 1993. Cell adhesion molecules as targets of autoantibodies in pemphigus and pemphigoid, bullous diseases due to defective epidermal cell adhesion. *Adv. Immunol.* 53:291–325.

Stappenbeck, T.S., and K.J. Green. 1992. The desmoplakin carboxyl terminus coaligns with and specifically disrupts intermediate filament networks when expressed in cultured cells. *J. Cell Biol.* 116:1197–1209.

Stappenbeck, T.S., J.A. Lamb, C.M. Corcoran, and K.J. Green. 1994. Phosphorylation of the desmoplakin COOH terminus negatively regulates its interaction with keratin intermediate filament networks. *J. Biol. Chem.* 269:29351–29354.

Sugi, T., T. Hashimoto, T. Hibi, and T. Nishikawa. 1989. Production of human monoclonal anti-basement membrane zone (BMZ) antibodies from a patient with bullous pemphigoid (BP) by Epstein-Barr virus transformation. *J. Clin. Invest.* 84:1050–1055.

Tamai, K., D. Sawamura, H.C. Do, Y. Tamai, K. Li, and J. Uitto. 1993. The human 230-kD bullous pemphigoid antigen gene (BPAG1): exon-intron organization and identification of regulatory tissue specific elements in the promoter region. *J. Clin. Invest.* 92:814–822.

Vassar R., P.A. Coulombe, L. Degenstein, K. Albers, and E. Fuchs. 1991. Mutant keratin expression in transgenic mice causes marked abnormalities resembling a human genetic skin disease. *Cell.* 64:365–380.

Wiche, G., D. Gromov, A. Donovan, M.J. Castanon, and E. Fuchs. 1993. Expression of plectin mutant cDNA in cultured cells indicates a role of COOH-terminal domain in intermediate filament association. *J. Cell Biol.* 121:607–619.

Yang, Y., J. Dowling, Q.-C. Yu, P. Kouklis, D.W. Cleveland, and E. Fuchs. 1996. An essential cytoskeletal linker protein connecting actin microfilaments to intermediate filaments. *Cell.* 86:655–665.

Functional Studies of the Protein 4.1 Family of Junctional Proteins in *Drosophila*

Richard G. Fehon, Dennis LaJeunesse, Rebecca Lamb, Brooke M. McCartney, Liang Schweizer, and Robert E. Ward

Developmental, Cell, and Molecular Biology Group, Department of Zoology, Duke University, Durham, North Carolina 27708-1000

Introduction

Cell–cell interactions play an important role in defining the fates of individual cells throughout the development of virtually all higher organisms. Based on classical signal transduction paradigms, one might expect three general classes of gene products to be involved in such interactions: (1) receptor-ligand proteins that act at the cell surface to mediate interactions between cells and to transduce signals from the outside to the inside of the cell; (2) proteins in the cytoplasm that serve to carry signals between the cell surface and the nucleus; and (3) transcription factors and related proteins that act within the nucleus to alter gene expression in response to extracellular signals. Recent genetic studies using model systems such as *Drosophila* and *C. elegans* have identified key players in all three categories. In particular, the use of genetic interaction screens has proven to be a highly efficient means to identify novel members of signaling pathways (Greenwald and Rubin, 1992; Artavanis-Tsakonas et al., 1995). Furthermore, genetic mutations in these pathway members have provided insights into functions that range from embryonic pattern formation to cytoskeletal assembly.

In addition to providing a superb genetic system with which to dissect molecular constituents of signaling pathways, the small size, relative ease of immunostaining, and highly polarized, single-layered epithelia of *Drosophila* have facilitated detailed studies of the cellular and subcellular localizations of signaling molecules. One of the most interesting outcomes of these studies has been the discovery that many transmembrane and cytoplasmic signaling proteins are tightly associated with cytologically defined cellular junctions. For example, studies of Notch,[1] a transmembrane receptor that functions to retain cells in an undifferentiated state, show that it is localized to the adherens junction in a variety of epithelia (Fehon et al., 1991). Likewise the Notch ligands, Delta and Serrate, are found in the same junctional region (Thomas et al., 1991; Kooh et al., 1993). Sevenless and the EGF-receptor, both of which are transmembrane receptor tyrosine-kinases, are also apically localized in the region of the adherens junction (Tomlinson et al., 1987; Zak and Shilo, 1992). Even more striking is evidence regarding *Drosophila* proteins whose vertebrate homologues were originally identified as junctional components. The *armadillo* gene encodes a homologue of β-catenin, a cytoplasmic protein that binds to and regulates the functions of junctional cadherins (Peifer and Wieschaus, 1990).

---

[1]In keeping with the convention for *Drosophila* genes, gene names are in italics, and their protein products are in roman characters.

Cytoskeletal Regulation of Membrane Function © 1997 by The Rockefeller University Press

Based on genetic studies, *armadillo* functions as a member of the *wingless* signaling pathway. Subsequent molecular and biochemical studies have shown that Armadillo has at least two separable functional domains, one that maintains junctional structure and another that functions, via a poorly understood mechanism, in *wingless* signal transduction (Orsulic and Peifer, 1996). Recently a receptor for Wingless with seven transmembrane segments, *Dfz2*, has been identified (Bhanot et al., 1996), although its subcellular localization has not yet been determined. These results indicate that most, if not all, cell to cell signaling mechanisms in *Drosophila* are restricted to specific domains of the cell membrane, specifically the apical junctional complex, rather than being evenly distributed over the cell surface.

To better understand the mechanisms by which receptors and related proteins are localized to cellular junctions and the functions of this localization in regulating signaling mechanisms, we are studying *Drosophila* members of a family of junctionally associated cytoplasmic proteins, the Protein 4.1 superfamily. Vertebrate members of this family include Protein 4.1, Talin, Ezrin, Moesin, Radixin, Merlin (the NF2 tumor suppressor), and several protein tyrosine phosphatases. In *Drosophila*, where genetic tools can be applied to dissect the functions of these genes, we have identified three Protein 4.1 family members. One other highly divergent member of this family, *expanded*, was previously identified in *Drosophila* (Boedigheimer and Laughon, 1993). In this review, we will summarize our findings to date regarding the expression, subcellular localization, and functions of these evolutionarily-conserved junctional proteins.

Cell Junctions in *Drosophila* Epithelia

In *Drosophila*, as in other cells, epithelial cells display a highly characteristic apical–basal polarity of the cell surface, including a distinct apical junctional complex. Within this region, two types of junctions are apparent in electron micrographs; the adherens junction and the septate junction (Tepass and Hartenstein, 1994). In contrast to vertebrate cells in which the tight junctions are apical to the adherens junctions, in *Drosophila* epithelia the adherens junctions are found in the most apical region of cell contact, while the septate junctions lie just basal to this region. The adherens junctions are both structurally and functionally analogous in *Drosophila* and vertebrates. In these and other animals, the adherens junctions appear to be major sites of cytoskeletal-membrane interaction and have high concentrations of cadherins and associated molecules. On the other hand, the septate junction, characterized by a ladder-like array of septa that connect adjacent cells, has not been observed in vertebrates. Instead, vertebrate epithelial cells display tight junctions that form both a transepithelial barrier and a barrier within the plane of the membrane that separates the apical membrane from the basolateral membrane (Mandel, 1993). While septate and tight junctions are morphologically quite distinct, previous studies have suggested that they may be functionally analogous, and recent molecular data seem to support this notion (Willott et al., 1993).

The Protein 4.1 Superfamily

Protein 4.1 has been extensively studied for its structural role in the erythrocyte membrane skeleton. From these studies, a model for the function of Protein 4.1 and the rest of this gene family has been derived. In brief, an NH$_2$-terminal, 35-kD proteolytic fragment from erythrocyte Protein 4.1 has been shown to interact with the cytoplasmic tail of glycophorin C, a transmembrane protein abundantly expressed in erythrocytes (Anderson and Marchesi, 1985; Marfatia et al., 1995). This NH$_2$-terminal region is well-conserved in all members of the family, and in fact is the defining structural feature of the Protein 4.1 superfamily. A smaller region located close to the COOH terminus seems to mediate interactions with spectrin and actin (Correas et al., 1986). Thus, Protein 4.1 maintains the biconcave shape of the erythrocyte by stabilizing interactions between the plasma membrane and the spectrin-actin cytoskeleton. Consistent with this notion, inherited defects in Protein 4.1 result in hemolytic anemia due to destabilization of the erythrocyte membrane (Marchesi et al., 1990).

Although clearly instructive, in some ways studies of erythrocyte Protein 4.1 may have confused our views of the function of Protein 4.1 and the other family members in non-erythroid cells. Expression studies have shown that the spectrin/actin-binding domain found in erythrocyte Protein 4.1 is composed in part by an alternatively spliced region that is almost unique to the erythrocyte (Tang et al., 1990; Conboy et al., 1991). Thus, the Protein 4.1 isoforms expressed in non-erythroid cells may have very different cellular functions than erythroid Protein 4.1. In addition, immunofluorescent studies have shown that Protein 4.1 is primarily junctionally localized in non-erythroid cells, suggesting a function that would not be relevant to erythrocytes that display neither cellular junctions nor stable intercellular contacts.

Interest in the Protein 4.1 superfamily, and in the so-called Ezrin-Radixin-Moesin (ERM) family within the superfamily, has increased recently due to the discovery that the neurofibromatosis 2 tumor suppressor Merlin (Schwannomin) is an ERM-related protein (Rouleau et al., 1993; Trofatter et al., 1993). Sequence analysis shows that Ezrin, Radixin, and Moesin are highly conserved, being ~85% identical over their entire lengths. Ezrin was originally identified as a component of the brush border (Bretscher, 1983), but subsequent studies of this and the other family members indicate that they show partially overlapping domains of expression that include the adherens junction and cleavage furrows (Sato et al., 1991, 1992; Franck et al., 1993). The high degree of structural similarity and functional studies using antisense oligonucleotides have led to the suggestion that the ERM proteins are partially redundant (Takeuchi et al., 1994). Though similar to the ERM proteins in its NH$_2$-terminal half, Merlin is clearly distinct from the ERM sequences, particularly in the COOH-terminal half of the protein. Despite this structural similarity, at present the cellular function of Merlin is not known, nor is the mechanism by which loss of Merlin function results in tumor formation understood.

Drosophila Coracle

Using degenerate oligonucleotide primers and PCR, we have identified a *Drosophila* homologue of Protein 4.1 termed Coracle (Fehon et al., 1994). *Drosophila* Coracle is nearly 55% identical to the human and *Xenopus* Protein 4.1 over a stretch of ~350 amino acids at the NH$_2$-terminus of these proteins. This region corresponds approximately to the 30-kD fragment that has been shown to bind the cytoplasmic tail of glycophorin C. In addition, Coracle and human Protein 4.1 are 40% identical over the COOH-terminal 70 aa, a region that is not conserved in any of the other family members. Thus, Coracle is a structural homologue of Protein 4.1 rather than a more distantly related family member. However, Coracle seems to lack the spectrin/actin binding domain found in the erythrocyte isoform of Protein 4.1, suggesting that it may be functionally more similar to the non-erythroid isoforms.

To examine the cellular functions of Coracle, we generated polyclonal and monoclonal antibodies and used them to determine the localization of Coracle in embryonic and developing adult (imaginal) tissues (Fehon et al., 1994). Several features of Coracle expression are unique. First, Coracle is not expressed maternally, and in the embryo is expressed relatively late, after the major morphogenetic events associated with gastrulation have been completed. Second, Coracle is expressed only in the epithelial derivatives of the ectodermal germ layer, and thus is not expressed in the midgut, the muscles, or the central nervous system. Third, Coracle is tightly associated with the cell membrane and in particular is found primarily in the region of the septate junction. In fact, the tissue specificity and developmental timing of Coracle expression correlate precisely with the appearance of "pleated" septate junctions ("smooth" septate junctions have few or no cross-septa and are found primarily in the midgut) during embryonic development (Tepass and Hartenstein, 1994). Finally, confocal optical sections of tissues stained with fluorescently labeled phalloidin, which binds to filamentous actin, show that actin filaments are predominantly found in the most apical junctional region, corresponding with the adherens junctions and clearly distinct from the septate junctions where Coracle is found. This observation is consistent with the finding that the Coracle sequence lacks a spectrin/actin binding domain, and suggests that unlike erythroid Protein 4.1, Coracle does not associate directly with filamentous actin.

To better understand the functions of Coracle in developmental processes, we have isolated mutations in the *coracle* gene. Severe loss of function *coracle* mutations are recessive embryonic lethal, resulting in a failure in a morphogenetic process termed dorsal closure. During dorsal closure, the lateral embryonic epidermal cells extend dorsally to cover the embryonic gut and overlying amnioserosa. Studies have revealed a large number of gene products are involved in this process, ranging from components of a signaling mechanism that is proposed to regulate interactions between the lateral epidermis and the amnioserosa (Glise et al., 1995; Riesgo-Escovar et al., 1996), to structural components such as the nonmuscle myosin, Zipper, that is localized to the leading edge of the advancing epidermal cells and is thought to generate the force required to pull the epidermis over the amnioserosa (Young et al., 1993). At the moment, either a structural role at cell junctions or a role facilitating cellular interactions (see below) seem equally possible for Coracle.

Although *coracle* mutations have an embryonic lethal phenotype, mutations in this gene were originally discovered in a screen for dominant modifiers of a gain-of-function mutation in the *Drosophila* epidermal growth factor receptor gene (*Egfr*) (Fehon et al., 1994). This allele of the *Egfr* gene, termed *Ellipse* (*Egfr^{Elp}*), appears to result in a hyperfunctional form of the receptor that is either constitutively active or hypersensitive to its ligand (Baker and Rubin, 1992). The most noticeable phenotype of the *Egfr^{Elp}* mutation is a rough and reduced eye that is due to a combination of misspecification of precursor cells and alterations in patterns of cell division in the developing eye primordium. In addition, *Drosophila Egfr* is believed to be involved in cellular interactions that determine cell fate in a variety of tissues, including the oocyte, the ventral midline of the developing embryo, and the developing wing (Clifford and Schupbach, 1994).

We have shown that loss-of-function mutations in the *coracle* gene dominantly suppress the rough eye phenotype of the *Egfr^{Elp}* mutation. Thus, removal by mutation of one of two normal doses of *coracle* partially corrects a phenotypic defect caused by the hyperfunctional *Egfr^{Elp}* receptor. To understand the significance of this result, it is important to note that *coracle* mutations are normally completely recessive. That is, loss of a single dose of the *coracle* gene normally has no visible phenotype, while in the background of the *Egfr^{Elp}* mutation, *coracle* mutations dominantly suppress the rough eye that is characteristic of this mutation. A variety of studies, especially on signaling pathways, have shown that such genetic interactions often reflect underlying functional interactions between the proteins that the interacting genes encode. In this case, the suppression of a phenotype caused by a hyperactivated form of the EGF receptor by mutations in the *coracle* gene implies that Coracle protein is necessary for the EGF receptor to transduce signals from the cell surface into the cytoplasm.

How might Coracle be involved in EGF receptor function? Previous studies of the subcellular localization of the EGF receptor protein in *Drosophila* epithelia have shown that it is highly localized to the apical membrane domain and the junctional complex. Likewise, our studies of Coracle have shown that it localizes to the apical junctional complex, particularly in the region of the septate junction, and is partially overlapping with the EGF-receptor (R. Fehon, unpublished observations). Given this similarity in subcellular localization, the proposed role of Protein 4.1 family members to bind to the cytoplasmic tail of transmembrane proteins, and the genetic interactions we have observed, it is tempting to speculate that Coracle may interact with the cytoplasmic tail of the EGF receptor or a closely linked component of the EGF-receptor pathway (such as Rhomboid or Star, two transmembrane proteins that have been shown to function in EGF-receptor–mediated signals), and thereby play a direct role in this signal transduction mechanism. Alternatively, it is possible that Coracle plays a more general role in organizing the apical cell surface, perhaps by contributing to the "fence" function of the septate junction that has been proposed to maintain separation of the apical and basolateral membrane domains in epithelial cells (Mandel et al., 1993). We cannot distinguish yet between these alternatives, though preliminary experiments have not shown any disruption of junctional integrity or apical-basal polarity in *coracle* mutant cells (R. Lamb and R. Fehon, unpublished observations). In either case, elucidation of the molecular framework within which Coracle functions, and in particular the identification of its binding partners, should clarify the exact role of the Coracle protein in these processes.

Drosophila Members of the ERM Family

Using a similar approach to that described for Coracle, we have identified two *Drosophila* members of the ERM group within the Protein 4.1 gene family (McCartney and Fehon, 1996). One gene encodes a protein that is equally similar to Ezrin, Radixin, and Moesin, but lacks the proline-rich region that is found in the former two, and hence has been termed *Moesin*. A partial clone from the same gene was identified previously in a screen for sequences that alter cell shape when overexpressed in the fission yeast *Schizosaccharomyces pombe* (Edwards et al., 1994). The second *Drosophila* gene is clearly more similar to human Merlin than to any of the other ERM family members (Fig. 1). Further comparisons between these sequences reveal interesting regions of conservation and divergence (McCartney and Fehon, 1996). Specifically, the NH_2-terminal 330 aa of all ERM family members is highly conserved, being ~45% identical between the *Drosophila* genes and their two human counterparts. However, alignment of the *Drosophila* and human Merlin sequences reveals other regions that are well conserved between these sequences but are divergent from the other ERM family members, especially in the COOH-terminal tail. This region, which is extremely well conserved in Ezrin, Radixin, and Moesin as well as in *Drosophila* Moesin, has been shown to serve as an F-actin binding domain. In marked contrast, the COOH-terminal tail of the Merlin sequences, while similar to one another, are highly divergent from the other ERM proteins, suggesting that Merlin does not have a COOH-terminal actin-binding domain.

Despite extensive screening of the amplification products using degenerate primers, we have so far failed to detect any other *Drosophila* members of the ERM family. Thus, it appears that in *Drosophila* there is only one ERM family member, rather than the three closely related ERM proteins that are found in higher vertebrates. In addition, the alignment tree shown in Fig. 1 suggests that these three vertebrate proteins arose from duplication of a single progenitor sequence somewhere in the vertebrate lineage, because the three vertebrate ERM proteins are much more similar to one another (~85%) than any one is to *Drosophila* Moesin (~55%). In contrast, human Merlin is more similar to the *Drosophila* Merlin sequence than it is to either Ezrin, Radixin, or Moesin (Fig. 1). This conclusion has important implications for the study of these proteins. First, it suggests that *Drosophila* Moesin may serve most or all of the functions that the three ERM proteins combined serve in vertebrate cells. Observations of the subcellular localization of Moesin in vivo lead to a similar conclusion (McCartney and Fehon, 1996). Second, if there is only a single *Drosophila* ERM family member, then it should be possible to study Moesin function in *Drosophila* without the confounding effects of antibody cross-reactivity and functional redundancy that have been seen previously in other

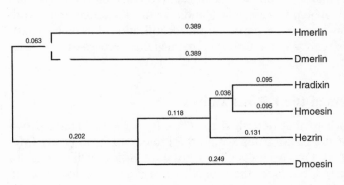

Figure 1. Sequence alignment tree (Unweighted Pair Group Method; Nei, 1987) of the ERM and Merlin proteins from humans and *Drosophila*. Genetic distances between pairs are shown numerically. Abbreviations: *Hmerlin*, human Merlin; *Dmerlin*, *Drosophila* merlin; *Hmoesin*, human Moesin; *Hradixin*, human Radixin; *Hezrin*, human Ezrin; *Dmoesin*, *Drosophila* Moesin.

systems (Takeuchi et al., 1994). In the absence of such redundancy mutations in *Drosophila Moesin* should be particularly useful for studying possible functional interactions with the *Merlin* tumor suppressor.

The similarities in the NH_2-terminal half of Merlin and the other ERM proteins has led previous studies to propose that these proteins all function as a link between the membrane and the cytoskeleton (Kinzler and Vogelstein, 1993; Rouleau et al., 1993; Trofatter et al., 1993). However, the observed structural differences between Merlin and the other ERM family members in a known actin-binding domain, and the observation that these differences are conserved in both humans and flies, raises questions about whether or not these proteins have similar functions. To investigate this question further, we examined expression of Merlin and Moesin throughout embryonic and adult development of the fly (McCartney and Fehon, 1996). In particular, the subcellular localization of these proteins was examined in detail in a variety of tissues, including epithelial cells and neuronal tissues. In short, although both proteins are found in association with the plasma membrane, there are striking differences in aspects of their subcellular localization. *Drosophila* Moesin is always found in close association with the cell membrane, either in the region of the adherens junction or on the apical membrane in association with microvilli. This pattern is essentially identical to that seen for ERM family members as a group in mammalian cells. In contrast, while Merlin is associated with the apical membrane and in particular the adherens junction region in *Drosophila* epithelial cells, this protein also is observed in punctate structures within the cytoplasm.

Similar differences in subcellular localization are also readily apparent when Moesin and Merlin are expressed in cultured cells. As expected, Moesin is concentrated at the cell membrane in cultured Schneider's 2 cells, even when overexpressed under an inducible promoter. Merlin on the other hand shows a much more dynamic pattern of expression in these cells. Initially Merlin appears to concentrate at the plasma membrane, just as we observed for Moesin. However, within 3 h after induction Merlin is found in punctate cytoplasmic structures that seem to progress to membrane bound vesicles within 5–7 h after induction. This progression from the plasma membrane to cytoplasmic vesicular structures over time implies that Merlin may be trafficking from the membrane to the cytoplasm, perhaps in association with an endocytic process. To test this hypothesis, we labeled endocytic compartments within the cell by adding a fluorescent dextran to the medium in cells that expressed Merlin, and then examined the correlation between Merlin and endocytosis by confocal microscopy. The results clearly showed that dextran-labeled endocytic structures are often surrounded by Merlin protein expressed in this cultured cell system. However it is important to note that similar experiments have not yet been performed using tissues that express Merlin endogenously to determine if this association can also be observed in vivo. Nonetheless, the subcellular pattern that we observe in vitro is consistent with that observed in vivo.

While it is still too early to rigorously define a cellular function for Merlin, it may be worthwhile to speculate on possibilities that on one hand are consistent with a role in tumor suppression, and on the other hand relate to endocytic processes. One possibility that seems consistent with both roles, and with what we know about the Protein 4.1 superfamily of proteins, is that Merlin may function in the endocytosis and downregulation of a transmembrane receptor involved in growth control. For example, peptide hormone receptors, such as the EGF- and

PDGF-receptors, are downregulated by endocytosis in response to ligand binding, and mutant forms of the EGF-receptor that are not properly endocytosed can lead to a transformed phenotype (Wells et al., 1990). Thus if Merlin is necessary for endocytosis of a transmembrane receptor involved in growth regulation, perhaps by binding directly to its cytoplasmic tail, then loss of Merlin function might be sufficient to cause overproliferation. In this regard, we can envision Merlin playing either an active role as an "adapter" protein linking the receptor to endocytic machinery, or perhaps a more indirect role, for example by clustering or localizing receptors to a particular region of the membrane for subsequent endocytosis. Further experiments will be required to test these intriguing possibilities.

Future Directions

Studies of Merlin and the other Protein 4.1 family members thus far have been informative, but we are still far from understanding their functions in complex cellular processes such as growth regulation and morphogenesis. To better understand the functions of these proteins, we need to learn the molecular framework within which they function, and in particular, the proteins with which they interact. Biochemical studies of human erythrocytes and the intestinal brush border have been successful in identifying binding partners for some of these proteins. However, it is clear that we do not yet have a complete understanding all of the interactions in which these proteins are involved. For example, although Protein 4.1 binds to glycophorin C in erythrocytes, its binding partners in non-erythroid cells, which do not express glycophorin C, have not yet been identified.

Further characterization of these interactions and the elucidation of their cellular functions will likely require a multifaceted approach to this problem. As a complement to ongoing biochemical experiments in other systems, the *Drosophila* experimental system with its array of genetic and molecular genetic tools provides powerful means to understand the functions of these proteins and to identify their interacting partners. In particular, genetic studies can be especially enlightening because they allow a simple experiment—examining the effect of removing a single gene product in a whole organism. One advantage of this approach is that it does not require any preconception of what functions a particular gene may possess. For example, our phenotypic analysis of *coracle* mutations suggests interactions between Coracle and the EGF-receptor that had not previously been considered (Fehon et al., 1994). Likewise, phenotypic analysis of *Drosophila Merlin* mutations, which is currently underway (B. McCartney and R. Fehon, unpublished data), should provide information regarding its cellular and developmental functions. Comparison of these phenotypes with those of known genes may also may help identify other genes that have similar functions. Therefore studies of *Drosophila* members of this well-conserved gene family should continue to provide insights into their functions.

Acknowledgments

Work in our laboratory has been supported by grants from the National Institutes of Health (NS34783) and the National Science Foundation (IBN-92066555).

References

Anderson, R.A., and V.T. Marchesi. 1985. Regulation of the association of membrane skeletal protein 4.1 with glycophorin by a polyphosphoinositide. *Nature.* 318:295–298.

Artavanis-Tsakonas, S., K. Matsuno, and M.E. Fortini. 1995. Notch signaling. *Science.* 268:225–232.

Baker, N.E., and G.M. Rubin. 1992. Ellipse mutations in the *Drosophila* homologue of the EGF receptor affect pattern formation, cell division, and cell death in eye imaginal discs. *Dev. Biol.* 150:381–396.

Bhanot, P., M. Brink, C. Harryman-Samos, J.C. Hsieh, Y. Wang, J. Macke, D. Andrew, J. Nathans, and R. Nusse. 1996. A new member of the *frizzled* family from *Drosophila* functions as a Wingless receptor. *Nature.* 382:225–230.

Boedigheimer, M., and A. Laughon. 1993. *expanded*: a gene involved in the control of cell proliferation in imaginal discs. *Development.* 118:1291–1301.

Bretscher, A. 1983. Purification of an 80,000-dalton protein that is a component of the isolated microvillus cytoskeleton, and its localization in nonmuscle cells. *J. Cell Biol.* 97:425–432.

Clifford, R., and T. Schupbach. 1994. Molecular analysis of the *Drosophila* EGF receptor homolog reveals that several genetically defined classes of alleles cluster in subdomains of the receptor protein. *Genetics.* 137:531–550.

Conboy, J., J. Chan, J. Chasis, Y. Kan, and N. Mohandas. 1991. Tissue- and development-specific alternative RNA splicing regulates expression of multiple isoforms of erythroid membrane protein 4.1. *J. Biol. Chem.* 266:8273–8280.

Correas, I., T.L. Leto, D.W. Speicher, and V.T. Marchesi. 1986. Identification of the functional site of erythrocyte protein 4.1 involved in spectrin-actin associations. *J. Biol. Chem.* 261:3310–3315.

Edwards, K.A., R.A. Montague, S. Shepard, B.A. Edgar, R.L. Erikson, and D.P. Kiehart. 1994. Identification of *Drosophila* cytoskeletal proteins by induction of abnormal cell shape in fission yeast. *Proc. Natl. Acad. Sci. USA.* 91:4589–4593.

Fehon, R.G., I.A. Dawson, and S. Artavanis-Tsakonas. 1994. A *Drosophila* homologue of membrane-skeleton protein 4.1 is associated with septate junctions and is encoded by the *coracle* gene. *Development.* 120:545–557.

Fehon, R.G., K. Johansen, I. Rebay, and S. Artavanis-Tsakonas. 1991. Complex cellular and subcellular regulation of Notch expression during embryonic and imaginal development of *Drosophila*: implications for Notch function. *J. Cell Biol.* 113:657–669.

Franck, Z., R. Gary, and A. Bretscher. 1993. Moesin, like ezrin, colocalizes with actin in the cortical cytoskeleton in cultured cells, but its expression is more variable. *J. Cell Sci.* 105:219–231.

Glise, B., H. Bourbon, and S. Noselli. 1995. *hemipterous* encodes a novel Drosophila MAP kinase kinase, required for epithelial cell sheet movement. *Cell.* 83:451–461.

Greenwald, I., and G.M. Rubin. 1992. Making a difference: the role of cell-cell interactions in establishing separate identities for equivalent cells. *Cell.* 68:271–281.

Kinzler, K.W., and B. Vogelstein. 1993. A gene for neurofibromatosis 2. *Nature.* 363:495–496.

Kooh, P.J., R.G. Fehon, and M.A. Muskavitch. 1993. Implications of dynamic patterns of Delta and Notch expression for cellular interactions during *Drosophila* development. *Development.* 117:493–507.

Mandel, L.J., R. Bacallao, and G. Zampighi. 1993. Uncoupling of the molecular 'fence' and paracellular 'gate' functions in epithelial tight junctions. *Nature.* 361:552–555.

Marchesi, S., J. Conboy, P. Agre, J. Letsinger, V. Marchesi, D. Speicher, and N. Mohandas. 1990. Molecular analysis of insertion/deletion mutations in protein 4.1 in elliptocytosis. I. Biochemical identification of rearrangements in the spectrin/actin binding domain and functional characterizations. *J. Clin. Invest.* 86:516–523.

Marfatia, S.M., R.A. Leu, D. Branton, and A.H. Chishti. 1995. Identification of the protein 4.1 binding interface on glycophorin C and p55, a homologue of the Drosophila discs-large tumor suppressor protein. *J. Biol. Chem.* 270:715–719.

McCartney, B.M., and R.G. Fehon. 1996. Distinct cellular and subcellular patterns of expression imply distinct functions for the *Drosophila* homologues of moesin and the neurofibromatosis 2 tumor suppressor, merlin. *J. Cell Biol.* 133:843–852.

Nei, M. 1987. Molecular Evolutionary Genetics. Columbia University Press, New York. 293–298.

Orsulic, S., and M. Peifer. 1996. An in vivo structure-function study of Armadillo, the β-catenin homologue, reveals both separate and overlapping regions of the protein required for cell adhesion and for Wingless signaling. *J. Cell Biol.* 134:1283–1300.

Peifer, M., and E. Wieschaus. 1990. The segment polarity gene *armadillo* encodes a functionally modular protein that is the Drosophila homolog of human plakoglobin. *Cell.* 63:1167–1178.

Riesgo-Escovar, J.R., M. Jenni, A. Fritz, and E. Hafen. 1996. The Drosophila Jun-N-terminal kinase is required for cell morphogenesis but not for DJun-dependent cell fate specification in the eye. *Genes Dev.* 10:2759–2768.

Rouleau, G.A., P. Merel, M. Lutchman, M. Sanson, J. Zucman, C. Marineau, K. Hoang-Suan, S. Demczuk, C. Desmaze, B. Plougastel, et al. 1993. Alteration in a new gene encoding, a putative membrane-organizing protein causes neurofibromatosis type 2. *Nature.* 363:515–521.

Sato, N., N. Funayama, A. Nagafuchi, S. Yonemura, S. Tsukita, and S. Tsukita. 1992. A gene family consisting of ezrin, radixin and moesin. Its specific localization at actin filament/plasma membrane association sites. *J. Cell Sci.* 103:131–143.

Sato, N., S. Yonemura, T. Obinata, S. Tsukita, and S. Tsukita. 1991. Radixin, a barbed end-capping actin-modulating protein, is concentrated at the cleavage furrow during cytokinesis. *J. Cell Biol.* 113:321–330.

Takeuchi, K., N. Sato, H. Kasahara, N. Funayama, A. Nagafuchi, S. Yonemura, S. Tsukita, and S. Tsukita. 1994. Perturbation of cell adhesion and microvilli formation by antisense oligonucleotides to ERM family members. *J. Cell Biol.* 125:1371–1384.

Tang, T., Z. Qin, T. Leto, V. Marchesi, and E.J. Benz. 1990. Heterogeneity of mRNA and protein products arising from the protein 4.1 gene in erythroid and nonerythroid tissues. *J. Cell Biol.* 110:617–624.

Tepass, U., and V. Hartenstein. 1994. The development of cellular junctions in the *Drosophila* embryo. *Dev. Biol.* 161:563–596.

Thomas, U., S.A. Speicher, and E. Knust. 1991. The *Drosophila* gene *Serrate* encodes an EGF-like transmembrane protein with a complex expression pattern in embryos and wing discs. *Development.* 111:749–761.

Tomlinson, A., D.L. Bowtell, E. Hafen, and G.M. Rubin. 1987. Localization of the sevenless protein, a putative receptor for positional information in the eye imaginal disc of Drosophila. *Cell.* 51:143–150.

Trofatter, J.A., M.M. MacCollin, J.L. Rutter, R. Eldridge, N. Kley, A.G. Menon, K. Pulaski, H. Haase, C.M. Ambrose, D. Munroe, et al. 1993. A novel moesin-, ezrin-, radixin-like gene is a candidate for the neurofibromatosis 2 tumor suppressor. *Cell.* 72:791–800.

Wells, A., J.B. Welsh, C.S. Lazar, H.S. Wiley, G.N. Gill, and M.G. Rosenfeld. 1990. Ligand-induced transformation by a noninternalizing epidermal growth factor receptor. *Science.* 247:962–964.

Willott, E., M.S. Balda, A.S. Fanning, B. Jameson, C. Van Itallie, and J.M. Anderson. 1993. The tight junction protein ZO-1 is homologous to the Drosophila *discs-large* tumor suppressor protein of septate junctions. *Proc. Natl. Acad. Sci. USA.* 90:7834–7838.

Young, P.I., A.M. Richman, A.S. Ketchum, and D.P. Kiehart. 1993. Morphogenesis in *Drosophila* requires nonmuscle myosin heavy chain function. *Genes Dev.* 7:29–41.

Zak, N.B., and B.Z. Shilo. 1992. Localization of DER and the pattern of cell divisions in wild-type and *Ellipse* eye imaginal discs. *Dev. Biol.* 149:448–456.

Chapter 5

Cytoskeletal Interaction with Ion Channels

Cytoskeletal Interactions with Glutamate Receptors at Central Synapses

Gary L. Westbrook,* Johannes J. Krupp,* and Bryce Vissel‡

*Vollum Institute, Oregon Health Sciences University, Portland, Oregon 97201; and
‡Salk Institute, La Jolla, California 92186

Introduction

Release of glutamate at central synapses results in activation of two ligand-gated channels, AMPA (α-amino-3-hydroxy-5-methyl-4-isoxazolepropionic) and NMDA (N-methyl-D-aspartate), that can colocalize at individual synaptic sites (Bekkers and Stevens, 1989). Despite exposure to the same glutamate transient ($\sim$1 mM for 1 ms, Clements et al., 1992), these channels differ markedly in their properties and effect on synaptic transmission. In hippocampal pyramidal cells, AMPA receptors mediate a brief synaptic current ($\sim$3 ms) due to opening of monovalent-selective cation channels, whereas the kinetics of the Na- and Ca-permeable NMDA receptors are much slower (rise 10 ms, duration 100–500 ms, Lester et al., 1990). These receptor characteristics imply that AMPA receptors act primarily to relay action potentials while NMDA receptors act in a neuromodulatory manner either by increasing membrane excitability due to their voltage dependence or acting as a source of calcium influx into the neuron due to their high calcium permeability (Mayer and Westbrook, 1987).

AMPA and NMDA receptors are imbedded in the postsynaptic density (PSD), a specialized compartment of the submembrane cytoskeleton (Kennedy, 1993). Although some of the proteins in the PSD have been identified, relatively little is known about protein–protein interactions between PSD components and glutamate receptors. It might be expected, by analogy with the neuromuscular junction, that cytoskeletal elements are involved in the clustering and anchoring of receptors at central synapses (see Froehner et al., 1997, in this volume). For example, at the neuromuscular junction (NMJ), both tyrosine phosphorylation and links to cytoskeletal proteins such as the 43K protein are important in receptor clustering (Froehner, 1993). Glutamate receptors on hippocampal pyramidal cells are also clustered on dendritic spines (Harris and Stevens, 1989; Harris and Kater, 1994). Circumstantial biochemical evidence has begun to reveal parallels with the NMJ, implicating certain glutamate receptor subunits in clustering. For example, the NMDA receptor subunit, NR2B that is expressed early in development in vivo and in vitro (e.g., Williams et al., 1993; Sheng et al., 1994; Zhong et al., 1995) is tyrosine phosphorylated, and binds novel cytoskeletal proteins such as PSD95 (Moon et al., 1994; Kornau et al., 1995; Hunt et al., 1996). These properties suggest that NMDA receptors containing the NR2B are crucial to the early formation and organization of central excitatory synapses. In addition, the NR1 subunit has been shown to cluster in vitro by a phosphorylation-dependent mechanism (Ehlers et al., 1995) and the COOH terminus of NR2A is also tyrosine phosphorylated (Köhr and Seeburg, 1996). Interactions of other synaptic-associated proteins with cytoskeletal proteins such as actin or PSD95 have been recently recognized (Kim et al., 1995; Brenman et al., 1996; Lyford et al., 1995). However the functional role of PSD proteins and their protein–protein interactions is still at a very early stage (Kennedy, 1993).

In our studies of the electrophysiological properties of glutamate receptors in cultured hippocampal neurons, we have found evidence for two different types of cytoskeletal interactions that directly alter the gating of AMPA and NMDA channels. Although details of the molecular interactions responsible for these physiological effects remain to be determined, our results provide evidence for a dynamic role of cytoskeletal elements in the function of membrane proteins in specialized domains such as synapses.

Anchoring of Protein Kinase A and Regulation of AMPA Receptors

AMPA receptors mediate the fast component of the postsynaptic excitatory response at central synapses. These channels are multimers composed of four subunits, GluR1-4, and are characterized by their low agonist affinity and rapid desensitization. Most AMPA receptors contain one or more copies of the GluR2 subunit and have a low permeability to calcium, whereas AMPA receptors lacking the GluR2 subunit have a measurable calcium permeability (Seeburg, 1993; Hollman and Heinemann, 1994). Functionally, AMPA receptors are the mediators of the fast relay of information from presynaptic to postsynaptic elements, although the calcium permeability may in some cases also provide a biochemical signalling mechanism. Thus their regulation may have a significant impact on the shape of individual excitatory postsynaptic currents (EPSCs), as well as on the ability of the postsynaptic cell to follow high frequency stimuli from the presynaptic cell.

Protein kinase A (PKA) is necessary for the maintenance of AMPA/kainate channel activity in hippocampal neurons (Greengard et al., 1991; Wang et al., 1991). Endogenous PKA appears to be compartmentalized in neurons by a family of homologous proteins that bind the type II regulatory subunit (RII) of the A kinase holoenzyme. These proteins were initially dubbed A kinase anchoring proteins (AKAPs) based on the hypothesis that they serve to localize the kinase near its substrate. AKAPs have a conserved RII binding domain and are thought to have a unique targeting domain as AKAPs can be found in various cellular compartments including the PSD (see Scott, 1997, in this volume). To test whether AKAPs might play a functional role in the regulation of AMPA channels, we generated a series of peptides based on the conserved RII binding domain of two different AKAPs, AKAP79 and HT31. HT31 was initially identified in thyroid cells whereas AKAP79 is present in the central nervous system. We first confirmed that several AKAPs are present in extracts of cultured hippocampal neurons using overlay assays to detect RII binding proteins.

We hypothesized that if AKAPs serve to localize the kinase near its substrate, in this case the AMPA receptor, competition with the peptides should disrupt this interaction. The specificity of the peptides was confirmed by overlay assays prepared from hippocampal cell extracts. Intracellular perfusion of cultured hippocampal neurons with the peptides prevented the PKA-mediated regulation of AMPA/

kainate currents as well as fast excitatory synaptic currents (Rosenmund et al.; 1994). This effect was mimicked by the PKA substrate inhibitor peptide PKI, but developed with a slower time course consistent with the idea that the peptide only became effective after unbinding of RII from the AKAP. Control peptides that did not bind RII in the overlay assay had no effect on AMPA responses in perfused cells. This effect could be overcome by adding the catalytic subunit of PKA, suggesting that the peptides affected the availability of catalytic subunit rather than directly interfering with PKA-substrate interactions. These results provide the first evidence that compartmentalization of kinases may be important in the regulation of postsynaptic receptors. A proposed scheme for this interaction is shown in Fig. 1. It has now been demonstrated that calcineurin (CaN) also binds to AKAPs (Coghlan et al., 1995; Scott, 1997, in this volume) suggesting a remarkably intricate bi-directional control of phosphorylation at synapses. The potential importance of tight temporal and spatial control of phosphorylation may extend to other synaptic proteins in the PSD, but this question has yet to be fully explored.

Calcium Regulation of NMDA Receptors: A Clue to Protein–Protein Interactions in the Postsynaptic Density (PSD)

The regulation of NMDA receptors is complex. They are regulated by a wide variety of molecules including divalent cations, protons, redox state, fatty acids, histamine, polyamines, neurosteroids, kinases, and phosphatases (McBain and Mayer, 1994). Native NMDA receptor activity is also controlled by three phenomenologically distinct forms of desensitization. Increases in intracellular Ca are responsible for one form of NMDA channel desensitization. As the NMDA channel is highly calcium permeable, this provides a feedback mechanism at excitatory synapses. In fact, this was the first form of NMDA receptor desensitization to be recognized

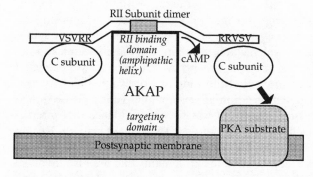

Figure 1. Role of compartmentalization in kinase function in the postsynaptic membrane. Peptides derived from the RII binding domain of AKAP79 disrupt the PKA-dependent activity of AMPA channels in cultured hippocampal neurons, suggesting that PKA localization is critical to its function at excitatory synapses. This schematic depicts how A kinase anchoring proteins (AKAPs) can affect regulation of AMPA/kainate channels. The RII subunit dimer binds to an AKAP placing the PKA holoenzyme in close proximity to substrates such as AMPA receptors in the postsynaptic membrane. Activation of cAMP leads to dissociation of the catalytic subunit (*C*) and results in preferential phosphorylation (*arrowhead*) of nearby substrates, either the channel or associated proteins.

(Mayer and Westbrook, 1985; Zorumski et al., 1989; Clark et al., 1990), and such Ca dependence has now been reported by several groups (Vyklicky, 1993; Zilberter et al., 1991; Kyrosis et al., 1995; Medina et al., 1995). We called this form of desensitization, Ca-dependent inactivation, as gating of the channel is not required to reduce subsequent channel activity (Legendre et al., 1993), suggesting that the regulatory site is accessible in the closed state. Ca-dependent proteins including CaM and calcineurin can affect NMDA channel activity (Lieberman and Mody, 1994; Tong and Jahr, 1995; Ehlers et al., 1996) but do not appear to be directly involved in Ca-dependent inactivation (Rosenmund and Westbrook, 1993*b*). Two other forms of NMDA receptor desensitization exist: a glycine-dependent form due to an allosteric interaction between glutamate and glycine binding sites (Mayer et al., 1989); and glycine-independent desensitization (Sather et al., 1990) that is extremely prominent in cell-free patches. Glycine-independent desensitization accelerates with time of recording, suggesting that dialysis enhances entry into the desensitized state (Tong and Jahr, 1994; Rosenmund et al., 1995*b*). The mechanism of glycine-independent desensitization is not well understood but can be modulated by dephosphorylation (Tong and Jahr, 1994). Our preliminary evidence suggests that glycine-independent desensitization requires specific domains in NR2A (not shown). NMDA receptor desensitization is rather slow (requiring hundreds of milliseconds to develop) and thus can alter either the EPSCs in a stimulus train or the duration of the NMDA EPSC (Mennerick and Zorumski, 1995; Tong and Jahr, 1995; Rosenmund et al., 1995*a*). This latter mechanism is analogous to Ca inactivation of voltage-gated calcium channels (Imredy and Yue, 1994).

The action of calcium on NMDA receptors has two kinetic components, a relatively rapid ($t \approx 1$ s) reduction in channel open probability (P_o) and a much slower reduction in P_o that evolves over minutes in whole-cell recording. The second component occludes the first, suggesting a final common pathway leading to a reduction in the open probability of NMDA channels, perhaps with some common biochemical steps. The slower component of calcium regulation can be prevented by adding ATP to the patch pipette, but does not appear to directly involve phosphorylation (Rosenmund and Westbrook, 1993*a*). Because we were unable to demonstrate that these effects were dependent on kinases, phosphatases, or the inhibition of pumps/exchangers, we investigated the possibility that interactions with the actin cytoskeleton were responsible. This approach was motivated by the known opposing actions of calcium and ATP on actin polymerization, and the fact that dendritic spines are packed with actin. Phalloidin, which stabilizes actin, completely mimicked the action of ATP whereas actin depolymerization with cytochalasins caused channel rundown. In addition, phalloidin prevented the rundown of channel activity suggesting that the state of actin polymerization affects channel gating (Rosenmund and Westbrook, 1993*b*). This surprising result was the first evidence that interactions with cytoskeletal proteins are important in NMDA channel activity. Recent biochemical studies suggest that CaM and cytoskeletal proteins can interact with intracellular domains of NMDA receptor subunits at domains that are also involved in Ca-dependent inactivation (Wyszynski et al. 1997), suggesting an intriguing link between dynamic regulation of channel activity (e.g., by compartmentalized regulatory proteins) and structural features such as channel anchoring and clustering. We outline here our functional studies directed at the molecular domains involved in Ca-dependent inactivation.

Molecular Analysis of Domains Involved in Calcium Regulation

Glutamate subunits differ from the nicotinic receptor superfamily in being much larger proteins with 50% of the protein ($\approx$500 aa) in the NH_2-terminal domain preceding the predicted TM1 domain. Although it was initially assumed that glutamate channels might follow the 4 TM model of AChRs, elegant work from several labs now suggests that glutamate channels have a unique 3 TM topology with the "TM2" domain forming a re-entrant membrane loop similar to the S5-6 P loop of voltage-gated channels (Hollmann et al., 1994; Wo and Oswald, 1995; Bennett and Dingledine, 1995; Wood et al., 1995). There is agreement that the P region contributes to the pore, but TM1 may also contribute to the outer vestibule (see, e.g., Wo and Oswald, 1995). Nakanishi et al. (1990) noted a homology between the NH_2-terminal domain of AMPA subunits and bacterial amino acid binding proteins. This led O'Hara et al. (1993) to develop a model of the glutamate binding site based on homology with two of these bacterial proteins including LIVBP (leucine, isoleucine, valine binding protein) and LAOBP (lysine/arginine/ornithine binding protein). Based on this model, glutamate binds to a LIVBP-like domain in metabotropic glutamate receptors and to a LAOBP-like domain in ionotropic glutamate receptors. Two regions of the LAOPB-like domain, labeled S1 and S2 in Fig. 2, seemingly operate in a clamshell manner to bind ligands in AMPA and NMDA receptors (Stern-Bach et al., 1994; Kuryatov et al., 1994). Based on the crystal structure with and without ligand, the two lobes of LAOBP, corresponding to S1 and S2, are hinged with the first lobe undergoing a large (50 degree) rotation; ligand binding is proposed to stabilize the closed conformation.

Native NMDA receptors are heteromers of NR1 and NR2 subunits (Meguro et al., 1992; Monyer et al., 1992). NR2 subunits are coded by four different genes whereas the NR1 subunit is encoded by a single gene that can be spliced to eight different variants. NR1 is expressed ubiquitously (Brose et al., 1993) with NR1-1a being the most common splice variant. The COOH-terminal domain of the most common NR1 splice variant (NR1-1a) includes three regions C0, C1, and C2. C1 contains PKC phosphorylation sites and C0/C1 can bind calmodulin (CaM) and

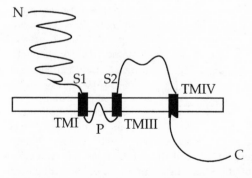

Figure 2. The three transmembrane (*TM*) domain model of glutamate receptor/channel subunits as proposed by several groups. The proximal portion of the NH_2 terminus and the TMIII-IV extracellular domains (labeled *S1*, *S2*) have been shown to be involved in agonist binding to AMPA and NMDA receptors and show homology with the bacterial periplasmic binding protein LAOBP (see text). The P region (formerly called TM2) appears to be a reentrant membrane loop similar to K channels and contributes to the pore of the channel. The intracellular COOH terminus of NR1 and NR22 subunits binds several proteins including CaM and PSD95. The COOH-terminal domains also have phosphorylation sites for multiple kinases.

perhaps other regulatory proteins. The NR2 subunits are distinguished by long COOH-terminal domains which are likely to be important in receptor regulation. It now appears that 2 NR1 subunits are required per channel (Behe et al., 1995), and pharmacological and immunoprecipitation studies suggest that NR2A/2B and NR2A/NR2C also can coexist, implying that there are at least 2 NR2 subunits per channel (Wafford et al., 1993; Chazot et al., 1994; Sheng et al., 1994).

To examine the molecular basis of inactivation, we expressed NMDA receptor subunits in HEK 293 cells in 4 combinations: NR1-1a with each of the NR2 subunits (A-D). We found that NR2A and NR2D-containing receptors demonstrate Ca-dependent inactivation that is very similar to hippocampal neurons (Fig. 3). As in native receptors, inactivation of NR1-1a/NR2A receptors was unaffected by inhibitors of calcineurin or serine/threonine kinases. Although CaM can affect gating of recombinant NR1/2A receptors in excised patches (Ehlers et al., 1996), CaM inhibitors (calmidazolium, CaM inhibitory peptide) had no affect on inactivation of NMDA receptors in hippocampal neurons. More interestingly, the recombinant NR1-1a/NR2A heteromers inactivated when 6 mM barium was substituted for calcium. As barium is a very weak activator of CaM (Chao et al., 1984), this provides further evidence against the idea that CaM binding to the COOH terminus of NR1-1a is the sole cause of inactivation. Strikingly NR1a/NR2C showed no inactivation whatsoever. The lack of inactivation was not due to lower calcium influx as NR2C-containing receptors produced similar intracellular calcium transients (Krupp et al., 1996).

The difference between NR2A- and NR2C-containing receptors provided a natural starting point to examine the domains responsible for inactivation. We first made chimeras with NR2A and NR2C in which the large COOH-terminal domains distal to TM4 were exchanged, called C4A and A4C, respectively. Given that inactivation was specific to NR2A-containing heteromers, the large COOH-terminal domain initially seemed a good candidate for the essential domain. If the COOH terminus of NR2A was critical to inactivation, the A4C chimera containing the COOH terminus of NR2C would not be expected to show inactivation. However we were surprised to find that the A4C chimera inactivated while the C4A chimera

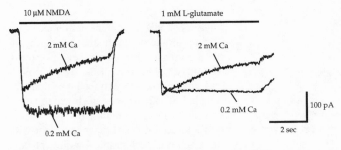

Figure 3. Calcium-dependent inactivation in HEK 293 cell expressing NR1-1a/2D heteromeric NMDA channels. (*A*) 5 second application of NMDA (10 μM) evoked a sustained inward current in low calcium solutions (0.2 mM Ca). However, in higher extracellular Ca (2 mM Ca), the current slowly decayed (inactivated) during the agonist application. (*B*) In another 293 cell, the natural agonist L-glutamate produced a similar degree of inactivation. Note that even in the presence of saturating concentrations of agonist, there was no desensitization in low calcium solutions. Thus the 2D-containing receptors show inactivation but not other forms of NMDA desensitization, suggesting that the distinct forms of desensitization involve separate receptor domains.

did not. Thus the COOH-terminal cytoplasmic domain of NR2A is not essential for inactivation.

We also examined the role of NR1 in inactivation. The cytoplasmic COOH-terminal domain of NR1-1a splice variant contains all three regions: C0, C1, and C2. C0 and C1 contain low and high affinity binding sites for calmodulin, respectively (Ehlers et al., 1996), and C1 also contains PKC phosphorylation sites (Roche et al., 1994). We tested the NR1-4a splice variant that lacks the C1 insert. Inactivation was still observed (not shown) consistent with our prior studies in hippocampal neurons that PKC-dependent phosphorylation/dephosphorylation or CaM binding is not responsible for inactivation. We then made NR1 deletion mutants that lacked either the entire COOH terminus or contained only the C0 region. A deletion mutant that contained only 5 aa following TM4 (NR1/stop838) showed no inactivation when expressed with NR2A whereas the deletion mutant containing C0 (NR1/stop863) showed normal inactivation (not shown). Thus the C0 region of NR1 is required for inactivation.

Based on the available data, several models are possible for inactivation. The first utilizes the recent evidence that the C0 domain binds CaM and α-actinin, and that binding of one displaces binding of the other (Wyszynski et al., 1997). In this model, calcium entry through the NMDA receptor binds to CaM which then competes with α-actinin for C0 binding. This model does not account for our inability to block inactivation with CaM inhibitors or why NR2 subunits are required. We propose an alternative hypothesis in which two intracellular domains of the assembled channel may interact to cause inactivation. This is conceptually similar to the

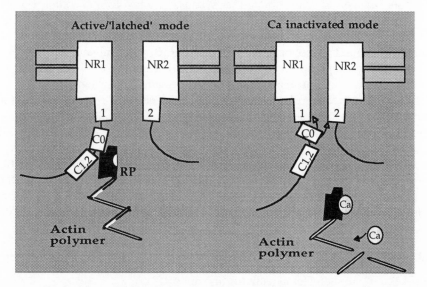

Figure 4. "Latching" model of calcium regulation of NMDA receptors. Single NR1 and NR2 subunits are shown with their cytoplasmic COOH-terminal domains. Our data suggests that the C0 domain of the NR1 COOH terminus and a region within NR2A between TM1 and the end of TM4 are necessary for inactivation. When unlatched by calcium influx, C0 interacts with either NR1 or NR2A (labeled *1, 2*). See text for details. Note that the COOH terminus of NR1 is "latched" by interaction with RP (a hypothetical Ca-dependent actin-binding protein) and actin filaments to prevent inactivation. In this model, C0 is the "ball" analogous to ball-and-chain inactivation of potassium channels.

ball-and-chain model of N-type voltage-gated channel inactivation proposed by Armstrong and Bezanilla (1977) and subsequently proven for N-type inactivation of Shaker potassium channels by Aldrich and colleagues (Hoshi et al., 1990; Zagotta et al., 1990). A working model is shown in Fig. 4. The model incorporates our evidence that the C0 region of the NR1 COOH terminus and a region of NR2A are required for inactivation. A ball-and-chain(s) model is also suggested by the aa similarity between the Shaker ball peptide, consisting of 10 or more hydrophobic residues adjacent to a string of charged residues (Zagotta et al., 1990), and portions of the NR1 COOH terminus. The sequence of C0 has a number of charged residues separated by a series of hydrophobic residues: EIAYKRHKDARRKQMQLAFA-AVNVWRKNLQ. The ball and chain model suggests two alternative mechanisms for the role of NR1 and NR2. Firstly, Ca-dependent release of a protein bound to C0 could allow C0 to interact/bind to another domain on NR1. In this case, NR2A/D would be permissive whereas NR2C would prevent such an interaction. Alternatively C0 could interact directly with a domain on NR2A/D. We are currently testing these hypotheses.

Summary

The data presented here are clearly just the beginning of any comprehensive understanding of the set of regulatory and cytoskeletal proteins that interact with membrane receptors in the postsynaptic density. They do, however, indicate that both glutamate channels at central excitatory synapses are involved in complex protein–protein interactions. For example, while NR2A is important for Ca-dependent inactivation of NMDA receptors, studies in several systems suggest that the other major NR2 subunit in hippocampal neurons, NR2B, predominates at critical times during synapse formation. In addition, the COOH terminus of NR2B binds to several novel cytoskeletal proteins. These results provide circumstantial evidence that NR2B may play specific roles in function and localization of receptors at excitatory synapses. The possible role of NR2B in early synaptic function gains additional support from functional data suggesting that NMDA receptors have specific roles during development (Komuro and Rakic, 1993; Rabacchi et al., 1992; Yen et al., 1993). The essential role of NR1 and NR2B in development is graphically demonstrated by the neonatal death of transgenic mice lacking either of these two subunits (Forrest et al., 1994; Kutsuwada et al., 1996) whereas NR2A and NR2C-deficient mice are less severely affected (Sakimura et al., 1995; Ebralidze et al., 1996).

Acknowledgments

We thank Dr. Stephen Heinemann for his support.

This work was supported by NIH grant MH46613 (G.L. Westbrook) and by fellowships from the Human Frontiers Science Program (J.J. Krupp and B. Vissel).

References

Armstrong, C.M., and F. Bezanilla. 1977. Inactivation of the sodium channel II. Gating current experiments. *J. Gen. Physiol.* 70:567–590.

Behe, P., P. Stern, D. Wyllie, M. Nassar, R. Schoepfer, and D. Colquhoun. 1995. Determination of NMDA NR1 subunit copy number in recombinant NMDA receptors. *Proc. Roy. B.* 262:205–213.

Bekkers, J., and C.F. Stevens. 1989. NMDA and non-NMDA receptors are co-localized at individual excitatory synapses in cultured rat hippocampus. *Nature.* 341:230–233.

Bennett, J.A., and R. Dingledine. 1995. Topology profile for a glutamate receptor: three transmembrane domains and a channel-lining re-entrant membrane loop. *Neuron.* 14:373–384.

Brenman, J.E., D.S. Chao, S.H. Gee, A.W. McGee, S.E. Craven, D.R. Santillano, Z. Wu, F. Huang , H. Xia, M.F. Peters, S.C. Froehner, and D.S. Bredt. 1996. Interaction of nitric oxide synthase with the postsynaptic density protein PSD-95 and a1-syntrophin mediated by PDZ domains. *Cell.* 84:757–767.

Brose, N., G.P. Gasic, D.E. Vetter, J.M. Sullivan, and S.F. Heinemann. 1993. Protein chemical characterization and immunocytochemical localization of the NMDA receptor subunit NMDA R1. *J. Biol. Chem.* 268:22663–22671.

Chao, S.-H., Y. Suzuki , J.R. Zysk, and W.Y. Cheung. 1984. Activation of calmodulin by various metal cations as a function of ionic radius. *Mol. Pharmacol.* 26:75–82.

Chazot, P.L., S.K. Coleman, M. Cik, and F.A. Stephenson. 1994. Molecular characterization of N-methyl-D-aspartate receptors expressed in mammalian cells yields evidence for the coexistence of three subunit types within a discrete receptor molecule. *J. Biol. Chem.* 269:24403–24409.

Clark, G.D., D.B. Clifford, and C.F. Zorumski. 1990. The effect of agonist concentration, membrane voltage and calcium on N-methyl-D-aspartate receptor desensitization. *Neuroscience.* 39:787–797.

Clements, J.D., R.A.J. Lester, G. Tong, C.E. Jahr, and G.L. Westbrook. 1992. The time course of glutamate in the synaptic cleft. *Science.* 258:1498–1501.

Coghlan, V.M., B. Perrino, M. Howard, L.K. Langeberg, J. Hicks, W.M. Gallatin, and J.D. Scott. 1995. Association of protein kinase A and protein phosphatase 2B with a common anchoring protein. *Science.* 267:108–111.

Ebralidze, A.K., D.J. Rossi, S. Tonegawa, and N.T. Slater. 1996. Modification of NMDA receptor-channels and synaptic transmission by targeted disruption of the NR2C gene. *J. Neurosci.* 16:5014–5025.

Ehlers, M.D., W.G. Tingley, and R.L. Huganir. 1995. Regulated subcellular distribution of the NR1 subunit of the NMDA receptor. *Science.* 269:1734–1737.

Ehlers, M.D., S. Zhang, J.P. Bernhardt, and R.L. Huganir. 1996. Inactivation of NMDA receptors by direct interaction of calmodulin with the NR1 subunit. *Cell.* 84:745–755.

Forrest, D., M. Yuzaki, H. Soares, L. Ng, D. Luk, M. Sheng, C. Stewart, J. Morgan, J. Connor, and T. Curran. 1994. Targeted disruption of NMDA receptor 1 gene abolishes NMDA response and results in neonatal death. *Neuron.* 13:325–338.

Froehner, S. 1993. Regulation of ion channel distribution at synapses. *Annu. Rev. Neurosci.* 16:347–368.

Froehner, S.C., M.E. Adams, M.F. Peters, and S.H. Gee. 1997. Syntyrophins: modular adapter proteins at the neuromuscular junction and the sarcolemma. *In* Cytoskeletal Regulation of Membrane Function. S.C. Froehner and G.V. Bennett, editors. The Rockefeller University Press, New York. 197–207.

Greengard, P., J. Jen, A.C. Nairn, and C.F. Stevens. 1991. Enhancement of the glutamate response by cAMP-dependent protein kinase in hippocampal neurons. *Science.* 253:1135–1138.

Harris, K., and J. Stevens. 1989. Dendritic spines of CA1 pyramidal cells in the rat hippocampus: serial electron microscopy with reference to their biophysical characteristics. *J. Neurosci.* 9:2982–2997.

Harris, K.M., and S.B. Kater. 1994. Dendritic spines: cellular specializations imparting both stability and flexibility to synaptic function. *Annu. Rev. Neurosci.* 17:341–371.

Hollmann, M., and S. Heinemann. 1994. Cloned glutamate receptors. *Annu. Rev. Neurosci.* 17:31–108.

Hollmann, M. , C. Maron, and S.F. Heinemann. 1994. N-glycosylation site tagging suggests a three transmembrane domain topology for the glutamate receptor GluR1. *Neuron.* 13: 1331–1343.

Hoshi, T., W.N. Zagotta, and R.W. Aldrich. 1990. Biophysical and molecular mechanisms of Shaker potassium channel inactivation. *Science.* 250:533–538.

Hunt, C.A., L.J. Schenker, and M.B. Kennedy. 1996. PSD-95 is associated with the postsynaptic density and not with the presynaptic membrane at forebrain synapses. *J. Neurosci.* 16:1380–1388.

Imredy, J., and D. Yue. 1994. Mechanism of Ca^{2+}-sensitive inactivation of L-type Ca^{2+} channels. *Neuron.* 12:1301–1318.

Kennedy, M.B. 1993. The postsynaptic density. *Curr. Opin. Neurobiol.* 3:732–737.

Kim, E., M. Niethammer, A. Rothschild, Y.N. Jan, and M. Sheng. 1995. Clustering of shaker-type K^+ channels by interaction with a family of membrane-associated guanylate kinases. *Nature.* 378:85–88.

Köhr, G., and P.H. Seeburg. 1996. Subtype-specific regulation of recombinant NMDA receptor-channels by protein kinases of the src family. *J. Physiol.* 492:445–452.

Komuro, H., and P. Rakic. 1993. Modulation of neuronal migration by NMDA receptors. *Science.* 260:95–97.

Kornau, H.-C., L.T. Schenker, M.B. Kennedy, and P.H. Seeburg. 1995. Domain interaction between NMDA receptor subunits and the postsynaptic density protein PSD-95. *Science.* 269:1737–1740.

Krupp, J., B. Vissel, S.F. Heinemann, and G.L. Westbrook. 1996. Calcium-dependent inactivation of recombinant NMDA receptors is NR2 subunit-specific. *Mol. Pharmacol.* 50:1680–1688.

Kuryatov, A., B. Laube, H. Bez, and J. Kuhse. 1994. Mutational analysis of the glycine-binding site of the NMDA receptor: structural similarity with bacterial amino acid-binding proteins. *Neuron.* 12:1291.

Kutsuwada, T., K. Sakimura, T. Manabe, C. Takayama, N. Katakura, E. Kushiya, R. Natsume, M. Watanabe, Y. Inoue, T. Yagi, S. Aizawa, M. Arakawa, T. Takahashi, Y. Nakamura, H. Mori, and M. Mishina. 1996. Impairment of suckling response, trigeminal neuronal pattern formation and hippocampal LTD in NMDA receptor e2 subunit mutant mice. *Neuron.* 16:333–344.

Kyrozis, A., P.A. Goldstein, M.J. Heath, and A.B. MacDermott. 1995. Calcium entry through a subpopulation of AMPA receptors desensitized neighboring NMDA receptors in rat dorsal horn neurons. *J. Physiol.* 485:373–381.

Legendre, P., C. Rosenmund, and G.L. Westbrook. 1993. Inactivation of NMDA channels in cultured hippocampal neurons by intracellular calcium. *J. Neurosci.* 13:674–684.

Lester, R.A.J., J.D. Clements, G.L. Westbrook, and C.E. Jahr. 1990. Channel kinetics determine the time course of NMDA receptor-mediated synaptic currents. *Nature.* 346:565–567.

Lieberman, D.N., and I. Mody. 1994. Regulation of NMDA channel function by endogenous Ca^{2+}-dependent phosphatase. *Nature.* 369:235–239.

Lyford, G.L., K. Yamagata, W.E. Kaufmann, C.A. Barnes, L.K. Sanders, N.G. Copeland, D.J. Gilbert, N.A. Jenkins, A.A. Lanahan, and P.F. Worley. 1995. Arc, a growth factor and activity-regulated gene, encodes a novel cytoskeleton-associated protein that is enriched in neuronal dendrites. *Neuron.* 14:433–445.

Mayer, M.L., L. Vyklicky, Jr., and J. Clements. 1989. Regulation of NMDA receptor desensitization in mouse hippocampal neurons by glycine. *Nature.* 338:425–427.

Mayer, M.L., and G.L. Westbrook. 1985. The action of N-methyl-D-aspartic acid on mouse spinal cord neurons in culture. *J. Physiol.* 354:29–53.

Mayer, M.L., and G.L. Westbrook. 1987. The physiology of excitatory amino acids in the mammalian central nervous system. *Prog. Neurobiol.* 28:197–276.

McBain, C., and M. Mayer. 1994. N-methyl-D-aspartic acid receptor structure and function. *Physiol. Rev.* 74:723–759.

Medina, I., N. Filippova, G. Charton, S. Rougeole, Y. Ben-Ari, M. Khrestchatisky, and P. Bregestovski. 1995. Calcium-dependent inactivation of heteromeric NMDA receptor-channels expressed in human embryonic kidney cells. *J. Physiol.* 482.3:567–573.

Meguro, H., H. Mori, K. Araki, E. Kushiya, T. Kutsuwada, M. Yamazaki, T. Kumanishi, M. Arakawa, K. Sakimura, and M. Mishina. 1992. Functional characterization of a heteromeric NMDA receptor channel expressed from cloned cDNAs. *Nature.* 357:70–74.

Mennerick, S., and C.F. Zorumski. 1995. Paired-pulse modulation of fast excitatory synaptic currents in microcultures of rat hippocampal neurons. *J. Physiol.* 488:85–101.

Monyer, H., R. Sprengel, R. Schoepfer, A. Herb, M. Higuchi, H. Lomeli, N. Burnashev, B. Sakmann, and P.H. Seeburg. 1992. Heteromeric NMDA receptors: molecular and functional distinction of subtypes. *Science.* 256:1217–1221.

Moon, I.S., M.L. Apperson, and M.B. Kennedy. 1994. The major tyrosine-phosphorylated protein in the postsynaptic density fraction is N-methyl-D-aspartate receptor subunit 2B. *Proc. Natl. Acad. Sci. USA.* 91:3954–3958.

Nakanishi, N., N.A. Schneider, and R. Axel. 1990. A family of glutamate receptor genes: evidence for the formation of heteromultimeric receptors with distinct channel properties. *Neuron.* 5:569–581.

O'Hara, P.J., P.O. Sheppard, H. Thogersen, D. Venezia, B.A. Haldeman, V. McGrane, K.M. Houamed, C. Thomsen, T.L. Gilbert, and E.R. Mulvihill. 1993. The ligand-binding domain in metabotropic glutamate receptors is related to bacterial periplasmic binding proteins. *Neuron.* 11:41–52.

Rabacchi, S., Y. Bailly, N. Delhaye-Bouchaud, and J. Mariani. 1992. Involvement of the N-methyl-D-aspartate NMDA receptor in synapse elimination during cerebellar development. *Science.* 256:1823–1825.

Roche, K.W., W.G. Tingley, and R.L. Huganir. 1994. Glutamate receptor phosphorylation and synaptic plasticity. *Curr. Opin. Neurobiol.* 4:383–388.

Rosenmund, C., and G.L. Westbrook. 1993*a*. Calcium-induced actin depolymerization reduces NMDA channel activity. *Neuron.* 10:805–814.

Rosenmund, C., and G.L. Westbrook. 1993*b*. Rundown of NMDA channels during prolonged whole-cell recording in cultured hippocampal neurons: Role of Ca^{2+} and ATP. *J. Physiol. (Lond.).* 470:705–729.

Rosenmund, C., D.W. Carr, S.E. Bergeson, G. Nilaver, J.D. Scott, and G.L. Westbrook. 1994. Anchoring of protein kinase A is required for modulation of AMPA/kainate receptors on hippocampal neurons. *Nature.* 368:853–855.

Rosenmund, C., A. Feltz, and G.L. Westbrook. 1995*a*. Calcium-dependent inactivation of synaptic NMDA receptors in hippocampal neurons. *J. Neurophysiol.* 73:427–430.

Rosenmund, C., A. Feltz, and G.L. Westbrook. 1995*b*. Synaptic NMDA receptors have a low open probability. *J. Neurosci.* 15:2788–2795.

Sakimura, K., T. Kutsuwada, I. Ito, T. Manabe, C. Takayama, E. Kushiya, T. Yagi, S. Aizawa, Y. Inoue, H. Sugiyama, and M. Mishina. 1995. Reduced hippocampal LTP and spatial learning in mice lacking NMDA receptor e1 subunit. *Nature.* 373:151–155.

Sather, W., J.W. Johnson, G. Henderson, and P. Ascher. 1990. Glycine-insensitive desensitization of NMDA responses in cultured mouse embryonic neurons. *Neuron.* 4:725–731.

Scott, J.D. 1997. Dissection of protein kinase and phosphatase targeting interactions. *In* Cytoskeletal Regulation of Membrane Function. S.C. Froehner and G.V. Bennett, editors. The Rockefeller University Press, New York. 227–239.

Seeburg, P.H. 1993. The molecular biology of mammalian glutamate receptor channels. *TINS.* 16:359–365.

Sheng, M., J. Cummings, L.A. Roldan, Y.N. Jan, and L.Y. Jan. 1994. Changing subunit composition of heteromeric NMDA receptors during development of rat cortex. *Nature.* 368:144–147.

Stern-Bach, Y., B. Bettler, M. Hartley, P.O. Sheppard, P.J. O'Hara, and S.F. Heinemann. 1994. Agonist selectivity of glutamate receptors is specified by two domains structurally related to bacterial amino acid-binding proteins. *Neuron.* 13:1345–1357.

Tong, G., and C. Jahr. 1995. Synaptic desensitization of NMDA receptors by calcineurin. *Science.* 267:1510–1512.

Tong, G., and C. Jahr. 1994. Regulation of glycine-insensitive desensitization of the NMDA receptor in outside-out patches. *J. Neurophysiol.* 72:754–761.

Vyklicky, L., Jr. 1993. Calcium-mediated modulation of N-methyl-D-aspartate NMDA responses in cultured rat hippocampal neurones. *J. Physiol.* 470:575–600.

Wafford, K.A., C.J. Bain, B. Le Bourdelles, P.J. Whiting, and J.A. Kemp. 1993. Preferential co-assembly of recombinant NMDA receptors composed of three different subunits. *Mol. Neurosci.* 4:1347–1349.

Wang, L.-Y., M.W. Salter, and J.F. MacDonald. 1991. Regulation of kainate receptors by cAMP-dependent protein kinase and phosphatases. *Science.* 253:1132–1135.

Williams, K., S.L. Russell, Y.M. Shen, and P.B. Molinoff. 1993. Developmental switch in the expression of NMDA receptors occurs in vivo and in vitro. *Neuron.* 10:267–278.

Wo, Z.G., and R.E. Oswald. 1995. Unraveling the modular design of glutamate-gated ion channels. *Trends Neurosci.* 18:161–168.

Wood, M., H. VanDongen, and A. VanDongen. 1995. Structural conservation of ion conduction pathways in K channels and glutamate receptors. *Proc. Natl. Acad. Sci. USA.* 92:4882–4886.

Wyszynski, M., J. Lin, A. Rao, E. Nigh, A.H. Beggs, A.M. Craig, and M. Sheng. 1997. Competitive binding of alpha-actinin and calmodulin to the NMDA receptor. *Nature.* 385:439–442.

Yen, L.-H., J.T. Sibley, and M. Constantine-Paton. 1993. Fine structural alterations and clustering of developing synapses after chronic treatments with low levels of NMDA. *J. Neurosci.* 13:4949–4960.

Zagotta, W.N., T. Hoshi, and R.W. Aldrich. 1990. Restoration of inactivation in mutants of Shaker potassium channels by a peptide derived from ShB. *Science.* 250:568–571.

Zhong, J., D. Carrozza, K. Williams, D. Pritchett, and P. Molinoff. 1995. Expression of mRNAs encoding subunits of the NMDA receptor in developing rat brain. *J. Neurochem.* 64:531–539.

Zilberter, Y., V. Uteshev, S. Sokolova, and B. Khodorov. 1991. Desensitization of N-methyl-D-aspartate receptors in neurons dissociated from adult rat hippocampus. *Mol. Pharmacol.* 40:337–341.

Zorumski, C.F., J. Yang, and G.D. Fischbach. 1989. Calcium-dependent, slow desensitization distinguishes different types of glutamate receptors. *Cell. Mol. Neurobiol.* 9:95–104.

A Membrane Skeleton that Clusters Nicotinic Acetylcholine Receptors in Muscle

Robert J. Bloch,* Gabriela Bezakova,* Jeanine A. Ursitti,* Daixing Zhou,* and David W. Pumplin‡

*Department of Physiology and ‡Department of Anatomy and Neurobiology, School of Medicine, University of Maryland at Baltimore, Baltimore, Maryland 21201

We have been investigating the structural proteins associated with nicotinic acetylcholine receptors (AChR) in skeletal muscle cells to learn how the receptors accumulate, or "cluster," in the sarcolemma in response to innervation by motor neurons. Muscle cells grown in tissue culture also cluster AChR in their plasma membranes in a way that resembles the earliest stages of postsynaptic differentiation in vivo. These clusters, formed in vitro, can be isolated and characterized to reveal the organization of structural proteins associated with AChR. Our results suggest that AChR are immobilized in clusters by virtue of their ability to attach to a membrane skeleton that resembles the skeleton of the human erythrocyte. An unusual form of spectrin is present at the cytoplasmic surface of AChR clusters, together with utrophin, dystrophin, syntrophin, dystroglycan, and the receptor-associated 43K protein, rapsyn. Semiquantitative fluorescence measurements indicate that AChR, rapsyn and syntrophin are present in clusters in approximately stoichiometric amounts, whereas spectrin is present in severalfold higher amounts. Dystrophin and utrophin together are present in smaller amounts. Selective extractions indicate that rapsyn is more tightly bound to the cluster membrane than spectrin. Actin, which is also present, is more easily removed from clusters than all the other receptor-associated proteins. Ultrastructural studies of platinum replicas indicate that the organization of the cytoplasmic face of AChR clusters is quantitatively similar to the spectrin-based skeleton of the human erythrocyte in terms of the diameters and lengths of filaments attached to the membrane, the number of intersections they make per unit area, and the number of filaments per intersection. Immunogold labeling of these structures shows that spectrin, dystrophin, utrophin, and syntrophin are present in the membrane skeletal lattice. Cytoplasmic actin filaments insert into the lattice. Decoration with anti-spectrin indicates that the spectrin is present as antiparallel homo-oligomers. We conclude that clustered AChR associate with a membrane skeleton resembling that of the erythrocyte but having unique components. To characterize this structure biochemically, we are trying to clone molecularly the cDNA that encodes the unusual, receptor-associated spectrin. We have also identified a region of rapsyn that is essential for clustering. This region, containing two zinc fingers (Scotland et al., 1993), was expressed as a GST-fusion protein and injected into muscle cells. AChR clusters in injected cells were rapidly and extensively disrupted. Control cells, injected with GST alone or with a double mutant peptide that failed to bind zinc, showed no significant effects. We obtained similar results with the wild-type and mutant zinc finger peptides that had

Cytoskeletal Regulation of Membrane Function © 1997 by The Rockefeller University Press

been cleaved from the fusion protein and purified by HPLC before injection. Our results suggest that the zinc finger domain has a binding activity required for AChR clustering. We are now trying to identify and characterize the ligand(s) of the zinc finger domain using affinity chromatography and immunoprecipitation.

Introduction

The formation of a synapse requires that several key events occur in a carefully co-ordinated fashion. The ingrowing neuronal process must find its way to the appropriate postsynaptic cell, the presynaptic and postsynaptic cells must exchange information that promotes further differentiation, the proteins essential to synaptic function must be assembled on the two sides of the synaptic cleft, and, finally, the structures formed initially must develop into the mature synapse. Many of these developmental events have been investigated at the vertebrate neuromuscular junction (NMJ) because it is so well characterized at the morphological, biochemical and pharmacological levels.

The mature NMJ has a distinctive morphology (e.g., Kelly and Zachs, 1969; Fertuck and Salpeter, 1976; Couteaux, 1981; reviewed in Salpeter, 1987). Presynaptically, small groups of vesicles filled with the neurotransmitter, acetylcholine, accumulate over specialized regions of membrane termed "active zones." Large intramembrane particles at active zones, probably equivalent to voltage-gated Ca^{2+} channels (Pumplin et al., 1981; Cohen et al., 1991; Sugiura et al., 1995), are concentrated adjacent to the sites of vesicle accumulation. The synaptic cleft is filled with a dense basal lamina that is attached via thin fibrils to both the presynaptic and postsynaptic membranes. The postsynaptic membrane is thrown into deep folds. The tops of the folds, closest to the nerve terminal, contain a paracrystalline array of acetylcholine receptors (AChR; e.g., Heuser and Salpeter, 1979; Fertuck and Salpeter, 1976; Cartaud et al., 1981) associated on the cytoplasmic surface with a highly specialized collection of peripheral membrane proteins. The binding of acetylcholine stimulates the flux of Na^+ (and, to a lesser extent, Ca^{2+}) through the receptor ion channels, depolarizing the postsynaptic region. The depths of the folds as well as the perisynaptic sarcolemma are enriched in voltage-gated Na^+ channels (Beam et al., 1985; Flucher and Daniels, 1989) that amplify this depolarization and initiate the action potential that eventually leads to muscle contraction.

Our laboratory has been studying the morphogenesis of the postsynaptic membrane of the mammalian neuromuscular junction in vivo and in vitro, with the object of understanding the mechanisms responsible for concentrating AChR there. The postsynaptic membrane contains ~70% of the total AChR present on muscle fibers, although it occupies only ~0.01% of the total fiber surface area. This is an extreme example of a cell membrane domain that is highly differentiated to serve a specialized function. Understanding how this membrane domain forms should reveal how similar domains assemble in other cells, especially neurons in the central nervous system, that also have high densities of postsynaptic receptors.

Here we focus on the results we obtained with acetylcholine receptor (AChR) clusters, a model of the embryonic postsynaptic membrane formed by rat myotubes grown in culture in the absence of neurons. We studied AChR clusters to understand the relationship between the receptor and intracellular structural (i.e., cytoskeletal) elements with which it is believed to interact. Our work has been facilitated by the fact that AChR clusters of rat myotubes form at sites of substrate adhesion, permitting their partial purification either by extraction with the cholesterol-specific detergent, saponin (Bloch, 1984), or by shearing with a stream of buffer (Avnur and Geiger, 1981). Shearing appears to leave relatively intact the cytoskeletal structures associated with the cytoplasmic surface of the clusters; detergent extraction yields a preparation that is more readily studied at the biochemical level. Either method yields a preparation of AChR clusters in which the cytoplasmic surface of the membrane is exposed to the solution and thus readily accessible to further extractions, immunolabeling, and ultrastructural examination. The extracellular surface is trapped between the membrane and the glass coverslip that serves as the tissue culture substrate and is thus partially protected from manipulations on the cytoplasmic surface.

A Spectrin-based Membrane Skeleton

The central hypothesis behind our research is that AChR binds to a cytoskeleton resembling the spectrin-based membrane skeleton of the human erythrocyte. To test this hypothesis, we looked for spectrin at AChR clusters and, finding it, we asked whether the membrane skeleton of clusters resembles the skeleton of the human erythrocyte in other ways. (i) Is the membrane skeleton of AChR clusters a layered, filamentous network resembling that found in erythrocytes? (ii) Are its components present in a stoichiometric complex, consistent with the association of a single AChR with an oligomer of spectrin? (iii) Are there proteins that play the role of ankyrin in the erythrocyte, lying between the spectrin network and the integral membrane protein, AChR? Although our answers to these questions are still somewhat tentative, they support our hypothesis that the overall organization of the membrane skeletons of AChR clusters and human erythrocytes is similar. We outline our key results below, then summarize other findings showing that, despite these basic similarities, the membrane skeleton of AChR clusters is in some ways quite distinct.

Spectrin in a Filamentous Network

The association of clustered AChR with spectrin was originally demonstrated by immunolabeling with monoclonal antibodies to β-spectrin (Bloch and Morrow, 1989; Daniels, 1990). A β-spectrin–like protein is present in regions of isolated clusters that are also rich in AChR, but it is absent from nearby membrane regions (Fig. 1 *B*). Immunoprecipitation experiments confirmed the specificity of the antibodies for a β-spectrin–like protein in isolated AChR clusters (Bloch and Morrow, 1989). The identity of this protein remains elusive, however, due in part to its unusual antigenic properties. We discuss these further below.

We applied ultrastructural techniques to determine whether the receptor-associated, spectrin-like molecule was arranged in a filamentous network at the cytoplasmic face of the membrane, as predicted from our earlier studies of the erythrocyte membrane (Ursitti et al., 1991). Those studies determined the size and connectivity of the spectrin network in intact red cell membranes. Previous studies, which examined spectrins after the membrane was dissolved, had determined the dimensions of the fully extended spectrin molecule to be ~100 nm for the αβ heterodimer and ~200 nm for the (αβ)$_2$ heterotetramer (Shotton et al., 1979; Shen et al., 1986; Liu et al., 1987; Byers and Branton, 1985). In situ, however, spectrin appeared as ~30-nm rods with a diameter of ~7.8 nm (value not corrected for platinum decoration; Ursitti et al., 1991; Fig. 2 *A*). Calculations suggested that these structures represented αβ heterodimers in which the 106 amino acid, triple-helical

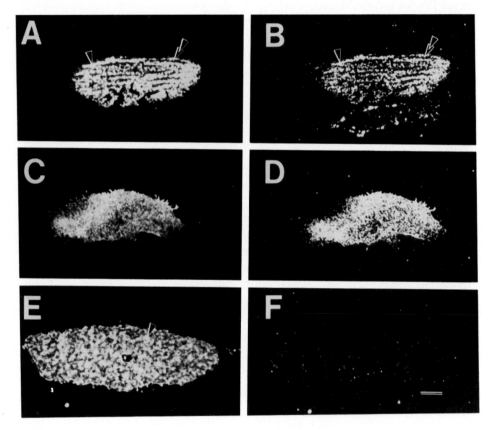

Figure 1. Spectrin at AChR clusters: normal distribution and selective removal. AChR clusters were labeled with R-BT and isolated by extraction with saponin (Bloch, 1984). After fixation, clusters were labeled with VIIF7 monoclonal antibodies to erythrocyte β-spectrin followed by FGAM. These samples showed spectrin present at AChR-rich domains (*A* and *B, arrowheads*) and absent at AChR-poor domains (*A* and *B, double arrowheads*). Alternatively, clusters were extracted at low ionic strength (*C* and *D*) or treated with chymotrypsin (*E* and *F*) before fixation and labeling with VIIF7. After extraction at low ionic strength (*C* and *D*) both spectrin and AChR move from their distinctive domains into a more uniform pattern, probably because actin associated with AChR clusters is removed (Bloch, 1986; see Table II). After treatment with chymotrypsin, all the spectrin detectable at AChR-rich membrane is removed (*F*) and AChR redistribute in the membrane (*E*), where they sometimes form distinctive aggregates (e.g., *E, arrowhead*). Bar, 10 μm. Figure reproduced from *The Journal of Cell Biology*, 1989, 108:481–493 (Bloch and Morrow, 1989), by copyright permission of The Rockefeller University Press.

repeats were organized compactly, like the folded pleats in an accordion (Bloch and Pumplin, 1992). Upon dissolution of the membrane, the molecule could expand as the pleats opened. On the basis of these calculations, we predicted that the folded triple helical repeat would have a length of ~4.5 nm, a value that was subsequently confirmed by crystallographic studies (Yan et al., 1993). In addition to revealing the likely shape and size of membrane-bound spectrin, our studies showed further that there were an average of ~400 filament intersections per square micron in the skeleton, and that each intersection was formed by an average of 3.4 filaments (Ursitti et al., 1991; see Table I).

We compared these values to those obtained from parallel experiments performed on isolated AChR clusters (Pumplin, 1995). We found clustered AChR to be associated with a filamentous network that is very similar to that of the spectrin-based membrane skeleton of the human erythrocyte (reviewed in Pumplin and Bloch, 1993), with respect to the lengths of the filaments (~30 nm), their widths (~8.6 nm, uncorrected for platinum decoration), and the average number of 500 filament intersections/μm^2 with each intersection containing 3.1–3.3 filaments (Pumplin, 1995; Table I). Thus, at the ultrastructural level, the spectrin-based membrane skeleton of AChR clusters is very similar indeed to that of the human erythrocyte.

These ultrastructural studies, as well as immunofluorescence experiments (Bloch, 1986), indicate still another similarity between AChR clusters and erythrocyte membranes—the presence of actin. In the spread erythrocyte skeleton, actin appears primarily as short oligomers that crosslink spectrin oligomers (Shotton et al., 1979; Shen et al., 1986; Liu et al., 1987; Pinder and Gratzer, 1983), but we did not readily observe actin oligomers in intact preparations of erythrocytes (Ursitti et al., 1991). Similarly, we could not detect short actin oligomers in intact AChR clusters (spread preparations have not yet been achieved), although longer actin filaments arising from deeper levels of the cytoplasm were frequently seen to insert into the filamentous network (Dmytrenko et al., 1993). Results summarized below suggest that much of the actin present in AChR clusters can be released under conditions that leave most of the spectrin and other receptor-associated proteins bound to cluster membrane.

Stoichiometry and Association with the Membrane

The membrane skeleton of the erythrocyte contains ~4 spectrin subunits for each integral membrane protein it anchors, and this anchoring is mediated by a single

TABLE I
Characteristic Parameters of Membrane Skeletons of Erythrocytes and AChR-rich Membrane*

	Erythrocyte	AChR-rich membrane
Filament length (nm)	28.8 ± 8.7	31.7 ± 10.9
Filament Diameter (nm)[‡]	7.8 ± 2.0	8.6 ± 2.0
Filaments/intersection	3.4	3.1–3.3
Intersections/μm^2	411	500

*Adapted from Pumplin (1995), based in part on data from Ursitti et al. (1991). [‡]The values for filament diameter are uncorrected for the thickness of the platinum layer deposited on the samples, which is estimated at ~2.5-nm.

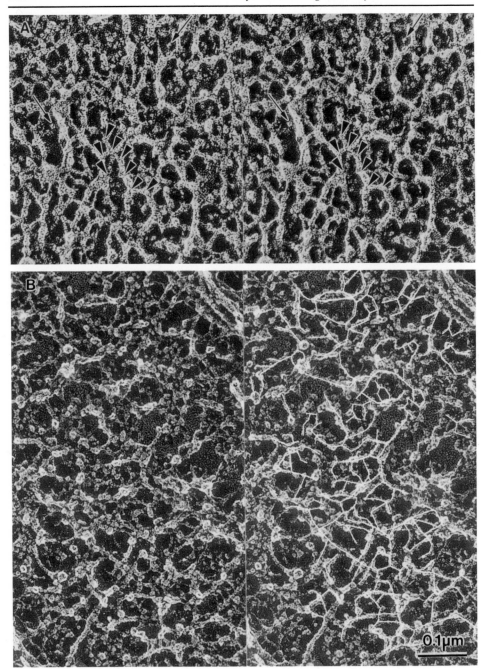

Figure 2. Stereo views of spectrin-based membrane skeletons visualized by quick-freeze, deep-etch, and rotary-replication with platinum and carbon. Membrane skeletons are from (*A*) erythrocyte and (*B*) acetylcholine receptor-rich regions of myotube membrane. Both samples display a dense, irregular array of interconnecting filaments. The filaments vary widely in length, but appear to have similar morphology. In *A*, junctions between filaments are indicated with arrowheads. The arrows indicate occasional particles that project from the

molecule of the peripheral membrane protein, ankyrin (reviewed in Bennett, 1992). We developed a semiquantitative immunofluorescence procedure to determine the relative amounts of spectrin, AChR, and other receptor-associated proteins in AChR clusters. The approach involved double labeling for AChR with monotet-ramethylrhodamine-α-bungarotoxin (R-BT) and for other proteins with monoclonal IgG antibodies followed by fluoresceinated goat anti–mouse IgG (FGAM). The fluorescence arising from the fluorescein and rhodamine labels within individual clusters was measured with a photomultiplier attached to the microscope, and, after correction for background values, a ratio was derived. Fortuitously, this ratio was 1 when we labeled clusters with saturating concentrations of anti-AChR antibodies (Bloch and Froehner, 1987). We also obtained a ratio of 1 with saturating concentrations of monoclonal antibodies to the AChR-associated 43K protein, rapsyn (Bloch and Froehner, 1987). This suggested that isolated clusters contain equal amounts of AChR and rapsyn, in agreement with studies of *Torpedo* electric organ (Cartaud et al., 1981; LaRochelle and Froehner, 1986; Sobel et al., 1978).

These experiments indicated that we could use the fluorescent signals arising from bound primary and secondary antibodies to compare the relative amounts of different proteins at AChR clusters. When we used this approach with the VIIF7 monoclonal antibodies to β-spectrin, we found values of ∼2 with clusters isolated with saponin and ∼4–7 with clusters isolated by shearing (Bloch and Morrow, 1989). The lower value was probably due to the fact that the spectrin epitope is lost from clusters as a function of time after isolation (Bloch and Morrow, 1989), which is more extended when saponin is used. We believe, therefore, that the higher values are more likely to reflect the stoichiometry in situ. These values are close to those expected for a spectrin-based membrane skeleton such as that of the erythrocyte.

These values are reliable only if several conditions are met. First, the FGAM used in these experiments must react equally well with each of the monoclonal antibodies used to probe AChR and receptor-associated proteins. We confirmed this by ELISA methods (unpublished data). Second, each of the monoclonal antibodies must label AChR clusters with equal valency. If one antibody binds to clusters through both its binding sites and others through only one, then the relative values we determine may be incorrect by a factor of two. Although we cannot rule out the possible effects of other factors, such as limited accessibility to antibodies, we believe they are unlikely, because selective extraction of proteins from AChR clusters did not increase the labeling of the proteins that remained.

The close association of AChR with the 43K protein, rapsyn, at the biochemical and ultrastructural levels (Burden et al., 1983; Bridgman et al., 1987; Toyoshima and Unwin, 1990; Mitra et al., 1989; Toyoshima and Unwin, 1988) has suggested to many investigators that rapsyn serves as a peripheral membrane protein to link AChR to cytoskeletal proteins involved in clustering (reviewed in Froehner, 1991). We have tried to demonstrate binding between rapsyn and erythroid spectrin, with

---

cytoplasmic face of the membrane between the elements of the membrane skeletal lattice. In *B*, filaments in the network were traced with a computer graphics program; the tracing is shown superimposed on the image (*right side* of *B*). The scale bar (0.1 μm) applies to both images. Panel *A* is reproduced from Ursitti et al., 1991, *Cell Motility and the Cytoskeleton*, 19:227–243, with permission from Wiley-Liss, Inc., a subsidiary of Wiley & Sons, Inc., and panel *B* is reproduced from Pumplin, 1995, *The Journal of Cell Science*, 108:3145–3154.

variable results. More indirectly, however, we have shown that rapsyn is considerably more difficult to remove from AChR clusters than spectrin (Table II). Unlike the spectrin of erythrocyte membranes, receptor-associated spectrin remained stably associated with cluster membrane upon extraction with solutions of low ionic strength (Bloch and Morrow, 1989; Table II). We were able to remove spectrin epitopes from clusters by mild digestion with chymotrypsin (Fig. 1, *E* and *F*; Table II; Bloch and Morrow, 1989), which has little effect on rapsyn (Bloch and Froehner, 1987). These results, as well as the slow loss of spectrin from isolated clusters under conditions in which rapsyn is almost completely retained (Bloch and Froehner, 1987; Krikorian and Bloch, 1992), suggest that rapsyn is more tightly associated with AChR-rich membrane than is spectrin.

Rapsyn was removed from AChR clusters by harsher extraction conditions, including solutions containing 6 M urea or 50 mM ethanolamine, pH 11 (Bloch and Froehner, 1987); these also remove the bulk of the spectrin (Table II; Bloch and Morrow, 1989). Removal of rapsyn doubled the binding to AChR of an antibody that reacts with both gamma and delta subunits of the receptor. This result was confirmed with antibodies to other cytoplasmic, but not to extracellular, epitopes of the AChR (Krikorian and Bloch, 1992). We interpret this to mean that additional cytoplasmic epitopes of the AChR become exposed when rapsyn is removed, consistent with a particularly close association between these proteins. We did not observe such an increase in immunolabeling of rapsyn, spectrin, or other receptor-associated proteins (see below) when more loosely bound proteins were removed by alterations in ionic strength or by proteolysis (Table II), consistent with the idea that, with the exception of the AChR itself, the access of antibodies to receptor-associated proteins is not restricted.

These results suggest that the network of spectrin filaments interacts with muscle membrane containing high concentrations of AChR and the peripheral membrane protein, rapsyn. This "layered" structure resembles the organization of the

TABLE II
Ratios Measured for Proteins in AChR Clusters

Protein	mAb*	PBS[‡]	Low[§]	CT[‖]	pH 11[¶]	LIS/urea**
AChR	88B[‡‡]	1	1	1	2	2
rapsyn	1579[‡‡]	1	1	1	0	0
syntrophin	1351[‡‡]	1	1	0	0.3	0
β-spectrin	VIIF7[§§]	4–7	1	0	0	0
dystrophin	1808[‡‡]	0.2	0.2	0	0.2	0.2
utrophin[‖‖]	Nova	∼0.3	∼0.3	0	∼0.3	∼0.2
actin	HP249	100%	35%	10%	—	—

*All monoclonal antibodies are IgG$_1$ and react equally well with FGAM, except the anti-actin antibody HP249, which is an IgM. [‡]Phosphate-buffered saline control. [§]Low ionic strength buffer (buffer A: 2 mM Tris-HCl, 0.2 mM ATP, pH 8.0). [‖]10 μg/ml chymotrypsin. [¶]50 mM ethanolamine, pH 11. **20 mM lithium diiodosalicylate in buffered saline, or 6 M urea. [‡‡]Provided by Dr. S. Froehner, Department of Physiology, University of North Carolina, Chapel Hill, NC. [§§]Provided by Dr. J. Morrow, Department of Pathology, Yale University School of Medicine, New Haven, CT. [‖‖]Values for utrophin are still preliminary. Antibody obtained from Novocastra, Newcastle upon Tyne, UK.

erythrocyte membrane, in which the spectrin network associates with an integral membrane protein, the anion transporter, largely through ankyrin. We still lack biochemical evidence to indicate whether rapsyn binds directly or indirectly to the membrane skeleton.

Differences between the Membrane Skeletons of AChR Clusters and Erythrocytes

The findings presented above are consistent with our hypothesis that AChRs cluster by virtue of their ability to associate with a spectrin-based membrane skeleton similar to that of the red cell, but there are significant biochemical and morphological differences between the two skeletons. Obviously, the key integral and peripheral membrane proteins are distinct in AChR clusters and erythrocyte membranes, with AChR replacing the anion exchanger in the membrane and rapsyn possibly replacing ankyrin. Furthermore, the receptor-associated form of spectrin is different from other known members of the spectrin superfamily. It does not appear to associate with any known α-spectrin (or α-fodrin) subunit, and it is immunologically distinct from other known spectrins (Bloch and Morrow, 1989; unpublished results). To date only one monoclonal antibody, VIIF7, generated against the β-subunit of erythrocyte spectrin, and a single polyclonal antibody, generated in rabbits against the same protein, have labeled AChR clusters (Bloch and Morrow, 1989). A set of other monoclonal and polyclonal antibodies to β-spectrin, β-fodrin (βII), and α-fodrin (αII), as well as antibodies generated in rabbits against the COOH-terminal peptide sequences of the erythrocyte (βIΣ1) and muscle (βIΣ2) isoforms of β-spectrin have failed to react with the receptor-associated spectrin. These results suggest that the spectrin associated with clustered AChR is not closely related to these three known subunits. Other studies suggest that it is probably also distinct from α-spectrin (Bloch and Morrow, 1989; unpublished results). Paradoxically, however, peptide maps of the major band of β-spectrin immunoprecipitated by VIIF7 resemble those of erythrocyte β-spectrin (βIΣ1), with $\sim$80% homology at the level of tyrosine-containing peptides (Bloch and Morrow, 1989). One possible explanation of this paradox is that βIΣ2 is the major protein present in immunoprecipitates, but that a different and presumably distant member of the spectrin family is the one that associates with AChR.

We are currently pursuing serial immunoprecipitation experiments to test this possibility and, if possible, to obtain enough of the receptor-associated spectrin to characterize it further. This task has been complicated by our recent finding that VIIF7 reacts with some forms of smooth muscle myosin as well as with β-spectrin. Antibodies that recognize both protein families have been described before (Wong et al., 1985), but the basis for cross-reactivity is still unclear. Comparisons of primary sequences do not show regions of significant homology, suggesting that the similarity may be structural. The ATPase region of myosin is not involved, however, as antibodies to this region do not cross-react with spectrins, nor do they label AChR-rich membrane. We are examining different anti-myosin and anti-spectrin antibodies with the hope of improving our chances of affinity purifying the receptor-associated spectrin. We are also pursuing PCR-based strategies to identify β-spectrin–like sequences that are expressed at high levels in myotubes and developing myofibers.

At the morphological level, the membrane skeleton of AChR clusters is also unusual. In the erythrocyte, and presumably in other spectrin-based membrane skeletons containing $\alpha\beta$ dimers and $(\alpha\beta)_2$ tetramers, the filaments that form the membrane-associated network are asymmetric. By contrast, the filaments at AChR clusters appear to be symmetric, as they label at both ends with VIIF7 (Pumplin, 1995). This, together with the absence of an identifiable α-subunit (Bloch and Morrow, 1989), suggests that the filaments are composed of anti-parallel homo-oligomers (Bloch and Morrow, 1989; Pumplin, 1995). Examples of homomeric spectrins associated with the plasma membrane are relatively rare (Woods and Lazarides, 1986; Bennett and Condeelis, 1988). The ability of a single spectrin species to assemble into filamentous networks would greatly expand the potential of this protein superfamily to organize distinct domains within the plasma membrane, and perhaps in internal membranes as well (e.g., Beck et al., 1994; Devarajan et al., 1996).

Other Membrane Skeletal Proteins at AChR Clusters

Elucidating the structure of the membrane skeleton of AChR clusters is further complicated by the fact that it contains other, more distant members of the spectrin superfamily. We demonstrated this using antibodies to dystrophin, utrophin (also known as dystrophin-related protein, or DRP) and two associated proteins, syntrophin and β-dystroglycan. We found that, like spectrin, both dystrophin and utrophin were present on the cytoplasmic face of AChR clusters, where they distributed into the network of filaments associated with AChR (Dmytrenko et al., 1993, and unpublished data; see Fig. 2 *B*). Both proteins appeared to be present in less than stoichiometric amounts, however. Using our semiquantitative immunofluorescence procedures, described above, we found that AChR exceeded dystrophin by $\sim$5-fold (Dmytrenko et al., 1993) and utrophin by $\sim$3-fold (Table II). If these values are correct, there is probably not enough of these two proteins together to associate with each AChR. However, although the monoclonal antibodies to utrophin and dystrophin react equally well with FGAM (Dmytrenko et al., 1993; unpublished results), we do not know the valency with which they react with the membrane, so our values may be subject to a twofold correction. In addition, these values are subject to considerable sampling error. Therefore, we cannot completely rule out the possibility that utrophin and dystrophin together form an equimolar complex with AChR.

In keeping with the close association of dystrophin and utrophin with AChR (Woodruff et al., 1987; Sealock et al., 1991; Dmytrenko et al., 1993; Cartaud et al., 1992; Ohlendieck et al., 1991; Phillips et al., 1993), we have also found syntrophin and β-dystroglycan at AChR clusters (Bloch et al., 1991, 1994; see also Froehner et al., 1987; Apel et al., 1995; Daniels et al., 1990; Peters et al., 1994). Syntrophin has been localized to the membrane skeletal lattice at the ultrastructural level (Bloch et al., 1991). Our studies of dystroglycan, though only qualitative to date, showed codistribution of this protein with AChR (not shown), in agreement with reports from other laboratories (Phillips et al., 1993; Apel et al., 1995; Cohen et al., 1995; see Fallon and Hall, 1994, for a review). Semiquantitative studies of syntrophin showed it to be present in amounts approximately equal to those of AChR, i.e., the ratio we determined from immunofluorescence measurements was $\sim$1 (Table II). If syntrophin associates with each dystrophin or utrophin molecule as a trimer (Yang et al.,

1995), then this value may be underestimated. Bearing in mind the caveats mentioned above, however, we believe that syntrophin is present in clusters in amounts equimolar with AChR.

We used selective extraction procedures to determine how tightly or loosely associated with cluster membrane these proteins are. In contrast to our results with spectrin and rapsyn, all the dystrophin and utrophin and ~30% of the syntrophin resisted extraction by alkaline pH or chaotropic reagents (Bloch et al., 1991). These results are consistent with the idea that these proteins remain associated with cluster membrane even when rapsyn and spectrin are quantitatively extracted. Thus, they are not likely to associate with the membrane or with the AChR through either of those proteins. We propose instead that they associate with the membrane through their ability to bind to dystroglycan, an integral membrane protein (Ibraghimov-Beskrovnaya et al., 1992; Gee et al., 1993).

Linkage to the Extracellular Matrix

The association of dystroglycan with clustered AChR, coupled with the ability of dystroglycan to interact with extracellular proteins such as laminin (Ibraghimov-Beskrovnaya et al., 1992; Gee et al., 1993, 1994), suggests a mechanism for linking AChR-rich membrane to the extracellular matrix. Immunofluorescent and ultra-structural observations have consistently shown that AChR-rich membrane domains have a specialized extracellular matrix (reviewed in Bloch and Pumplin, 1988), but the means of anchoring this matrix selectively at AChR-rich sites has remained unknown. We therefore designed an experiment to learn whether this anchoring was dependent on intracellular proteins present in the receptor-associated membrane skeleton.

AChR clusters were isolated and then exposed to insoluble preparations of chymotrypsin which was bound to cellulose or agarose particles. The insoluble chymotrypsin, like its soluble counterpart, digested the membrane skeleton of isolated clusters, and in the process removed the bulk of the spectrin and syntrophin from the cytoplasmic face of the membrane. The insoluble enzyme did not gain access to the extracellular proteins that were protected by overlying membrane, however, as we could detect no degradation of fibronectin and no loss of laminin, heparansulfate proteoglycan, or type IV collagen (Dmytrenko and Bloch, 1993). When we examined the extracellular proteins in treated clusters, however, they had redistributed from their normal place in AChR-rich domains (Dmytrenko et al., 1990) to nearby regions of the membrane, just as the AChR redistributed as the membrane skeleton was degraded (Dmytrenko and Bloch, 1993). These results suggest that elements exposed on the cytoplasmic face of the cluster membrane not only bind to and immobilize AChR, they also immobilize proteins of the extracellular matrix.

A Model of the Membrane Skeleton of AChR Clusters

We outlined six key results above. (i) Spectrin is present in a filamentous network at AChR clusters. (ii) This network also contains smaller amounts of dystrophin and utrophin, as well as syntrophin and dystroglycan. (iii) Rapsyn is closely associated with AChR but does not mediate the membrane binding of dystrophin, utrophin, or syntrophin. (iv) Rapsyn, AChR, and syntrophin are present in approximately equimolar amounts, while spectrin is present in severalfold excess. Dystrophin and utrophin, by contrast, are present in smaller amounts, although together they may be stoichiometric with AChR. (v) AChR clusters contain two transmembrane elements, the AChR itself, and dystroglycan. (vi) Extracellular matrix proteins associate with AChR by virtue of their ability to anchor to components on the cytoplasmic surface of the membrane, probably through dystroglycan. Taking all these observations into account, we created a model of the AChR cluster, as depicted in Fig. 3.

Several features of the model are noteworthy. First, it proposes that all associations among proteins of AChR-rich membrane are mediated by the spectrin-based membrane skeleton. Although this is consistent with our results, it does not readily account for the ability of rapsyn and dystroglycan to coaggregate when the two proteins are heterologously expressed in avian fibroblasts (Apel et al., 1995). Heterologous expression has been a powerful tool to study the proteins involved in AChR clustering, but its relationship to the process of AChR clustering in a skeletal muscle cell has never been fully established. If the coaggregation of rapsyn and dystroglycan is in fact physiological, then our model will have to be revised. Second, it proposes that spectrin, dystrophin and utrophin combine to form overlapping or interacting structural elements within the filamentous, membrane skeletal network. Evidence for this is indirect (Dmytrenko et al., 1993), and further studies will be needed to test the possible interactions among these proteins. Third, the model proposes that spectrin, or dystrophin and utrophin, can bind rapsyn and so anchor

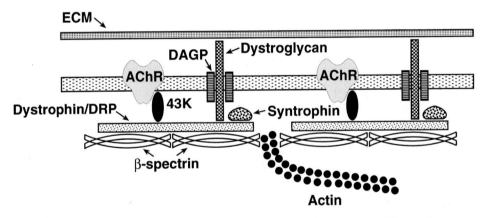

Figure 3. Model of membrane skeletal organization at AChR clusters. The model postulates that each AChR in a cluster is linked by a single rapsyn molecule (*43K*) to a network containing a 4-fold molar excess of the receptor-associated form of β-spectrin with small amounts of dystrophin and utrophin (*DRP*). It is not yet clear if the two latter proteins together are present in amounts that are equimolar with AChR and rapsyn (see text). The spectrin and dystrophin/utrophin together form the network of filaments pictured in Fig. 2 *B*. Actin filaments attach to this network, presumably by binding to the actin-binding domains present in spectrin, dystrophin, and utrophin. Each molecule of dystrophin and utrophin is associated with syntrophin and each is linked to the membrane independently through its binding to dystroglycan and other dystrophin-associated glycoproteins (*DAGP*). Dystroglycan in turn links these elements to the extracellular matrix (*ECM*) by virtue of its ability to bind to laminin.

AChR. Blot overlay experiments have so far failed to give consistent results (unpublished data). Definitive identification of rapsyn's ligands will require more rigorous experimentation.

Intracellular Injection Studies of Rapsyn

We have tried to identify rapsyn's ligands by first determining the domains of the protein that are essential for AChR clustering. Site-directed mutagenesis studies (Phillips et al., 1991; Scotland et al., 1993) have identified sites at the NH_2 and COOH termini and a leucine zipper motif in the middle of the molecule that are important for rapsyn function in heterologous expression systems. We have begun a series of experiments to test the roles of these regions of rapsyn in the clustering of AChR in cultured rat muscle cells.

Our initial experiments utilized intracellular injections of peptides encompassing the COOH-terminal sequence of the molecule, which has been shown to contain two zinc fingers (Scotland et al., 1993). Injection into myotubes of a glutathione S-transferase (GST) fusion protein containing this domain disrupted AChR clusters (Fig. 4, *w1* and *w2*), whereas injection of a fusion protein containing a double mutant that fails to bind zinc (Scotland et al., 1993) had no significant effect (Fig. 4, *m1* and *m2*). Injection of GST alone also had no effect (not shown). When we released the zinc finger domain from the GST fusion protein by proteolysis, purified it by HPLC, and injected it into myotubes, clusters were again disrupted. Parallel experiments with the double zinc finger domain of the human immunodeficiency virus (HIV-1) showed no effect. Thus, the zinc finger domain of rapsyn acts in a dominant negative fashion to destabilize AChR clusters, consistent with the idea that this domain contains a binding site important for clustering.

We are currently using affinity chromatography procedures to try to identify the ligand(s) of the zinc finger domain. Previous experiments have suggested that the zinc finger domain is probably not involved in rapsyn binding either to itself or to the AChR (Scotland et al., 1993; our unpublished results). If our model (Fig. 3) is correct, the zinc finger domain should bind to one or more of the proteins that form the filamentous membrane skeleton—spectrin, dystrophin, and utrophin.

Summary

We have presented ultrastructural and semiquantitative immunofluorescence evidence to support the idea that AChR are clustered in rat myotubes by virtue of their ability to associate with a spectrin-based membrane skeleton. Many of the interactions postulated to be involved in the formation of this skeleton, and in the anchoring of AChR to it, must still be examined at the biochemical level, but the overall similarity of this structure to that of the human erythrocyte is already clear.

The ability of different members of the spectrin superfamily to associate in various combinations to form distinct plasmalemmal domains provides some exciting hints as to how the surface membrane can be organized efficiently to subserve multiple purposes. One of the challenges of future research will be to learn how innervation regulates the assembly of the membrane skeleton at the developing NMJ, and how this structure is altered as the junction matures. Another will be to learn if the principles of neuromuscular synaptogenesis are relevant to interactions between neurons in the brain, where cells must distinguish between multiple synaptic inputs and assemble synaptic structures at thousands of distinct sites on the neurolemma. Members of the spectrin superfamily have been identified in synaptic structures in the central nervous system (e.g., Carlin et al., 1983; LeVine and Sahyoun, 1986; Malchiodi-Albedi et al., 1993), so much of what we have learned at the neuromuscular junction may be applicable to central synapses.

Acknowledgments

We thank Wendy Resneck, Andrea O'Neill, John Strong and Tessa Chaney for their expert assistance, and our collaborators at the University of Maryland and

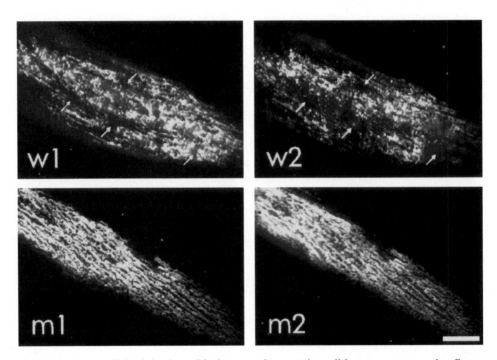

Figure 4. Intracellular injection of fusion proteins carrying wild type or mutant zinc finger domains. Rat myotubes were labeled with R-BT and then injected with solutions containing fusion protein consisting of GST linked to the wild type zinc finger domain of rapsyn (*w1* and *w2*) or to the double mutant that is unable to bind zinc (*m1* and *m2*). Fusion proteins were created in the pGEX-3x vector, which places the rapsyn sequence downstream of GST. The expressed proteins were purified on glutathione-Sepharose followed by HPLC and diluted to 1 μM in a high K^+ buffer before injection. The AChR cluster on the myotube injected with the wild type zinc finger construct, which in this case was a loosely organized structure immediately after injection (*w1*), lost significant areas of AChR-rich membrane over the next 1.5 h (*w2*: compare regions labeled with *small arrows*). The AChR cluster on the myotube injected with the mutant zinc finger construct was not affected between the time of injection (*m1*) and 1.5 h after injection (*m2*). Disrupted clusters like that shown in *w2* were observed in 47 of 96 cells (49%) injected with the wild type fusion protein, but with only 4 of 48 cells (8%) injected with the mutant. Bar, 10 μm.

elsewhere for their seminal contributions to the research summarized here. We are also grateful to many of our colleagues for their generous gifts of antibodies and other reagents used in the course of these studies.

Our research has been supported by grants (NS17282, NS15513, and HD16596) and NRSAs (HL08840 and NS07375) from the National Institutes of Health, the Muscular Dystrophy Association and the Myasthenia Gravis Foundation.

References

Apel, E.D., S.L. Roberds, K.P. Campbell, and J.P. Merlie. 1995. Rapsyn may function as a link between the acetylcholine receptor and the agrin-binding dystrophin-associated glyco-protein complex. *Neuron.* 15:115–126.

Avnur, Z., and B. Geiger. 1981. Substrate-attached membranes of cultured cells. Isolation and characterization of ventral cell membranes and the associated cytoskeleton. *J. Mol. Biol.* 153:361–379.

Beam, K.G., J.H. Caldwell, and D.T. Campbell. 1985. Na$^+$ channels in skeletal muscle concentrated near the neuromuscular junction. *Nature.* 313:588–590.

Beck, K.A., J.A. Buchanan, V. Malhotra, and W.J. Nelson. 1994. Golgi spectrin: identification of an erythroid β-spectrin homolog associated with the Golgi complex. *J. Cell Biol.* 127:707–723.

Bennett, V. 1992. Ankyrins. Adaptors between diverse plasma membrane proteins and the cytoplasm. *J. Biol. Chem.* 267:8703–8706.

Bennett, H., and J. Condeelis. 1988. Isolation of an immunoreactive analogue of brain fodrin that is associated with the cell cortex of Dictyostelium amoebae. *Cell Motil. Cytoskel.* 11:303–317.

Bloch, R.J. 1984. Isolation of acetylcholine receptor clusters in substrate-associated material from cultured rat myotubes using saponin. *J. Cell Biol.* 99:984–993.

Bloch, R.J. 1986. Actin at receptor-rich domains of isolated acetylcholine receptor clusters. *J. Cell Biol.* 102:1447–1458.

Bloch, R.J., and S.C. Froehner. 1987. The relationship of the postsynaptic 43K protein to acetylcholine receptors in receptor clusters isolated from cultured rat myotubes. *J. Cell Biol.* 104:645–654.

Bloch, R.J., and J.S. Morrow. 1989. An unusual β-spectrin associated with clustered acetylcholine receptors. *J. Cell Biol.* 108:481–493.

Bloch, R.J., and D.W. Pumplin. 1988. Molecular events in synaptogenesis: nerve-muscle adhesion and postsynaptic differentiation. *Am. J. Physiol.* 254:C345–C364.

Bloch, R.J., and D.W. Pumplin. 1992. A model of spectrin as a concertina in the erythrocyte membrane skeleton. *Trends Cell Biol.* 2:186–189.

Bloch, R.J., W.G. Resneck, A. O'Neill, J. Strong, and D.W. Pumplin. 1991. Cytoplasmic components of acetylcholine receptor clusters of cultured rat myotubes: the 58-kD protein. *J. Cell Biol.* 115:435–446.

Bloch, R.J., R. Sealock, P.W. Luther, D.W. Pumplin, and S.C. Froehner. 1994. Association of acetylcholine receptor with peripheral membrane proteins: evidence from antibody-induced coaggregation. *J. Membr. Biol.* 138:13–28.

Bridgman, P.C., C. Carr, S.E. Pedersen, and J.B. Cohen. 1987. Visualization of the cytoplasmic surface of Torpedo postsynaptic membranes by freeze-etch and immunoelectron microscopy. *J. Cell Biol.* 105:1829–1846.

Burden, S.J., R.L. DePalma, and G.S. Gottesman. 1983. Crosslinking of proteins in acetylcholine receptor-rich membranes: association between the β-subunit and the 43 kd subsynaptic protein. *Cell.* 35:687–692.

Byers, T.J., and D. Branton. 1985. Visualization of the protein associations in the erythrocyte membrane skeleton. *Proc. Natl. Acad. Sci. USA.* 82:6153–6157.

Carlin, R.K., D.C. Bartelt, and P. Siekevitz. 1983. Identification of fodrin as a major calmodulin-binding protein in postsynaptic density preparations. *J. Cell Biol.* 96:443–448.

Cartaud, A., M.A. Ludosky, F.M.S. Tome, H. Collin, F. Stetzkowski-Marden, T.S. Khurana, L.M. Kunkel, M. Fardeau, J.P. Changeux, and J. Cartaud. 1992. Localization of dystrophin and dystrophin-related protein at the electromotor synapse and neuromuscular junction in Torpedo marmorata. *Neuroscience.* 48:995–1003.

Cartaud, J., A. Sobel, A. Rousselet, P.F. Devaux, and J.-P. Changeux. 1981. Consequences of alkaline treatment for the ultrastructure of the acetylcholine-receptor-rich membranes from Torpedo marmorata electric organ. *J. Cell Biol.* 90:418–426.

Cohen, M.W., O.T. Jones, and K.J. Angelides. 1991. Distribution of Ca^{2+} channels on frog motor nerve terminals revealed by fluorescent ω-conotoxin. *J. Neurosci.* 11:1032–1039.

Cohen, M.W., C. Jacobson, E.W. Godfrey, K.P. Campbell, and S. Carbonetto. 1995. Distribution of α-dystroglycan during embryonic nerve-muscle synaptogenesis. *J. Cell Biol.* 129:1093–1101.

Couteaux, R. 1981. Structure of the subsynaptic sarcoplasm in the interfolds of the frog neuromuscular junction. *J. Neurocytol.* 10:947–962.

Daniels, M.P. 1990. Localization of actin, β-spectrin, 43×10^3 M_r and 58×10^3 M_r proteins to receptor-enriched domains of newly formed acetylcholine receptor aggregates in isolated myotube membranes. *J. Cell Sci.* 97:615–627.

Daniels, M.P., J.G. Krikorian, A.J. Olek, and R.J. Bloch. 1990. Association of cytoskeletal proteins with newly formed acetylcholine receptor aggregates induced by embryonic brain extract. *Exp. Cell Res.* 186:99–108.

Devarajan, P., P.R. Stabach, A.S. Mann, T. Ardito, M. Kashgarian, and J.S. Morrow. 1996. Identification of a small cytoplasmic ankyrin (Ank_{G119}) in the kidney and muscle that binds β1Σ* spectrin and associates with the Golgi apparatus. *J. Cell Biol.* 133:819–830.

Dmytrenko, G.M., and R.J. Bloch. 1993. Evidence for transmembrane anchoring of extracellular matrix at acetylcholine receptor clusters. *Exp. Cell Res.* 206:323–334.

Dmytrenko, G.M., D.W. Pumplin, and R.J. Bloch. 1993. Dystrophin in a membrane skeletal network: localization and comparison to other proteins. *J. Neurosci.* 13:547–558.

Dmytrenko, G.M., M.G. Scher, G. Poiana, M. Baetscher, and R.J. Bloch. 1990. Extracellular glycoproteins at acetylcholine receptor clusters of rat myotubes are organized into domains. *Exp. Cell Res.* 189:41–50.

Fallon, J.R., and Z.W. Hall. 1994. Building synapses: agrin and dystroglycan stick together. *Trends Neurosci.* 17:469–473.

Fertuck, H.C., and M.M. Salpeter. 1976. Quantitation of junctional and extrajunctional acetylcholine receptors by electron microscope autoradiography after ^{125}I-α-bungarotoxin binding at mouse neuromuscular junctions. *J. Cell Biol.* 69:144–158.

Flucher, B.E., and M.P. Daniels. 1989. Distribution of Na^+ channel and ankyrin in neuromuscular junction is complementary to acetylcholine receptor and 43-kD protein. *Neuron.* 3:163–175.

Froehner, S.C. 1991. The submembrane machinery for nicotinic acetylcholine receptor clustering. *J. Cell Biol.* 114:1–7.

Froehner, S.C., A.A. Murnane, M. Tobler, H.B. Peng, and R. Sealock. 1987. A postsynaptic M_r 58,000 (58K) protein concentrated at acetylcholine receptor-rich sites in Torpedo electroplaques and skeletal muscle. *J. Cell Biol.* 104:1633–1646.

Gee, S.H., R.W. Blacher, P.J. Douville, P.R. Provost, P.D. Yurchenco, and S. Carbonetto. 1993. Laminin-binding protein 120 from brain is closely related to the dystrophin-associated glycoprotein, dystroglycan, and binds with high affinity to the major heparin binding domain of laminin. *J. Biol. Chem.* 268:14972–14980.

Gee, S.H., F. Montanaro, M.H. Lindenbaum, and S. Carbonetto. 1994. Dystroglycan-alpha, a dystrophin-associated glycoprotein, is a functional agrin receptor. *Cell.* 77:675–686.

Heuser, J.E., and S.R. Salpeter. 1979. Organization of acetylcholine receptors in quick-frozen, deep-etched, and rotary-replicated Torpedo postsynaptic membrane. *J. Cell Biol.* 82:150–173.

Ibraghimov-Beskrovnaya, O., J.M. Ervasti, C.J. Leveille, C.A. Slaughter, S.W. Sernett, and K.P. Campbell. 1992. Primary structure of dystrophin-associated glycoproteins linking dystrophin to the extracellular matrix. *Nature.* 355:696–702.

Kelly, A.M., and S.I. Zachs. 1969. The fine structure of motor endplate morphogenesis. *J. Cell Biol.* 42:154–169.

Krikorian, J.G., and R.J. Bloch. 1992. Treatments that extract the 43K protein from acetylcholine receptor clusters modify the conformation of cytoplasmic domains of all subunits of the receptor. *J. Biol. Chem.* 267:9118–9128.

LaRochelle, W.J., and S.C. Froehner. 1986. Determination of the tissue distributions and relative concentrations of the postsynaptic 43-kDa protein and the acetylcholine receptor in Torpedo. *J. Biol. Chem.* 261:5270–5274.

LeVine, H., and N.E. Sahyoun. 1986. Involvement of fodrin-binding proteins in the structure of the neuronal postsynaptic density and regulation by phosphorylation. *Biochem. Biophys. Res. Comm.* 138:59–65.

Liu, S.-C., L.H. Derick, and J. Palek. 1987. Visualization of the hexagonal lattice in the erythrocyte membrane skeleton. *J. Cell Biol.* 104:527–536.

Malchiodi-Albedi, F., M. Ceccarini, J.C. Winkelmann, J.S. Morrow, and T.C. Petrucci. 1993. The 270 kDa splice variant of erythrocyte β-spectrin (βIΣ2) segregates in vivo and in vitro to specific domains of cerebellar neurons. *J. Cell Sci.* 106:67–78.

Mitra, A.K., M.P. McCarthy, and R.M. Stroud. 1989. Three-dimensional structure of the nicotinic acetylcholine receptor and location of the major associated 43-kD cytoskeletal protein, determined at 22 Å by low dose electron microscopy and X-ray diffraction to 12.5 Å. *J. Cell Biol.* 109:755–774.

Ohlendieck, K., J.M. Ervasti, K. Matsumura, S.D. Kahl, C.J. Leveille, and K.P. Campbell. 1991. Dystrophin-related protein is localized to neuromuscular junctions of adult skeletal muscle. *Neuron.* 7:499–508.

Peters, M.F., N.R. Kramarcy, R. Sealock, and S.C. Froehner. 1994. Beta 2-syntrophin: localization at the neuromuscular junction in skeletal muscle. *NeuroReport.* 5:1577–1580.

Phillips, W.D., M.M. Maimone, and J.P. Merlie. 1991. Mutagenesis of the 43-kD postsynaptic protein defines domains involved in plasma membrane targeting and AChR clustering. *J. Cell Biol.* 115:1713–1723.

Phillips, W.D., P.G. Noakes, S.L. Roberds, K.P. Campbell, and J.P. Merlie. 1993. Clustering and immobilization of acetylcholine receptors by the 43-kD protein: a possible role for dystrophin-related protein. *J. Cell Biol.* 123:729–740.

Pinder, J.C., and W.B. Gratzer. 1983. Structural and dynamic states of actin in the erythrocyte. *J. Cell Biol.* 96:768–775.

Pumplin, D.W. 1995. The membrane skeleton of acetylcholine receptor domains in rat myotubes contains antiparallel homodimers of β-spectrin in filaments quantitatively resembling those of erythrocytes. *J. Cell Sci.* 108:3145–3154.

Pumplin, D.W., and R.J. Bloch. 1993. The membrane skeleton. *Trends Cell Biol.* 3:113–117.

Pumplin, D.W., T.S. Reese, and R. Llinas. 1981. Are the presynaptic membrane particles the calcium channels? *Proc. Natl. Acad. Sci. USA.* 78:7210–7213.

Salpeter, M.M. 1987. The Vertebrate Neuromuscular Junction. Alan R. Liss, Inc. New York. 1–115.

Scotland, P.B., M. Colledge, I. Melnikova, Z. Dai, and S.C. Froehner. 1993. Clustering of the acetylcholine receptor by the 43-kD protein: involvement of the zinc finger domain. *J. Cell Biol.* 123:719–728.

Sealock, R., M.H. Butler, N.R. Kramarcy, K.-X. Gao, A.A. Murnane, K. Douville, and S.C. Froehner. 1991. Localization of dystrophin relative to acetylcholine receptor domains in electric tissue and adult and cultured skeletal muscle. *J. Cell Biol.* 113:1133–1144.

Shen, B.W., R. Josephs, and T.L. Steck. 1986. Ultrastructure of the intact skeleton of the human erythrocyte membrane. *J. Cell Biol.* 102:997–1006.

Shotton, D.M., B.E. Burke, and D. Branton. 1979. The molecular structure of human erythrocyte spectrin: biophysical and electron microscopic studies. *J. Mol. Biol.* 131:303–329.

Sobel, A., T. Heidmann, J. Hofler, and J.-P. Changeux. 1978. Distinct protein components from Torpedo marmorata membranes carry the acetylcholine receptor site and the binding site for local anesthetics and histrionicotoxin. *Proc. Natl. Acad. Sci. USA.* 75:510–514.

Sugiura, Y., A. Woppmann, G.P. Miljanich, and C.-P. Ko. 1995. A novel omega-conopeptide for the presynaptic localization of calcium channels at the mammalian neuromuscular junction. *J. Neurocytol.* 24:15–27.

Toyoshima, C., and N. Unwin. 1988. Ion channel of acetylcholine receptor reconstructed from images of postsynaptic membranes. *Nature.* 336:247–250.

Toyoshima, C., and N. Unwin. 1990. Three-dimensional structure of the acetylcholine receptor by cryoelectron microscopy and helical image reconstruction. *J. Cell Biol.* 111:2623–2635.

Ursitti, J.A., D.W. Pumplin, J.B. Wade, and R.J. Bloch. 1991. Ultrastructure of the human erythrocyte cytoskeleton and its attachment to the membrane. *Cell Motil. Cytoskel.* 19:227–243.

Wong, A.J., D.P. Kiehart, and T.D. Pollard. 1985. Myosin from human erythrocytes. *J. Biol. Chem.* 260:46–49.

Woodruff, M.L., J. Theriot, and S.J. Burden. 1987. 300-kD subsynaptic protein copurifies with acetylcholine receptor-rich membranes and is concentrated at neuromuscular synapses. *J. Cell Biol.* 104:939–946.

Woods, C.M., and E. Lazarides. 1986. Spectrin assembly in avian erythroid development is determined by competing reactions of subunit homo- and hetero-oligomerization. *Nature.* 321:85–89.

Yan, Y., E. Winograd, A. Viel, T. Cronin, S.C. Harrison, and D. Branton. 1993. Crystal structure of the repetitive segments of spectrin. *Science.* 262:2027–2030.

Yang, B., D. Jung, J.A. Rafael, J.S. Chamberlain, and K.P. Campbell. 1995. Identification of α-syntrophin binding to syntrophin triplet, dystrophin, and utrophin. *J. Biol. Chem.* 270:4975–4978.

Syntrophins: Modular Adapter Proteins at the Neuromuscular Junction and the Sarcolemma

Stanley C. Froehner, Marvin E. Adams, Matthew F. Peters, and Stephen H. Gee

Department of Physiology, University of North Carolina, Chapel Hill, North Carolina 27599-7545

The identification and isolation of the gene that when altered causes Duchenne muscular dystrophy (Hoffman et al., 1987) led to the characterization of dystrophin and the dystrophin complex, a new family of membrane-associated proteins involved in membrane specialization, signaling, and stability. The dystrophin protein family currently has four members: dystrophin, utrophin, dystrophin-related protein 2, and dystrobrevin (Blake et al., 1996; Carr et al., 1989; Hoffman et al., 1987; Roberts et al., 1996; Sadoulet-Puccio et al., 1996; Tinsley et al., 1992; Yoshida et al., 1995). Multiple forms of each member of this family are generated by alternative splicing and/or by utilization of alternative promoters that produce amino-terminally truncated forms (reviewed in Ahn and Kunkel, 1993). Dystrophin, and probably the other members of the family, are associated with a transmembrane complex of glycoproteins, which include β-dystroglycan and the sarcoglycans (α, β, γ, and δ)(Ervasti and Campbell, 1991; Matsumura et al., 1992; Yoshida and Ozawa, 1990). β-dystroglycan is in turn associated with α-dystroglycan, a heavily glycosylated extracellular protein that binds laminin and agrin (reviewed in Sealock and Froehner, 1994). On the cytoplasmic side of the membrane, dystrophin associates directly via its amino-terminal region with actin (Rybakaova et al., 1996; Way et al., 1992). Thus, one function of the dystrophin complex is to link the extracellular matrix to the membrane cytoskeleton.

In addition to the transmembrane and extracellular glycoproteins, the dystrophin complex also contains tightly associated cytoplasmic proteins. The functions of the cytoplasmic proteins are less obvious in comparison to the proteins involved in the extracellular matrix-cytoskeleton link. They are likely involved in membrane-associated signaling events, thus providing a regulatory function for the dystrophin complex in addition to its proposed structural role. This chapter reviews the current status of one cytoplasmic dystrophin-associated protein, syntrophin, and considers its role in targeting signaling proteins to the sarcolemma and neuromuscular postsynaptic membrane.

Syntrophin Diversity: Gene Structure, Chromosomal Localization, and Expression of Three Isoforms

Syntrophin was first discovered in Torpedo electric organ as a 58-kD protein associated with the postsynaptic membrane, a site of high nicotinic acetylcholine receptor density (Froehner, 1984; Froehner et al., 1987). The mammalian ortholog is highly

Cytoskeletal Regulation of Membrane Function © 1997 by The Rockefeller University Press

concentrated at the neuromuscular synapse and is also associated with the sarco-lemma (Froehner et al., 1987). Immunoaffinity purification studies showed that syntrophin is associated with dystrophin in the electric organ and in skeletal muscle, and with dystrophin short forms, dystrobrevin and utrophin in other tissues (Butler et al., 1992; Kramarcy et al., 1994). Because of these interactions, we named this protein from the Greek, syntrophos, meaning companion or associate.

Initial biochemical studies suggested that multiple syntrophins existed, al-though the basis for this multiplicity was not known (Yamamoto et al., 1993). This hypothesis was confirmed by molecular cloning studies, which revealed three syn-trophins encoded by separate genes (Adams et al., 1993, 1995; Ahn and Kunkel, 1995; Ahn et al., 1994). To date, three forms of human and mouse syntrophin, and one form of rabbit and Torpedo syntrophin have been cloned (Adams et al., 1993, 1995; Ahn et al., 1996; Ahn and Kunkel, 1995; Ahn et al., 1994; Yang et al., 1995). These results also confirmed that the 59-kD protein associated with the dystrophin complex purified from rabbit skeletal muscle was syntrophin (Yang et al., 1995). The nomenclature for the syntrophin family members is based on the observation that α1-syntrophin has a slightly acidic isoelectric point, whereas β1- and β2-syntro-phin are quite basic (Yoshida et al., 1995). The isoelectric points and other charac-teristics of the three syntrophins are compared in Table I.

Characterization of the mouse genes encoding α1- and β2-syntrophin (located on mouse chromosomes 2 and 8, respectively) revealed that the exon/intron struc-tures of these two genes are quite similar but not identical (Adams et al., 1995). The α1-syntrophin gene is comprised of 8 exons. The β2-syntrophin gene has introns lo-cated at all positions corresponding to the α1-syntrophin gene, with one exception. It lacks an intron corresponding to the first α1-syntrophin intron. Thus, the first β2-syntrophin exon corresponds to a fusion of the first two exons in the α1-syntrophin gene. Sequences that lie 5′ of the transcriptional start site contain numerous puta-tive transcriptional regulatory elements (Adams et al., 1995) which are likely to be responsible for the isoform specific expression patterns. The putative promoter re-gion of the β2-syntrophin gene contains an N box, a 12-bp sequence first found in

TABLE I
Characteristics of Three Syntrophin Isoforms

		α1-syntrophin	β1-syntrophin	β2-syntrophin
No. amino acids	mouse	503	537	520
	human	505	538	540
M_r (kD)	mouse	53.7	58.1	56.4
	human	54.0	58.0	57.9
Isoelectric	mouse	6.7	8.3	8.7
point	human	6.4	8.6	8.7
Chromosome	mouse	2	?	8
	human	20q11	8q23-24	16q23
Gene symbol		snta1	sntb1	sntb2
Tissue distribution		Skeletal and cardiac muscle	Ubiquitous	Ubiquitous
Localization within skeletal muscle		Sarcolemma conc. at NMJ	Sarcolemma conc. at NMJ	Only at NMJ

the acetylcholine receptor δ subunit gene promoter (Koike et al., 1995). This element is thought to confer preferential transcription in nuclei that lie just beneath the neuromuscular synapse, a feature that may be partly responsible for the restricted distribution of β2 syntrophin on the postsynaptic membrane (Peters et al., 1994; see below for discussion).

Northern blot analysis has shown that the tissue expression patterns of the syntrophins are quite different. α1-syntrophin is encoded by a single mRNA of 2.2 kb that is most highly expressed in skeletal muscle with moderate amounts in cardiac, kidney, and brain. Both β syntrophins are widely expressed, although the tissue distributions are different. In addition, multiple species of mRNA encode the β syntrophins (Adams et al., 1993; Ahn et al., 1994, 1996). The basis for this multiplicity of mRNAs is unknown, but it is possible that alternative splicing may produce different forms of the β syntrophins.

Interaction of Syntrophins and Dystrophin Family Members

Results from several laboratories have identified a major syntrophin binding site on dystrophin and related proteins (Fig. 1). Copurification studies showed that syntrophins are associated with full-length dystrophin and utrophin, and with truncated forms of these proteins (Kramarcy et al., 1994). The association of syntrophins with

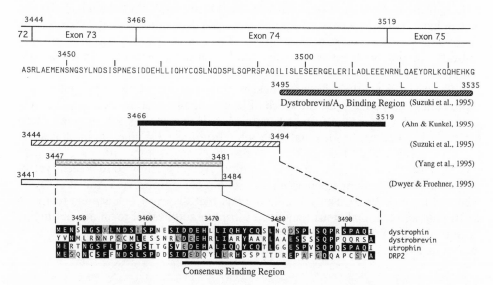

Figure 1. Location of the syntrophin binding region in the COOH-terminal domain of dystrophin. The region of dystrophin (Ahn et al., 1996; Ahn and Kunkel, 1995; Suzuki et al., 1995; Yang et al., 1995) and dystrobrevin (Dwyer and Froehner, 1995) important for binding syntrophin has been determined by four separate groups. The positions of these sequences aligned with human dystrophin and its exon borders are indicated by bars. The amino acid sequence of this region in each of the four dystrophin family members is shown. The sequence representing the overlap of the syntrophin binding region of all four groups is underlined. The dystrobrevin (A_0) binding region, which contains a leucine zipper, is indicated by the rounded bar. This region may also bind β1-syntrophin.

Dp 71 and dystrobrevin (formerly 87K protein), both of which lack the NH_2-terminal actin binding domain and all spectrin-like repeats of the rod domain, showed that a binding site is located in the cysteine-rich (CR) and carboxyterminus (CT) domains (Kramarcy et al., 1994). Following on this work, four laboratories used in vitro binding approaches to localize a binding site for α1-syntrophin in the region of dystrophin encoded by exon 74, and to the analogous region of utrophin and dystrobrevin (Ahn and Kunkel, 1995; Dwyer and Froehner, 1995; Suzuki et al., 1995; Yang et al., 1995). In addition, a second site immediately downstream of the exon 74-encoded region has been reported to bind β1-syntrophin (Suzuki et al., 1995). The copurification of syntrophins with an alternatively spliced form of dystrobrevin, in which part of this region is removed by alternative splicing, suggests that this syntrophin binding site can be even more precisely defined (Blake et al., 1996). Additional in vitro binding experiments are necessary, however, to test this possibility.

Exon 74 and the exons immediately surrounding it are subject to alternative splicing, such that some forms of dystrophin expressed in brain lack this syntrophin binding site (Bies et al., 1992; Feener et al., 1989). Although this would provide a mechanism by which syntrophin association with dystrophin could be regulated, in vivo experiments suggest that the situation is more complex than expected from the in vitro binding experiments. Chamberlain and colleagues have produced transgenic mice in which dystrophin is expressed in skeletal muscle of mdx mice. Expression of full-length dystrophin restores associated proteins, including the syntrophins, to the sarcolemma (Phelps et al., 1995). When this same approach was used with dystrophin lacking the sequences encoded by exons 71-74 (which includes the syntrophin binding domain; see Fig. 1), syntrophin association with the sarcolemma was still restored (Rafael et al., 1994, 1996). Thus, interactions between syntrophin and other members of the dystrophin complex must be important in forming this complex.

The discovery of three isoforms of syntrophin and their differential patterns of expression suggested a model in which different members of the dystrophin family are associated with distinct forms of syntrophin. However, in vitro binding experiments do not support the idea that syntrophin isoforms have inherent binding specificity for particular dystrophin-related proteins. Using coprecipitation of in vitro translated products as a measure of association, Ahn et al. showed that each syntrophin is capable of binding to utrophin, dystrophin or dystrobrevin (Ahn et al., 1996; Ahn and Kunkel, 1995). Thus, despite the fact that the exon 74–encoded region in dystrophin and the analogous regions in utrophin and dystrobrevin are not highly conserved, each binds to all three syntrophin isoforms. It is not known, however, if the binding affinity is the same for each combination.

Distribution of Syntrophin Isoforms in Skeletal Muscle

The binding studies described above showed that each syntrophin can associate with each dystrophin family member. In vivo, however, these interactions could be regulated so that defined complexes exist in different cells or under different physiological conditions. To begin to investigate this possibility, we produced peptide antibodies that recognized each of the mouse syntrophins specifically, and used them first to determine the localization of syntrophins in skeletal muscle. Studies of two syntrophins, α1 and β2, have been completed (Peters et al., 1994). α1-syntrophin is concentrated at the neuromuscular junction but is also present on the sarcolemma. In mdx skeletal muscle, which lacks dystrophin because of a mutation in the gene, α1-syntrophin is essentially absent from the sarcolemma but is retained at the neuromuscular synapse (Peters et al., 1994; Yang et al., 1995). This finding is consistent with the idea that α1-syntrophin is associated with dystrophin in the sarcolemma. Its retention at the synapse suggests an association with another member of the dystrophin family, possibly utrophin.

The localization and presumably the function of β2-syntrophin in skeletal muscle is more restricted. Staining with antibodies to β2-syntrophin is confined largely to the postsynaptic membrane with only very weak staining of the sarcolemma (Peters et al., 1994). As with α1-syntrophin, the synaptic localization of β2-syntrophin is preserved in mdx muscle (Peters and Froehner, unpublished results). These results are consistent with association of β2-syntrophin with utrophin. However, preliminary results from high resolution immunofluorescence analysis of the synaptic localization of these two proteins suggests that they do not share identical positions along the postsynaptic folds (Kramarcy and Sealock, unpublished results). Thus, additional studies are needed to establish the pairing of syntrophins and dystrophin family members at the postsynaptic site.

Domain Structure of Syntrophins: Implications for Function

Although the amino acid sequences of the three syntrophins are only about 50% identical, all three syntrophins have a common domain structure (Adams et al., 1995; Ahn et al., 1996). Approximately three-fourths of the sequence of syntrophins is comprised of domains found in membrane-associated signaling proteins and cytoskeletal proteins (Fig. 2). Two tandem pleckstrin homology (PH) domains occupy the amino-terminal part of syntrophin (Adams et al., 1995; Gibson et al., 1994). PH domains were first identified as ~100 amino acid repeats in pleckstrin, a major protein kinase C substrate in platelets. Since the discovery of PH domains approximately three years ago, more than 100 proteins have been found to contain this domain (Shaw, 1996). Several binding functions have been attributed to PH domains, including association with lipid bilayers containing phosphatidyl inositol-4,5-bisphosphate and interaction with the βγ subunits of G-proteins and protein kinase C (reviewed in Shaw, 1996). Thus, it seems likely that the PH domains of syntrophins provide a site for interaction with the membrane bilayer or with other proteins, but to date, no candidates have been identified.

Two features of syntrophin PH domains are unusual. First, proteins with two tandem PH domains are rare, with only seven proteins known to have this arrangement (Gibson et al., 1994). Second, the first PH domain in syntrophins is interrupted by an insert of approximately 150 amino acids. Interruptions in PH domains have been found in only a few proteins but always at precisely the same site within the domain. The PH domain of phospholipase Cγ, for example, contains an insert of approximately 340 amino acids consisting of two SH2 domains and one SH3 domain (Shaw, 1996). Although a large insert at this site in PH domains of phospholipase Cγ and syntrophin might be expected to disrupt the tertiary structure of the domain, recent analysis of the complete three-dimensional structure of several PH

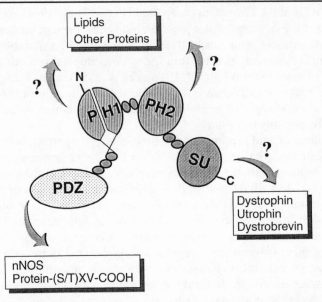

Figure 2. Schematic model of syntrophin domain structure showing known and hypothetical interactions of syntrophin domains. Syntrophin is comprised mainly of protein domains
(shown schematically as large ovals) found in membrane-associated signaling proteins. Two
tandem PH domains (PH1, PH2) occupy the NH_2-terminal half of syntrophin. The NH_2-ter-
minal PH domain (PH1) contains a 150 amino acid insert that divides it into two halves. By
analogy with other PH domain containing proteins, the PH domains of syntrophin may interact with membrane lipids (Harlan et al., 1994) or with other proteins, such as G-protein subunits (Touhara et al., 1994). The 150 aa insert in PH1 is comprised of linker sequence (*small
circles*) and a PDZ domain. The PDZ domain of α1-syntrophin binds to neuronal nitric oxide
synthase (nNOS) and targets it to the sarcolemmal membrane (Brenman et al., 1996). The
syntrophin PDZ domain may also bind to transmembrane proteins containing COOH-terminal S/TXV sequences. The last ~100 amino acids of syntrophin, which are highly conserved
among isoforms, form the syntrophin-unique (SU) domain. We have depicted the SU domain
binding to members of the dystrophin family based on the data of results of Ahn et al. (1994),
although we cannot rule out contributions by other domains.

domains suggests that inserted sequence can be easily accommodated (Ferguson et
al., 1994; Shaw, 1996; Timm et al., 1994; Yoon et al., 1994). PH domains are comprised of seven β-strands arranged in two β-sheets with a terminal α helix. The in-
sertions occur in the loop between the third and fourth β-strands, and thus can be
accommodated as a separate module.

     In the syntrophins, the insert in the first PH domain encodes a PDZ domain
(Adams et al., 1995; Ahn et al., 1996). PDZ domains (formerly called GLGF or
DHR domains) are named for three proteins in which they were first described
(PSD-95, Discs-large, and ZO-1) and are typically found in proteins that are localized at sites of membrane specialization, such as synapses, adherens junctions, and
tight junctions (Sheng, 1996). The family of membrane-associated guanylate kinases (MAGUKs), including PSD-95, SAP-97, Chapsyn 110, and hDlg (the human
homologue of Drosophila discs-large tumor suppresser), is perhaps the most nota-

ble group of PDZ-containing proteins (reviewed in Sheng, 1996). MAGUKs have three tandem PDZ domains followed by an SH3 domain and a guanylate kinase-like region. Recent studies have shown that some MAGUKs interact via their PDZ domains with the carboxy-terminal tail of transmembrane proteins (reviewed in Sheng, 1996). For example, PSD-95 and Chapsyn-110 bind directly to the NR2 subunits of NMDA type-glutamate receptor (Kornau et al., 1995) and to the Kv 1.4 voltage-regulated potassium channels (Kim et al., 1996; Kim et al., 1995). The feature in common between these two quite different types of ion channels that is responsible for binding to PDZ domains is the carboxy-terminal sequence, Ser/Thr-X-Val. Searches of protein databases reveal several interesting classes of transmembrane proteins with this terminal sequence, including cell adhesion molecules, receptor tyrosine kinases, receptor tyrosine phosphatases, and other voltage-regulated ion channels.

These results with MAGUKs suggest that the PDZ domains of syntrophins could serve as a linker between transmembrane proteins and the dystrophin complex. To date, this type of interaction has not been identified for any of the syntrophins. However, syntrophins are involved in a completely different type of interaction, PDZ heterodimerization, which targets an enzyme potentially involved in cell signaling to the membrane.

Interaction of α1-syntrophin with Neuronal Nitric Oxide Synthase

The neuronal form of nitric oxide synthase, nNOS, has a catalytic domain in the carboxy-terminal half of the molecule, and a PDZ domain at the amino-terminal end (Brenman et al., 1996). Despite its name, this enzyme, which catalyzes the formation of NO from arginine in a calcium/calmodulin dependent manner, is expressed in skeletal muscle, where it is associated with the sarcolemma (Kobzik et al., 1994). Bredt and colleagues noted that nNOS is absent from the sarcolemma in mdx and Duchenne skeletal muscle, as determined by immunofluorescence (Brenman et al., 1995; Chao et al., 1996). Furthermore, nNOS coimmunoprecipitated with the dystrophin complex (Brenman et al., 1995). These results suggested that nNOS is targeted to the sarcolemma via interaction with some component of the dystrophin complex.

A role for syntrophin in nNOS association with the dystrophin complex was demonstrated in two ways (Brenman et al., 1996). First, nNOS could be coimmunoprecipitated with α1-syntrophin from extracts of normal or mdx skeletal muscle. Thus, this association did not depend on the presence of dystrophin. Second, coprecipitation experiments show that a GST fusion protein corresponding to the amino-terminal region of nNOS (containing the PDZ domain) precipitates α1-syntrophin from muscle extracts. To test for a direct interaction between α1-syntrophin and the PDZ region of nNOS, we performed overlays with fusion proteins corresponding to the four syntrophin domains. Only the α1-syntrophin PDZ domain bound the nNOS PDZ fusion protein. No binding was observed with the other domains, nor did GST alone bind to any of the syntrophin domains. Thus, α1-syntrophin links nNOS to dystrophin and the sarcolemma, probably by PDZ heterodimerization. The importance of nNOS sequence downstream of the PDZ domain remains to be determined. It may be required for complete folding of the PDZ domain, or alternatively, it may participate directly in the interaction with α1-syntrophin PDZ.

PDZ domains are capable of two types of interactions. In one case, they bind the COOH-terminal tail of proteins with a particular sequence. Alternatively, they participate in interactions with PDZ domains of other proteins. Preliminary results suggest that these two binding modes are mutually exclusive. The second PDZ domain of PSD-95 also binds nNOS PDZ but a peptide corresponding to the COOH-terminal region of the NMDA NR2 subunit effectively inhibits this interaction (Brenman et al., 1996). Nevertheless, the ability of certain MAGUKs to form multimeric complexes with the cognate transmembrane protein suggests a model in which the PDZ-containing proteins link complexes of NMDA receptors to nNOS. Since NMDA receptors serve as ligand-gated ion channels for calcium ions, this model could account for the production of NO that occurs upon NMDA receptor activation.

The role of nNOS in skeletal muscle is largely unknown, although some results suggest that NO may inhibit contraction. Furthermore, the source of the calcium needed to activate nNOS is also not known. Our results showing that nNOS is targeted to the sarcolemma via binding with α1-syntrophin and dystrophin suggest a model in which this interaction is necessary to localize nNOS near a regulated calcium source at the membrane. Whether the source of calcium for this purpose is extracellular or intracellular is unknown.

The binding studies of nNOS were done with α1-syntrophin. Although the PDZ sequences of the three syntrophins are highly conserved, key differences among them suggest that the PDZ domains in α1-, β1-, and β2-syntrophin could have specificity for different PDZ domains and potentially for different carboxy-terminal sequences. In a more general scheme then, the dystrophin complex could be involved in targeting different PDZ-containing proteins to the sarcolemma, depending on the identity of the syntrophin (or combination of syntrophins) present in the complex. The syntrophin composition of dystrophin, utrophin, and dystrobrevin-containing complexes, the identity of PDZ-containing proteins that interact with syntrophin PDZ domains, and the associations mediated by the PH domains, all need to be examined in order to test this model.

Acknowledgments

This work was supported by grants from the National Institutes of Health and the Muscular Dystrophy Association to S.C. Froehner. M.E. Adams was a Muscular Dystrophy Association postdoctoral fellow and S.H. Gee holds a postdoctoral fellowship from the Human Frontier Science Program.

References

Adams, M.E., M.H. Butler, T.M. Dwyer, M.F. Peters, A.A. Murnane, and S.C. Froehner. 1993. Two forms of mouse syntrophin, a 58 kd dystrophin-associated protein, differ in primary structure and tissue distribution. *Neuron.* 11:531–540.

Adams, M.E., T.M. Dwyer, L.L. Dowler, R.A. White, and S.C. Froehner. 1995. Mouse α1- and β2-syntrophin gene structure, chromosome localization, and homology with a discs large domain. *J. Biol. Chem.* 270:25859–25865.

Ahn, A.H., C.A Freener, E. Gussoni, M. Yoshida, E. Ozawa, and L.M. Kunkel. 1996. The three human syntrophin genes are expressed in diverse tissues, have distinct chromosomal locations, and each bind to dystrophin and its relatives. *J. Biol. Chem.* 271:2724–2730.

Ahn, A.H., and L.M. Kunkel. 1993. The structural and functional diversity of dystrophin. *Nat. Genet.* 3283–3291.

Ahn, A.H., and L.M. Kunkel. 1995. Syntrophin binds to an alternatively spliced exon of dystrophin. *J. Cell Biol.* 128:363–371.

Ahn, A.H., M. Yoshida, M.S. Anderson, C.A. Feener, S. Selig, Y. Hagiwara, E. Ozawa, and L.M. Kunkel. 1994. Cloning of human basic A1, a distinct 59-kDa dystrophin-associated protein encoded on chromosome 8q23-24. *Proc. Natl. Acad. Sci. USA.* 91:4446–4450.

Bies, R.D., S.F. Phelps, M.D. Cortez, R. Roberts, C.T. Caskey, and J.S. Chamberlain. 1992. Human and murine dystrophin mRNA transcripts are differentially expressed during skeletal muscle, heart, and brain development. *Nucleic Acids Res.* 20:1725–1731.

Blake, D.J., R. Nawrotzki, M.F. Peters, S.C. Froehner, and K.E. Davies. 1996. Isoform diversity of dystrobrevin, the murine 87-kDa postsynaptic protein. *J. Biol. Chem.* 271:7802–7810.

Brenman, J.E., D.S. Chao, S.H. Gee, A.W. McGee, S.E. Craven, D.R. Santillano, Z. Wu, F. Huang, H. Xia, M.F. Peters, et al. 1996. Interaction of nitric oxide synthase with the postsynaptic density protein PSD-95 and α1-syntrophin mediated by PDZ domains. *Cell.* 84:757–767.

Brenman, J.E., D.S Chao, H. Xia, K. Aldape, and D.S. Bredt. 1995. Nitric oxide synthase complexed with dystrophin and absent from skeletal muscle sarcolemma in Duchenne muscular dystrophy. *Cell.* 82:743–752.

Butler, M.H., K. Douville, AA. Murnane, N.R. Kramarcy, J.B. Cohen, R. Sealock, and S.C. Froehner. 1992. Association of the Mr 58,000 postsynaptic protein of electric tissue with *Torpedo* dystrophin and the Mr 87,000 postsynaptic protein. *J. Biol. Chem.* 267:6213–6218.

Carr, C., G.D. Fischbach, and J.B. Cohen. 1989. A novel 87,000-Mr protein associated with acetylcholine receptors in *Torpedo* electric organ and vertebrate skeletal muscle. *J. Cell Biol.* 109:1753–1764.

Chao, D.S., J.R. Gorospe, J.E. Brenman, J.A. Rafael, M.F. Peters, S.C. Froehner, E.P. Hoffman, J.S. Chamberlain, and D.S. Bredt. 1996. Selective loss of sarcolemmal nitric oxide synthase in Becker muscular dystrophy. *J. Exp. Med.* 184:609–618.

Dwyer, T.M., and S.C. Froehner. 1995. Direct binding of *Torpedo* syntrophin to dystrophin and the 87 kDa dystrophin homologue. *FEBS Lett.* 375:91–94.

Ervasti, J.M., and K.P. Campbell. 1991. Membrane organization of the dystrophin-glycoprotein complex. *Cell.* 66:1121–1131.

Feener, C.A., M. Koenig, and L.M. Kunkel. 1989. Alternative splicing of human dystrophin mRNA generates isoforms at the carboxy terminus. *Nature.* 338:509–511.

Ferguson, K.M., M.A. Lemmon, J. Schlessinger, and P.B. Sigler. 1994. Crystal structure at 2.2 A resolution of the pleckstrin homology domain from human dynamin. *Cell.* 79:199–209.

Froehner, S.C. 1984. Peripheral proteins of postsynaptic membranes from Torpedo electric organ identified with monoclonal antibodies. *J. Cell Biol.* 99:88–96.

Froehner, S.C., A.A. Murnane, M. Tobler, H.B. Peng, and R. Sealock. 1987. A postsynaptic Mr 58,000 (58K) protein concentrated at acetylcholine receptor-rich sites in *Torpedo* electroplaques and skeletal muscle. *J. Cell Biol.* 104:1633–1646.

Gibson, T.J., M. Hyvonen, A. Musacchio, M. Saraste, and E. Birney. 1994. PH domain: the first anniversary. *Trends Biochem. Sci.* 19:349–353.

Harlan, J.E., P.J. Hajduk, H.S. Yoon, and S.W. Fesik. 1994. Pleckstrin homology domains bind to phosphatidylinositol-4,5-bisphosphate. *Nature.* 371:168–170.

Hoffman, E.P., R.H. Brown, Jr., and L.M. Kunkel. 1987. Dystrophin: the protein product of the Duchenne muscular dystrophy locus. *Cell.* 51:919–928.

Kim, E., K. Cho, A. Rothschild, and M. Sheng. 1996. Heteromultimerization and NMDA receptor-clustering activity of chapsyn-110, a member of the PSD-95 family of proteins. *Neuron.* 17:103–113.

Kim, E., M. Niethammer, A. Rothschild, Y.N. Jan, and M. Sheng. 1995. Clustering of Shaker-type K$^+$ channels by interaction with a family of membrane-associated guanylate kinases. *Nature.* 378:85–88.

Kobzik, L., M.B. Reid, D.S. Bredt, and J.S. Stamler. 1994. Nitric oxide in skeletal muscle. *Nature.* 372:546–548.

Koike, S., L. Schaeffer, and J.-P. Changeux. 1995. Identification of a DNA element determining synaptic expression of the mouse acetylcholine receptor δ-subunit gene. *Proc. Natl. Acad. Sci. USA.* 92:10624–10628.

Kornau, H.C., L.T. Schenker, M.B. Kennedy, and P.H. Seeburg. 1995. Domain interaction between NMDA receptor subunits and the postsynaptic density protein PSD-95. *Science.* 269:1737–1740.

Kramarcy, N.R., A. Vidal, S.C. Froehner, and R. Sealock. 1994. Association of utrophin and multiple dystrophin short forms with the mammalian M(r) 58,000 dystrophin-associated protein (syntrophin). *J. Biol. Chem.* 269:2870–2876.

Matsumura, K., J.M. Ervasti, K. Ohlendieck, S.D. Kahl, and K.P. Campbell. 1992. Association of dystrophin-related protein with dystrophin-associated proteins in *mdx* mouse muscle. *Nature.* 360:588–591.

Peters, M.F., N.R. Kramarcy, R. Sealock, and S.C. Froehner. 1994. β2-Syntrophin: localization at the neuromuscular junction in skeletal muscle. *Neuroreport.* 5:1577–1580.

Phelps, S.F., M.A. Hauser, N.M. Cole, J.A. Rafael, R.T. Hinkle, J.A. Faulkner, and J.S. Chamberlain. 1995. Expression of full-length and truncated dystrophin mini-genes in transgenic mdx mice. *Hum. Mol. Genet.* 4:1251–1258.

Rafael, J.A., G.A. Cox, K. Corrado, D. Jung, K.P. Campbell, and J.S. Chamberlain. 1996. Forced expression of dystrophin deletion constructs reveals structure-function correlations. *J. Cell Biol.* 134:93–102.

Rafael, J.A., Y. Sunada, N.M. Cole, K.P. Campbell, J.A. Faulkner, and J.S. Chamberlain. 1994. Prevention of dystrophic pathology in *mdx* mice by a truncated dystrophin isoform. *Hum. Mol. Genet.* 3:1725–1733.

Roberts, R.G., T.C. Freeman, E. Kendall, D.L. Vetrie, A.K. Dixon, C. Shaw-Smith, Q. Bone, and M. Bobrow. 1996. Characterization of DRP2, a novel human dystrophin homologue. *Nat. Genet.* 13:223–226.

Rybakaova, I.N., K.J. Amann, and J.M. Ervasti. 1996. A new model for the interaction of dystrophin with F-actin. *J. Cell Biol.* 135:661–672.

Sadoulet-Puccio, H.M., T.S. Khurana, J.B. Cohen, and L.M. Kunkel. 1996. Cloning and characterization of the human homologue of a dystrophin related phosphoprotein found at the *Torpedo* electric organ post-synaptic membrane. *Hum. Mol. Genet.* 5:489–496.

Sealock, R., and S.C. Froehner. 1994. Dystrophin-associated proteins and synapse formation: is α-dystroglycan the agrin receptor? *Cell.* 77:617–619.

Shaw, G. 1996. The pleckstrin homology domain: an intriguing multifunctional protein module. *Bioessays.* 18:35–46.

Sheng, M. 1996. PDZs and receptor/channel clustering: rounding up the latest suspects. *Neuron.* 17:575–578.

Suzuki, A., M. Yoshida, and E. Ozawa. 1995. Mammalian α1- and β1-syntrophin bind to the alternative splice-prone region of the dystrophin COOH terminus. *J. Cell Biol.* 128:373–381.

Timm, D., K. Salim, I. Gout, L. Guruprasad, M. Waterfield, and T. Blundell. 1994. Crystal structure of the pleckstrin homology domain from dynamin. *Nature Struct. Biol.* 1:782–788.

Tinsley, J.M., D.J. Blake, A. Roche, U. Fairbrother, J. Riss, B.C. Byth, A.E. Knight, J. Kendrick-Jones, G.K. Suthers, D.R. Love, et al. 1992. Primary structure of dystrophin-related protein. *Nature.* 360:591–593.

Touhara, K., J. Inglese, J.A. Pitcher, G. Shaw, and R.J. Lefkowtiz. 1994. Binding of G proteins βγ-subunits to pleckstrin homology domains. *J. Biol. Chem.* 269:10217–10220.

Way, M., B. Pope, R.A. Cross, J. Kendrick-Jones, and A.G. Weeds. 1992. Expression of the N-terminal domain of dystrophin in E. coli and demonstration of binding to F-actin. *FEBS Lett.* 301:243–245.

Yamamoto, H., Y. Hagiwara, Y. Mizuno, M. Yoshida, and E. Ozawa. 1993. Heterogeneity of dystrophin-associated proteins. *J. Biochem.* 114:132–139.

Yang, B., D. Jung, J.A. Rafael, J.S. Chamberlain, and K.P. Campbell. 1995. Identification of α-syntrophin binding to syntrophin triplet, dystrophin, and utrophin. *J. Biol. Chem.* 270:4975–4978.

Yoon, H.S., P.J. Hajduk, A.M. Petros, E.T. Olejniczak, R.P. Meadows, and S.W. Fesik. 1994. Solution structure of a pleckstrin-homology domain. *Nature.* 369:672–675.

Yoshida, M., and E. Ozawa. 1990. Glycoprotein complex anchoring dystrophin to sarcolemma. *J. Biochem.* 108:748–752.

Yoshida, M., H. Yamamoto, S. Noguchi, Y. Mizuno, Y. Hagiwara, and E. Ozawa. 1995. Dystrophin-associated protein A0 is a homologue of the *Torpedo* 87K protein. *FEBS Lett.* 367:311–314.

Mechanical Transduction by Ion Channels: How Forces Reach the Channel

Frederick Sachs

Biophysical Sciences, SUNY Buffalo, Buffalo, New York, 14214

Mechanically sensitive ion channels are ubiquitous transducers. In the higher animals they serve the neurological senses of touch, hearing, kinesthesis, and fullness of the hollow organs. Their function outside of specialized receptors remains to be determined. The study of these channels is difficult because of a lack of specific reagents, a lack of controlled stimulation, and in eukaryotes, the use of auxiliary structures to couple force to the channels. In prokaryotes, the situation is simpler. Mechanosensitive ion channels (MSCs) from *E. coli* have been be cloned and functionally reconstituted into lipid bilayers.

In eukaryotes, force is transmitted to the MSCs by linking elements other than the lipid bilayer—either the cytoskeleton, the extracellular matrix, or both. What does seem clear is that actin, tubulin, and dystrophin are not essential. The mechanical properties of the linking elements are not known, nor are the constitutive properties of the other elements lying in parallel and series with the channels. The consequence is that it is impossible to make a controlled stimulator for eukaryotic MSCs. One cannot build a "tension clamp" to produce defined stress in a particular component of the cortical membrane. In most cases, the best we can do is to say that the stimulus is monotonic. Because of time-dependent changes in properties of the linkers, the coupling of stress to MSCs is generally time dependent, making it even more difficult to create reliable stimulation protocols.

The dependence of eukaryotic MSC activity on the intracellular and extracellular matrix makes it tricky to test potential clones of MSCs. Clones cannot be reasonably reconstituted in artificial lipid bilayers since the natural cellular environment does not use bilayer stress for gating. Clones must be tested in whole cells and attention must then be paid to the possibility that a prospective clone is altering the properties of the linkers and causing activation of a previously unseen endogenous channel.

In a recent review of membrane mechanics, it was concluded that mechanosensitive channels (MSCs) cannot be activated by tension in the bilayer of eukaryotic cells since the tension is too small (Sheetz and Dai, 1996). There is little disagreement with this conclusion for it has been assumed since MSCs were discovered (Guharay and Sachs, 1984) that the forces which activate them come through the cytoskeleton. Although more than 10 years have elapsed since the discovery of these channels in eukaryotic cells, we still do not know how mechanical forces are transmitted to them, and we only know the components that are not involved. In contrast to eukaryotes, MSCs from prokaryotes can be activated in pure lipid bilayers. This brief review deals with what is known about how MSCs are activated and the consequent difficulties of testing putative clones. For the interested reader, more extensive general reviews of MSC channel properties are available (Sachs and Morris, 1996; Martinac, 1993; Yang and Sachs, 1993; Lecar and Morris, 1993; Martinac, 1993; Sackin, 1995; Hamill and McBride, 1996).

Mechanosensitive ion channels occur in two major classes, stretch-activated channels (SACs) (Guharay and Sachs, 1984) and stretch-inactivated channels (SICs) (Morris and Sigurdson, 1989). The former is the most common. A SAC variant exists that is sensitive to the sign of the membrane curvature (opening only with the patch pushed toward the cytoplasm) (Bowman et al., 1992).

The key property of an MSC that distinguishes it from other types of channels is that opening must be accompanied by a large change in dimension in the direction of the applied force. This permits the external force field to do work on the channel. A SAC must increase its dimensions with opening whereas a SIC must decrease its dimensions. The details of the change in shape are unclear. In cochlear hair cells, where the applied force seems to act along a linear string, or tip-link attaching adjacent cilia (Pickles et al., 1989; Pickles and Corey, 1992), a linear change seems appropriate. This change in dimensions between the open and closed states has been estimated to be about 2 nm for hair cells (Howard and Hudspeth, 1988). For channels activated by tension in a patch of membrane, an "in-plane" area change seems more appropriate. We do not know the symmetry of the forces at the molecular level so we also do not know how to characterize the relevant deformation. MSCs could be just as well be attached to pairs or arrays of cytoskeletal or extracellular "tip links" (Fig. 1).

When MSCs are active in pure lipid bilayers, tension is the appropriate unit. For these channels some measurements of the dimensional changes are available. Alamethicin is an antibiotic that forms a channel by sequentially adding monomers to form a ring with a conducting hole (Fig. 2). The addition of one monomer to a

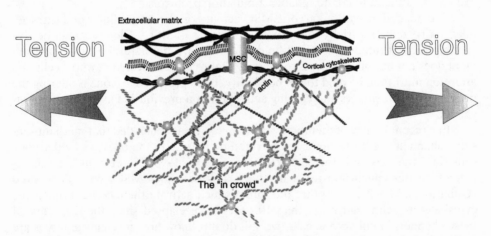

Figure 1. Cartoon of some structures that can affect eukaryotic MSC function. Forces can be transmitted to the channels through the extracellular matrix and various components of the cytoskeleton, many of which are reversibly cross-linked and introduce a time-dependent mechanical behavior. In mechanical terms, the membrane is not two dimensional, but three, so the cartoon represents a section. The "in-crowd" refers to the multitude of cytoskeletal components whose names end with "in."

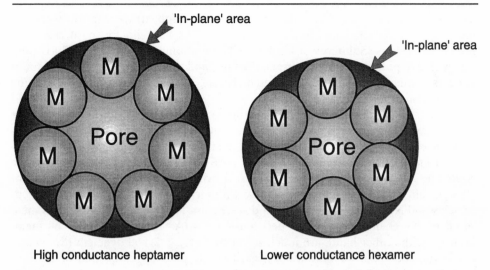

Figure 2. Cartoon of a gating transition in an oligomeric channel like almethicin where stability of the larger diameter complex is favored by the presence of tension in the surrounding bilayer (not shown). M indicates a monomer.

preexisting (conducting) oligomer increased the conductance and the in-plane area by about 1.2 nm^2 (Opsahl and Webb, 1994). The sensitivity was about 8 dyne/cm/ e-fold change in the probability of opening to a higher conductance state. For the cloned bacterial MSC (MscL) which is much more sensitive to tension (Sukharev et al., 1994), the area increase between closed and open appears to be about 3.5 nm^2, which is equivalent to a free energy difference of $\Delta G \approx 16\ k_B T$ at the tension required for half activation (Sigurdson et al., 1997).

In the simple case of a binary (closed-open) channel, the energy difference between open and closed states can be estimated from the slope of the Boltzmann curve of activation (Martinac, 1993):

$$P_1/P_2 = \exp(-\Delta G/k_B T)$$

where P_i is the probability of being in state i, ΔG is the energy difference between states 1 and 2 and $k_B T$ is Boltzmann's constant times temperature. For linear changes in dimensions, $\Delta G = F^* \Delta x$ where F is the applied force and Δx is the movement associated with opening. For channels that activate with a change in area, $\Delta G = T^* \Delta A$ where T is the membrane tension and ΔA is the change of in-plane area between the open and closed states. The tension can be calculated in certain circumstances from the law of Laplace, which for uniform curvature is given by: $T = \Delta Pr/2$, where ΔP is the hydrostatic pressure drop across a membrane and r is the radius of curvature. Thus, in a lipid bilayer it is possible to estimate T if one knows the geometry of the membrane and the pressure gradient (Fung, 1981). In a cell or a patch of membrane from a cell, the tension may be shared between many different components and the relevant forces are not clearly defined. And herein lies the rub.

The ability to reconstitute function in a lipid bilayer was essential for the strategy of expression cloning MscL (Sukharev et al., 1994). MscL is one of five different SACs present in *E. coli* (Berrier et al., 1996), and appears to bear only weak homology to the others, since low stringency screens have not been successful in finding the others (Sukharev et al., 1997). The monomer of MscL is a 17-kD peptide with two putative transmembrane domains and a large extracellular domain. It appears to function as a hexamer based on studies of concatamers and cross-linking (Blount et al., 1996).

Given this sturdy beginning of a prototype for structural studies of MSCs, why can't we extend the approach to eukaryotic MSCs? The answer is that in normal eukaryotic cells, the bilayer does not bear much stress. Many data support this conclusion. From pipette aspiration experiments, pure lipid membranes, particularly those containing cholesterol, are known to be stiff, with area elasticities in the range of 10^3 dyne/cm, and the membranes will lyse with increases in area over a few percent (Zhelev and Needham, 1993; Evans and Needham, 1987). The stiffness, K_a, is measured by the change in area per unit tension: $\Delta A/A = K_a T$. Patches of cell membranes, on the other hand, are much softer, with $K_a = 25$–60 dyne/cm (Sokabe et al., 1991), and these membranes will easily undergo changes in area of 20–50% (Sokabe et al., 1991; Sokabe and Sachs, 1990).

By measuring the patch capacitance, area, and SAC activity as a function of mean patch tension, Sokabe et al. (1991) showed that mechanical stress causes the lipids to flow through the cytoskeletal matrix so that in steady state the lipids are fluid and have little resting tension. This reasoning is shown in Fig. 3.

Different kinds of data illustrating the absence of tension in the lipid bilayer of eukaryotic cells came from experiments in which thin tubular threads (tethers) of the bilayer were pulled from cells (Dai and Sheetz, 1995). The force required to pull such a tether depends on the square root of the tension in the cell from which it was pulled (Heinrich and Waugh, 1996). Estimated resting tensions were 0.01–0.1 dyne/cm. The force should be linear with distance if there is a large buffer of lipids available (Heinrich and Waugh, 1996). If the cell had limited lipids, then as the length of the tether increased, the pool of lipids in the cell would be depleted, and the tension in the tether would increase super linearly with distance, but it does not. The absence of tension in the bilayer of normal cells obviously does not mean that such tensions cannot be created—cells can be lysed! Such catastrophic failure, however, follows disruption of the cytoskeleton. The reason for the striking difference in mechanical properties between pure lipid bilayers and the membrane of eukaryotic cells is that in cells forces applied to the cell cortex are carried by the cytoskeleton and the extracellular matrix rather than the bilayer (Fig. 1).

Since cortical forces exist outside the bilayer, MSCs have to be pulled by one or more of the remaining elements. In specialized mechanoreceptors the extracellular matrix may play a key role. There is good evidence that in hair cells, the force balance exists across the extracellular tip links, the channels, myosin, and intracellular actin bundles (Pickles and Corey, 1992). Recent genetic data from the nematode *C. elegans* suggests that forces are transmitted to MSCs through collagen (Garcia-Anoveros and Corey, 1996; Driscoll, 1996; Liu et al., 1996). It is still unknown whether the force balance (required to prevent the channels from being dragged through the membrane) is fulfilled by extracellular or by intracellular components. Data from *C. elegans* suggests the involvement of a specialized tubulin (Huang et al., 1995; Garcia-Anoveros and Corey, 1996).

As for acute experiments with the extracellular matrix, there is little data avail-

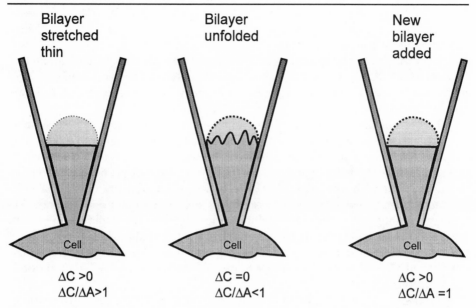

Figure 3. The effect of pipette suction on the change in capacitance (ΔC) and specific capacitance ($\Delta C_s = \Delta C/\Delta A$) for three different models of a patch. At zero pressure the patch is normal to the pipette axis. In the left panel, the membrane behaves like a rubber sheet, getting larger in area and getting thinner. In the center panel, the patch is wrinkled (at a resolution below the light microscope) and unfolds with suction. In the right panel, new membrane is brought in (at constant density) from extra material along the sides of the pipette and the cell. The experimental observations fit the model in the right panel so that for all tensions, $\Delta C_s = 0.7$ μF/cm^2, a typical value for phospholipids. Under stress, new lipids flow into the patch even though the attachment points of the patch to the pipette remain unchanged. In steady state, tension is not borne by the lipids, as they are available in excess and as liquids flow under tension. The gating of SACs in the patch is controlled by the tension calculated from the mean radius of curvature and the pressure gradient, so that force must be transmitted to the channels by a component other than the bilayer (Sokabe et al., 1991).

able. It is possible to reduce the extracellular matrix with reagents such as β-D-xyloside that inhibit polysaccharide extension (Hamati et al., 1989). This agent does not block SAC channel activation although it may improve the ability to form gigaseals (Izu and Sachs, 1991) and it alters the mechanical properties of cells and patches (Olesen, 1995).

The original suggestion that the cytoskeleton was the pathway by which force was transmitted to SACs (Guharay and Sachs, 1984) was not based on any data, but simply on the presence of more material inside the cell than out. Attempts to block MSC transduction by the use of cytoskeletal reagents have not been successful. Disruption of tubulin has no significant effect (Guharay and Sachs, 1984), although disruption of actin by cytochalasins can increase sensitivity (Guharay and Sachs, 1984). Since MSCs are observed in dystrophic muscle (Franco et al., 1991), dystrophin and its partner proteins (Matsumura et al., 1992) are not essential. What is left? The most obvious candidate is the spectrin/fodrin family (Yan et al., 1993; Bennett, 1990), a set of flexible linear polymers that, at least in red cells, forms a hexagonal array under the membrane (McGough and Josephs, 1990). Unfortunately, there are no reagents for spectrin, and no experiments have been published concerning MSCs in spectrin mutants.

It is probably presumptive to assume that there is only one specific protein that links to MSCs. As many gene knockout experiments have shown, cellular components have parallel functional partners. MSCs may be linked to several different components simultaneously.

The complexity of these auxiliary components suggests that the isolation of cells from tissues may destroy or modify the transduction mechanism by disrupting extracellular components. We have found that the mechanical sensitivity in acutely isolated cells varies with the preparation batch. We have not yet been able to track down the source of variability, but it is possible that the proteolytic enzymes needed to isolate the cells are also removing key extracellular components. Another possibility is that since cells exhibit a "use dependent" modification of their mechanical properties, the agitation associated with isolation affects responsiveness. It is known that flexing patches of cell membrane (Small and Morris, 1994; Hamill and McBride, 1992) or cells (White et al., 1993) may lead to an alteration in transduction.

A striking example of the time dependence of this process came from experiments in which fibroblasts were grown on a rubber sheet which could be stretched (Pender and McCulloch, 1991). Within 10 s of a stretch, the actin stress fibers of control cells nearly disappeared, whereas when the preparation was left to rest for 60 s, the stress fibers reformed to nearly twice the original density. We have observed similar dynamic changes in the stiffness of patches (Sigurdson and Sachs, 1994).

The complexity of the force coupling system raises the question of how we could assay a suspected clone of a eukaryotic MSC. The simplest technique would be to have a high affinity label whose binding could be taken as definitive. Unfortunately no such reagent yet exists, although promising new peptide ligands have been discovered (Chen et al., 1996).

Without a binding assay, we need a functional assay. Unfortunately this cannot be carried out in bilayers for two reasons. (1) Since there is no tension in the eukaryotic bilayer, MSC activity in a bilayer would appear to be a fluke or an artifact. (2) Planar bilayers are "tension clamped" by the excess lipids stored in the torus around the bilayer (Elliott et al., 1983). As one bends a planar bilayer by changing the water level on one side, the tension remains constant while the radius changes. The cloned epithelial Na channel, ENaC, was postulated to be mechanosensitive based on homology with some *mec* genes from *C. elegans.* When tested in a planar bilayer, ENaC appeared to be mechanosensitive (Awayda et al., 1995), but given the problems listed above, the conclusion is unsafe. In vivo, ENaC occasionally appears to be mechanically sensitive (Palmer et al., 1996).

Putative clones of eukaryotic MSCs must be tested in cells in which the necessary auxiliary components are present. Needless to say, the proper cells for expression must not have endogenous MSCs of similar selectivity and conductance. The *Xenopus* oocyte, the most common expression system, has a significant population of cationic SACs, and in our lab, attempts to express K-selective SACs from a variety of sources was unsuccessful. We have found that another common expression system, the HEK293 human embryonic kidney cell line, provided a rather quiet background, although other investigators have found the background significant (C.E. Morris, personal communication) suggesting that culture conditions or passage number may affect expression.

After having chosen a cell line for expression, our next problem is how to stimulate the channels. Traditionally, MSCs are activated by suction applied to a patch pipette. Unfortunately, patches have a small area, and MSCs have not been observed in high density in any preparation, and in a heterologous expression system the density may be far below the normal biological density of $\approx 1/\mu m^2$. At low density, verification of expression is a statistical measure relative to background. When the probability of finding an active channel in a patch is low, the background must be much lower if a reasonable number of patches are to suffice. It would be good if a reliable whole cell assay could be developed.

Whole cell recording of mechanosensitive currents has proved difficult. One common method has been to use osmotic stress. Unfortunately, osmotic stress and direct mechanical strain may activate different channel systems (Ackerman et al., 1994; Hu and Sachs, 1996). Volume stress usually activates a Cl^- conductance (Grunder et al., 1992) whereas direct strain activates a cationic system (Yang and Sachs, 1990, 1993; Hu and Sachs, 1996). Direct strain can be applied with stretched flexible substrates (Pender and McCulloch, 1991), but it is difficult to record electrically from attached cells. Fluorometric probes could prove useful for estimating potential changes (Haugland, 1994) in this preparation, but since the essential stimulus is a change in cell shape, brightness changes are ambiguous and some form of ratio recording is important. Cells can be distorted with one pipette while voltage clamped with another (Hu and Sachs, 1996). While this technique works, the stimulus cannot be easily quantitated to permit a comparison between cells. There appears to be no simple mechanical stimulation technique for whole cells.

If we had cloned what we thought was an MSC, and it exhibited mechanosensitive single channel activity in a patch, how could we be sure that we really had cloned the channel? How could we be sure we did not clone, for example, a cofactor or modulator of the auxiliary stress applying network that altered the stress distribution on an endogenous channel? Probably the only convincing demonstration would be site-directed mutagenesis that would alter the conductivity or selectivity of the channel.

The presence of auxiliary structures attached to eukaryotic MSCs has advantages and disadvantages. On the negative side, it makes reconstitution difficult. On the positive side, we can view the MSCs as an intrinsic probe of stress in part of the cell cortex. Once we understand that component, we have a unique probe to study the mechanical properties of cells.

It is likely that most of the cellular processes that involve MSCs have yet to be determined since MSCs occur in cells that are not obviously specialized for mechanical transduction. No matter how difficult it is to study MSCs, they are key to a great deal of fundamental physiology ranging from sex to gas pain.

Acknowledgments

Supported by the NIH, USARO, and MDA.

References

Ackerman, M.J., K.D. Wickman, and D.E. Clapham. 1994. Hypotonicity activates a native chloride current in *Xenopus* oocytes. *J. Gen. Physiol.* 103:153–179.

Awayda, M.S., I.I. Ismailov, B.K. Berdiev, and D.J. Benos. 1995. A cloned renal epithelial Na^+ channel protein displays stretch activation in planar lipid bilayers. *Am. J. Physiol.* 268:C1450–C1459.

Bennett, V. 1990. Spectrin-based membrane skeleton: a multipotential adaptor between plasma membrane and cytoplasm. *Physiol. Rev.* 70:1029–1065.

Berrier, C., M. Besnard, B. Ajouz, A. Coulombe, and A. Ghazi. 1996. Multiple mechanosensitive ion channels from *Escherichia coli*, activated at different thresholds of applied pressure. *J. Membr. Biol.* In press.

Blount, P., S.I. Sukharev, P.C. Moe, M.J. Schroeder, R.H. Guy, and C. Kung. 1996. Membrane topology and multimeric structure of a mechanosensitive channel protein of *Escherichia coli. EMBO J.* 15:4798–4805.

Bowman, C.L., J.P. Ding, F. Sachs, and M. Sokabe. 1992. Mechanotransducing ion channels in astrocytes. *Brain Res.* 584:272–286.

Chen, Y., S.M. Simasko, J. Niggel, W.J. Sigurdson, and F. Sachs. 1996. Ca^{2+} uptake in GH3 cells during hypotonic swelling: the sensory role of stretch-activated ion channels. *Am. J. Physiol.* 270:C1790–C1798.

Dai, J., and M.P. Sheetz. 1995. Mechanical properties of neuronal growth cone membranes studied by tether formation with laser optical tweezers. *Biophys. J.* 68:988–996.

Driscoll, M. 1996. Molecular genetics of touch sensation in *C. Elegans*: mechanotransduction and mechano-destruction. *Biophys. J.* 70:A1 (Abstr.).

Elliott, J.R., D. Needham, J.P. Dilger, and D.A. Haydon. 1983. The effects of bilayer thickness and tension on gramicidin single-channel lifetime. *Biochim. Biophys. Acta.* 735:95–103.

Evans, E., and D. Needham. 1987. Physical properties of surfactant bilayer membranes: thermal transitions, elasticity, rigidity, cohesion, and colloidal interactions. *J. Phys. Chem.* 91:4219–4228.

Franco, A., Jr., B.D. Winegar, and J.B. Lansman. 1991. Open channel block by gadolinium ion of the stretch-inactivated ion channel in mdx myotubes. *Biophys. J.* 59:1164–1170.

Fung, Y.C. 1981. Biomechanics. Springer Verlag, New York.

Garcia-Anoveros, J., and D.P. Corey. 1996. Mechanosensation: touch at the molecular level. *Curr. Opin. Biol.* 6:541–543.

Grunder, S., A. Thiemann, M. Pusch, and T.J. Jentsch. 1992. Regions involved in the opening of CIC-2 chloride channels by voltage and cell volume. *Nature.* 360:759–762.

Guharay, F., and F. Sachs. 1984. Stretch-activated single ion channel currents in tissue-cultured embryonic chick skeletal muscle. *J. Physiol. (Lond.).* 352:685–701.

Hamati, H.F., E.L. Britton, and D.J. Carey. 1989. Inhibition of proteoglycan synthesis alters extracellular matrix deposition, proliferation, and cytoskeletal organization of rat aortic smooth muscle cells in culture. *J. Cell Biol.* 108:2495–2505.

Hamill, O.P., and D.W. McBride. 1996. The pharmacology of mechanogated membrane ion channels. *Pharmacol. Rev.* 48:231–252.

Hamill, O.P., and D.W. McBride, Jr. 1992. Rapid adaptation of single mechanosensitive channels in *Xenopus* oocytes. *Proc. Natl. Acad. Sci. USA.* 89:7462–7466.

Haugland, R.P. 1994. Molecular probes. Molecular Probes, Eugene, OR.

Heinrich, V., and R.E. Waugh. 1996. A piconewton force transducer and its application to measure the bending stiffness of phospholipid membranes. *Ann. Biomed. Eng.* 24:595–605.

Howard, J., and A.J. Hudspeth. 1988. Compliance of the hair bundle associated with gating of mechanoelectrical transduction channels in the bullfrog's saccular hair cell. *Neuron.* 1:189–199.

Hu, H., and F. Sachs. 1996. Mechanically activated currents in chick heart cells. *J. Membr. Biol.* 154:205–216.

Huang, M., G. Gu, E.L. Ferguson, and M. Chalfie. 1995. A stomatin-like protein necessary for mechanosensation in C. elegans. *Nature.* 378:292–295.

Izu, Y.C., and F. Sachs. 1991. β-D-Xyloside treatment improves patch clamp seal formation. *Pflug. Arch.* 419:218–220.

Lecar, H., and C. Morris. 1993. Biophysics of mechanotransduction. *In* Mechanoreception by the Vascular Wall. G.M. Rubany, editor. Futura Publishing, Mount Kisco, NY. 1–11.

Liu, J., B. Schrank, and R.H. Waterston. 1996. Interaction between a putative mechanosensory membrane channel and a collagen. *Science.* 273:361–364.

Martinac, B. 1993. Mechanosensitive ion channels: biophysics and physiology. *In* Thermodynamics of Cell Surface Receptors. M. Jackson, editor. CRC Press, Boca Raton, FL. 327–351.

Matsumura, K., J.M. Ervasti, K. Ohlendieck, S.D. Kahl, and K.P. Campbell. 1992. Association of dystrophin-related protein with dystrophin-associated protiens in mdx mouse muscle. *Nature.* 360:588–594.

McGough, A.M., and R. Josephs. 1990. On the structure of erythrocyte spectrin in partially expanded membrane skeletons. *Proc. Natl. Acad. Sci. USA.* 87:5208–5212.

Morris, C.E., and W.J. Sigurdson. 1989. Stretch-inactivated ion channels coexist with stretch-activated ion channels. *Science.* 243:807–809.

Olesen, S.-P. 1995. Cell membrane patches are supported by proteoglycans. *J. Membr. Biol.* 144:245–248.

Opsahl, L.R., and W.W. Webb. 1994. Transduction of membrane tension by the ion channel alamethicin. *Biophys. J.* 66:71–74.

Palmer, R.E., A.J. Brady, and K.P. Roos. 1996. Mechanical measurements from isolated cardiac myocytes using a pipette attachment system. *Am. J. Physiol.* 270:C697–C704.

Pender, N., and C.A. McCulloch. 1991. Quantitation of actin polymerization in two human fibroblast sub-types responding to mechanical stretching. *J. Cell Sci.* 100:187–193.

Pickles, J.O., J. Brix, S.D. Comis, O. Gleich, C. Koppl, G.A. Manley, and M.P. Osborne. 1989. The organization of tip links and stereocilia on hair cells of bird and lizard basilar papillae. *Hearing Research.* 41:31–42.

Pickles, J.O., and D.P. Corey. 1992. Mechanoelectrical transduction by hair cells. [Review]. *TINS.* 15:254–259.

Sachs, F., and C. Morris. 1997. Mechanosensitive ion channels in non-specialized cells. *Rev. Physiol. Biochem. Pharmacol.* In press.

Sackin, H. 1995. Mechanosensitive channels. *Annu. Rev. Physiol.* 57:333–353.

Sheetz, M.P., and J. Dai. 1996. Modulation of membrane dynamics and cell motility by membrane tension. *Trends Cell Biol.* 6:85–89.

Sigurdson, W.J., and F. Sachs. 1994. Sarcolemmal mechanical properties in mouse myoblasts and muscle fibers. *Biophys. J.* 66:A171 (Abstr.).

Sigurdson, W.J., S. Sukharev, C. Kung, and F. Sachs. 1997. Energetic parameters for gating of the *E. coli* large conductance mechanosensitive channel (MSCL). *Biophys. J.* 72:A267 (Abstr.).

Small, D.L., and C.E. Morris. 1994. Delayed activation of single mechanosensitive channels in Lymnaea neurons. *Am. J. Physiol.* 267:C598–C606.

Sokabe, M., F. Sachs, and Z. Jing. 1991. Quantitative video microscopy of patch clamped membranes: stress, strain, capacitance and stretch channel activation. *Biophys. J.* 59:722–728.

Sokabe, M., and F. Sachs. 1990. The structure and dynamics of patch-clamped membranes: a study by differential interference microscopy. *J. Cell Biol.* 111:599–606.

Sukharev, S.I., P. Blount, B. Martinac, F.R. Blattner, and C. Kung. 1994. A large conductance mechanosensitive channel in *E. coli* encoded by mscL alone. *Nature.* 368:265–268.

Sukharev, S.I., P. Blount, B. Martinac, and C. Kung. 1997. Mechanosensitive channels of *Escherichia coli*: the MscL gene, protein and activities. *Annu. Rev. Physiol.* 59:633–657.

White, E., J.Y. Le Guennec, J.M. Nigretto, F. Gannier, J.A. Argibay, and D. Garnier. 1993. The effects of increasing cell length on auxotonic contractions: membrane potential and intracellular calcium transients in single guinea-pig ventricular myocytes. *Exp. Physiol.* 78:65–78.

Yan, Y., E. Winograd, A. Viel, T. Cronin, S.C. Harrison, and D. Branton. 1993. Crystal structure of the repetitive segments of spectrin. *Science.* 262:2027–2030.

Yang, X.C., and F. Sachs. 1990. Characterization of stretch-activated ion channels in *Xenopus* oocytes. *J. Physiol. (Lond.).* 431:103–122.

Yang, X.C., and F. Sachs. 1993. Mechanically sensitive, non-selective, cation channels. *In* Non-selective Ion Channels. D. Siemen and J. Hescheler, editors. Springer-Verlag, Heidelberg. 79–92.

Zhelev, D.V., and D. Needham. 1993. Tension stabilized pores in giant vesicles: determination of pore size and pore line tension. *Biochim. Biophys. Acta.* 1147:89–104.

Chapter 6

Signaling via the Membrane Cytoskeleton

Cell Signalling and CAM-mediated Neurite Outgrowth

Frank S. Walsh, Karina Meiri, and Patrick Doherty

Department of Experimental Pathology, UMDS, Guy's Hospital, London SE1 9RT, United Kingdom

Summary

A wide range of molecules promote nerve growth, and these include cell adhesion molecules (CAMs), NCAM, N-cadherin, and the L1 glycoprotein are CAMs that are normally found on both the advancing growth cone and also on cellular substrates, and in general operate via a homophilic binding mechanism. In recent years it has become clear that nerve growth stimulated by these CAMs does not rely on the adhesion function of these molecules, but instead requires that the CAMs activate second messenger cascades in neurons. A large body of evidence supports the hypothesis that homophilic binding of the CAM in the substrate to the CAM in the neuron leads to activation of the neuronal FGF receptor, possibly via a direct interaction in cis between the CAM and the FGF receptor. The consequential activation of PLC γ is both necessary and sufficient to account for the neurite outgrowth response stimulated by the above three CAMs.

Based on the above model, we reasoned that soluble CAMs might also be able to stimulate neurite outgrowth and that such agents might be developed as potential therapeutic agents for stimulating nerve regeneration. To this end we have made soluble chimeric molecules consisting of the extracellular domain of NCAM or L1 fused to the Fc region of human IgG 1. We have found that these molecules can stimulate neurite outgrowth from rat and mouse cerebellar granule cells cultured on a variety of tissue culture substrates and that they do so by activating the FGF receptor signal transduction cascade in the neurons. Consistent with this model, we find that neurons that have their FGF receptor function ablated as a consequence of the expression of dominant negative FGF receptors, no longer respond to the soluble CAM. Downstream targets of CAM function have also been studied. Addition of soluble CAMs to isolated growth cone preparations from mouse or rat brain leads to enhanced phosphorylation of the GAP-43 protein providing a link between the cell surface and the cytoskeleton.

Introduction

The growth cones of axons can navigate to target tissues during development and, in doing so, sense and respond to specific guidance cues. Highly specific but transient interactions occur between growth cone receptors for guidance molecules and components of the extracellular matrix and surrounding cells (Kater and Redher, 1995). It is also becoming increasingly clear that the final pattern of growth is a compromise between positive growth cone inductive cues and also negative or inhibitory influences. Growth and guidance molecules are a heterogeneous group of proteins and can be either soluble factors such as nerve growth factor and fibroblast growth factor, chemoattractants such as the netrins (Tessier-Levigne, 1994), or inhibitory proteins such as the collapsin-semaphorin family (Goshima et al., 1995; Luo et al., 1995). Other macromolecules which are important include members of the extracellular matrix group of proteins such as laminin (Reichardt and Tomaselli, 1991) or cell adhesion molecules (CAMs) which are cell surface glycoproteins belonging to specific superfamilies of recognition molecules (Bixby and Harris, 1991).

We have studied three CAMs which are believed to be particularly important in axon growth. These are L1, NCAM, and N-cadherin. These CAMs bind predominantly in a homophilic manner and are involved in cell to cell interactions. Their expression on the cell surface of the growing axon and also on other nonneuronal cells with which growth cones interact is consistent with a role in axonal growth. L1 and NCAM are members of the immunoglobulin superfamily (Brummendorf and Rathjen, 1995) whereas N-cadherin is a member of the cadherin superfamily (Takeichi, 1995). Results from a number of model systems but in particular antibody perturbation methods have shown that these CAMs are important in the promotion of axonal growth. Two additional systems have verified these studies, namely purification and coating of adhesion molecules to tissue culture substrates and secondly a transfection-based model which involves expressing CAMs in cells such as fibroblasts which do not express such molecules. For all three CAMs (L1, N-CAM, and N-cadherin) it has been shown that expression of physiological levels in fibroblast cells leads to robust induction of neurite outgrowth (Matsunaga et al., 1988; Doherty et al., 1990; Williams et al., 1992). A major question has been whether the neurite outgrowth response directly relates to the adhesive properties of the CAMs. Recent results utilizing soluble chimeric forms of CAMs have effectively separated the adhesive properties of such CAMs from their ability to stimulate neurite outgrowth (Doherty et al., 1995a). It is becoming increasingly clear that CAMs can activate intracellular second messenger pathways and that this is the main mechanism by which they function in neurite outgrowth. A number of nonreceptor tyrosine kinases such as pp60[c-src] and pp59[fyn] might be involved in CAM function (Beggs et al., 1994). Also the serine kinase p90[rsk] co-precipitates with the L1 CAM (Wong et al., 1996). It is however unclear at present whether these kinases are specific components of a signalling cascade involved in neurite outgrowth or whether they may regulate the ligand binding properties of CAMs via inside out signalling mechanisms, as have been identified for integrin receptors.

A large number of pharmacological studies of CAM-mediated neurite outgrowth have identified some key components of an intracellular signalling cascade. Inhibitors of tyrosine kinases and a specific inhibitor of the enzyme PLCγ blocked CAM function (Doherty et al., 1995b). Also pharmacological inhibitors of the enzyme diacylglycerol lipase and calcium channel antagonists acted in a highly specific manner. A high degree of specificity was found in that integrin-dependent neurite outgrowth was unaffected by a large number of pharmacological agents that specifically inhibited the CAM response (Doherty and Walsh, 1996).

The fibroblast growth factor (FGFR) receptor is a key component of the CAM signalling cascade. FGFRs are a diverse group of receptor tyrosine kinases (Green et al., 1996). They are widely distributed in the nervous system and are believed to be specifically involved in processes of neuronal growth, development, and regeneration. The extracellular region of the FGFR contains a number of immunoglobulin domains followed by a single membrane spanning region and an intracellular catalytic domain. It is generally believed that dimerization of receptors activates the intracellular tyrosine kinase domain resulting in autophosphorylation and subsequent binding of effector molecules such as PLCγ. FGF is a ligand for the FGFR and this agent can stimulate neurite outgrowth from neurons. The PLCγ binding site on FGFR1 is important for the neurite outgrowth response mediated by FGF (Hall et al., 1996). Also a kinase-deleted mutant of FGFR1 has been found in experimental systems, such as transfection of PC12 cells, to act as a dominant negative receptor both for FGF function and CAM function (Doherty and Walsh, 1996). It is likely that the FGF and CAM signalling pathways are common in many ways in that agents which specifically block CAM function also appear to block FGF function. A number of other pieces of data point to the FGFR as being required for CAM function. Pretreatment of neurons with antibodies to FGFR block both FGF and CAM neurite outgrowth (Doherty et al., 1995*b*). Finally, FGF and chimeric forms of the L1 adhesion molecule induce changes in tyrosine phosphorylation of similar neuronal proteins (Williams et al., 1994). Also, transgenic animals have been produced which stably express a kinase-deficient variant of the FGFR under the control of a neuronal-specific promoter called neuron specific enolase. Cerebellar neurons cultured from such transgenic animals appear to have lost their ability to respond to CAMs but retain integrin-dependent neurite outgrowth responses. One of the main reasons for believing that the FGFR is involved in CAM action was based on the observation of a region of sequence similarity to adhesion molecules on the FGFR and the theory that this region may act as a docking or binding site for CAMs. This region has been called CAM homology domain (CHD) and is composed of a string of 20 conserved amino acids that show sequence similarity to L1, NCAM, and N-cadherin. The N-cadherin motif on the FGFR contains the sequence HAV, which has been found to be important in the homophilic binding properties of cadherins. With the recent crystal structure of the first domain of N-cadherin being published (Shapiro et al., 1995) some mechanistic possibilities have been identified. The HAV binding motif on N-cadherin has been identified by Shapiro et al. (1995), and this region is duplicated in the N-cadherin protein (Doherty and Walsh, 1996). Synthetic peptides that correspond to the CHD or the FGFR binding motifs act as effective competitors of CAM-FGFR interactions (Doherty et al., 1995*b*). Although CAMs appear to be able to mimic some FGF responses it is unlikely that they will give a totally overlapping pattern of responses. This is because FGF is generally believed to be translocated to the nucleus as part of the mitogenic response of cells to this trophic factor. It is also possible that CAMs interact only with a subset of FGFRs, which could lead to a restriction on the activation of specific second messenger cascades. Some FGFRs can activate both the PLCγ second messenger cascade and also the mitogen-activated protein kinase (MAP kinase) cascade whereas others can only activate MAP kinase. There is no evidence at present to suggest that CAMs can actually activate the MAP kinase second messenger cascade.

Downstream Targets of CAM Action

The functional state of the growth cone is regulated by specific posttranslational modifications in this subcellular compartment. A large body of data suggests that proteins such as GAP-43 are important during axonal growth and also in synaptic plasticity. Its expression is intimately related with the process of axonogenesis (Skene, 1989), and mice that have had the GAP-43 gene deleted display pathfinding errors (Strittmatter et al., 1995). Phosphorylation of GAP-43 on a specific serine residue (serine 41) occurs in a highly specific manner and can be stimulated by specific growth factors and contact with a variety of cell types (Meiri et al., 1991). It is also highly likely that the state of the phosphorylation of GAP-43 may be specifically involved in the response of growth cones to extracellular signals. One attractive possibility is that CAMs might modulate the phosphorylation state of GAP-43 in growth cones. One experimental model that could be used to test this hypothesis is to analyze isolated growth cones from neonatal animals. Growth cone preparations retain both structural and physiological characteristics of normal growth cones and, under certain circumstances, can be shown to extend filopodia on appropriate substances. As mentioned earlier, one versatile tool for studying NCAM-mediated neurite outgrowth is bivalent NCAM-Fc chimeras. We have found that these chimeras can stimulate neurite outgrowth in a dose-dependent manner and that this can be blocked specifically by antibodies against neuronal NCAM. Our previous studies (Doherty et al., 1995*a*) showed that an L1-Fc chimera stimulated neurite outgrowth and that this also required the presence of L1 in the neuron. It is therefore likely that homophilic binding of the CAM-Fc chimeras to their homologue on neurons stimulates neurite outgrowth.

To assess the involvement of downstream target molecules such as GAP-43, we have utilized isolated growth cone preparations. Treatment of these growth cone preparations with soluble NCAM or L1-Fc chimeras resulted in a steady increase in the level of phosphorylation of GAP-43. The phosphorylation of GAP-43 was analyzed using antibody 2G12, which is only reactive when serine 41 is phosphorylated. Interestingly, the soluble neurotrophic factor FGF also induced phosphorylation of GAP-43. The dose-response curve for GAP-43 phosphorylation and neurite outgrowth are also remarkably similar. With doses of NCAM-Fc chimera as low as 0.5 μg per ml there was significant increase in GAP-43 phosphorylation with a maximum response at 5 μg per ml. This dose-response curve is similar to that found for neurite outgrowth with the chimeras. Also, we have found using structure function approaches that the first three immunoglobulin domains of NCAM are sufficient to elicit the above response. A series of domain deletion constructs of NCAM have been subcloned into the pIg 1 expression vector, and recombinant protein has been produced. We have found that it is possible to delete significant portions of the extracellular domain of NCAM without leading to an abrogation of the response. The first three immunoglobulin domains are sufficient to elicit the maximum response. This concurs with previous data showing that sites important for homophilic binding are contained within this region. We have also found that the FGF receptor–PLCγ signalling cascade is important in the phosphorylation of GAP-43 and that inhibition of the PLCγ activation cascade via addition of a cell-permeable phosphopeptide that inhibits the ability of the FGF receptor to activate PLCγ (Hall et al., 1996) blocks the ability of NCAM-Fc chimeras to induce GAP-43 phosphorylation.

These experiments suggest that phosphorylation of GAP-43 is a key event in the FGFR-PLCγ cascade and that CAMs can also activate phosphorylation via this pathway. Clear correlations have been found between the ability of a variety of agents to induce neurite outgrowth and their ability to phosphorylate GAP-43. The results suggest that GAP-43 phosphorylation may be a key limiting step in CAM-mediated neurite outgrowth. Dephosphorylated GAP-43 has been shown to bind calmodulin specifically, which results in inhibition of substrate attachment and axon outgrowth (Widner and Caroni, 1993; Meiri et al., 1996). One potential model for this is that phosphorylated and calmodulin-bound GAP-43 can interact with actin filaments. It is possible that phosphorylated GAP-43 can stabilize long filaments whereas calmodulin-bound GAP-43 may act as a capping protein and inhibit polymerization of actin filaments. The CAM signalling cascade may therefore allow changes in GAP-43 phosphorylation which could promote axon growth by inducing GAP-43 into a form which does not continue to bind calmodulin and could therefore stabilize actin filaments. Further work remains to be done on the molecular basis of GAP-43 phosphorylation. We do not know at present whether CAMs and FGF are activating kinases, inhibiting phosphatases, or altering the susceptibility of GAP-43 to the action of such proteins. Nevertheless, our recent studies provide direct evidence that one important downstream target of CAM action in growth cones is the protein GAP-43. The results also reinforce the idea that CAMs are acting in a highly dynamic manner to alter signalling cascades in growth cones.

References

Beggs, H.E., P. Soriano, and P.F. Maness. 1994. NCAM dependent neurite outgrowth is inhibited in neurons from fyn-minus mice. *J. Cell Biol.* 127:825–833.

Bixby, J.L., and W.A. Harris. 1991. Molecular mechanisms of axon growth and guidance. *Annu. Rev. Cell Biol.* 7:117–159.

Brummendorf, T., and G. Rathjen. 1995. Cell adhesion molecules 1: immunoglobulin superfamily. *Protein Profile.* 2:963–1058.

Doherty, P., M. Fruns, P. Seaton, G. Dickson, C.H. Barton, T.A. Sears, and F.S. Walsh. 1990. A threshold effect of the major isoforms of NCAM on neurite outgrowth. *Nature.* 343:464–466.

Doherty, P., E.J. Williams, and F.S. Walsh. 1995a. A soluble chimeric form of the L1 glycoprotein stimulates neurite outgrowth. *Neuron.* 14:1–20.

Doherty, P., M.S. Fazeli, and F.S. Walsh. 1995b. The neural cell adhesion molecule and synaptic plasticity. *J. Neurobiol.* 26:437–446.

Doherty, P., and F.S. Walsh. 1996. CAM-FGF receptor interactions: a model for axonal growth. *Mol. Cell. Neurosci.* 8:99–112.

Goshima, Y., F. Nakamura, P. Strittmatter, and S.M. Strittmatter. 1995. Collapsin induced growth cone collapse mediated by an intracellular protein related to UNC-33. *Nature.* 376:509–514.

Green, P.J., F.S. Walsh, and P. Doherty. 1996. Promiscuity of fibroblast growth factor receptors. *Bioessays.* 18:639–646.

Hall, H., E.J. Williams, S.E. Moore, F.S. Walsh, A. Prochiantz, and P. Doherty. 1996. Inhibition of FGF-stimulated phosphatidylinositol hydrolysis and neurite outgrowth by a cell-membrane permeable phosphopeptide. *Curr. Biol.* 6:580–587.

Kater, S.B., and V. Rehder. 1995. The sensory motor role of growth cone filopodia. *Curr. Opin. Neurobiol.* 5:68–74.

Luo, Y., O. Shepherd, J. Li, M.J. Renzi, L.S. Chang, and J.A. Raper. 1995. A family of molecules related to collapsin in the embryonic chick nervous system. *Neuron.* 14:1131–1140.

Matsunaga, J., K. Hatta, A. Nagafuchi, and M. Takeichi. 1988. Guidance of optic nerve fibers by N-cadherin adhesion molecules. *Nature.* 334:62–64.

Meiri, K.F., and D.A. Burdick. 1991. Nerve growth factor stimulation of GAP-43 phosphorylation in intact isolated growth cones. *J. Neurosci.* 11:3155–3164.

Meiri, K.F., J.P. Hammang, E.W. Dent, and E.E. Baetge. 1996. Mutagenesis of ser41 to ala inhibits the association of GAP-43 with the membrane skeleton of GAP-43-deficient PC12B cells: effects on cell adhesion and the composition of neurite cytoskeleton and membrane. *J. Neurobiol.* 29:213–232.

Reichardt, L.F., and K.J. Tomaselli. 1991. Extracellular matrix molecules and their receptors: functions in neural development. *Annu. Rev. Neurosci.* 14:531–570.

Shapiro, L., A.M. Fannon, P.D. Kwong, A. Thompson, M.S. Lehmann, G. Grubel, J.F. Legrand, J. Als-Nielsen, D.R. Colman, and W.A. Hendrickson. 1995. Structural basis of cell-cell adhesion by cadherins. *Nature.* 374:327–337.

Skene, J.H.P. 1989. Axonal growth-associated proteins. *Annu. Rev. Neurosci.* 12:127–156.

Strittmatter, S.M., T. Vartanian, and M.C. Fishman. 1995. GAP-43 as a plasticity protein in neuronal form and repair. *J. Neurobiol.* 23:507–520.

Takeichi, M. 1995. Morphogenetic roles of classic cadherins. *Curr. Opin. Cell Biol.* 7:619–627.

Tessier-Levigne, M. 1994. Axon guidance by diffusible repellants and attractants. *Curr. Opin. Genet. Dev.* 4:596–601.

Widmer, F., and P. Caroni. 1993. Phosphorylation-site mutagenesis of the growth associated protein GAP-43 modulates its effects on cell spreading and morphology. *J. Cell Biol.* 120:503–512.

Williams, E.J., P. Doherty, G. Turner, R.A. Reid, J.J. Hemperly, and F.S. Walsh. 1992. Calcium influx into neurons can solely account for cell-contact dependent neurite outgrowth stimulated by transfected L1. *J. Cell Biol.* 119:885–892.

Williams, E.J., J. Furness, F.S. Walsh, and P. Doherty. 1994. Activation of the FGF receptor underlies neurite outgrowth stimulated by L1, N-CAM and N-cadherin. *Neuron.* 13:593–594.

Wong, E.W., A.W. Schaefer, G. Landreth, and V. Lemmon. 1996. Casein kinase II phosphorylates the neural cell adhesion molecule L1. *J. Neurochemistry.* 66:779–786.

Dissection of Protein Kinase and Phosphatase Targeting Interactions

John D. Scott

Vollum Institute, Portland, Oregon 97201-3098

Protein phosphorylation is a primary means of mediating signal transduction events that control cellular processes. Accordingly, the activities of protein kinases and phosphatases are highly regulated. One level of regulation is that the subcellular distribution of several kinases and phosphatases is restricted by association with targeting proteins or subunits. This mechanism promotes rapid and preferential modulation of specific targets within a defined microenvironment in response to diffusible second messengers. The type II cAMP-dependent protein kinase (PKA) is targeted by association of its regulatory subunit (RII) with A-kinase anchoring proteins (AKAPs). To date, 36 unique AKAPs have been identified. Each of these proteins contains a conserved amphipathic helix responsible for AKAP association with cellular structures. Disruption of PKA/AKAP interaction with peptides patterned after the amphipathic helix region blocks certain cAMP responses, including the modulation of glutamate receptor ion-channel activity in neurons and transcription of cAMP-responsive genes. Yeast two-hybrid screening methods have identified neuronal specific AKAP79-binding proteins including the β isoform of the phosphatase 2B, calcineurin. Biochemical and immunological studies have confirmed the two-hybrid results and identified additional members of this multienzyme signaling complex, including certain protein kinase C isoforms. These findings are consistent with colocalization of CaN, PKC, and type II PKA by AKAP79 and suggest a novel model for reversible phosphorylation in which the opposing kinase and phosphatase actions are colocalized in a signal transduction complex by association with a common anchor protein.

Introduction

Phosphorylation of protein substrates by kinases and phosphatases is a major process in the control of cellular function (Krebs and Beavo, 1979). Accordingly, the mechanisms involved in the regulation of kinase and phosphatase activity have been the subject of intense investigation since glycogen phosphorylase was first recognized to be regulated by phosphorylation (Fischer and Krebs, 1955; Krebs et al., 1959). One level of regulation involves control of enzyme activity, whereas the subcellular localization of kinases and phosphatases is restricted through interactions with targeting proteins (Kemp and Pearson, 1990; Faux and Scott, 1996a). The focus of this chapter will be to describe the use of synthetic peptides in identifying functional domains on a multivalent kinase and phosphatase targeting protein, the characterization of bioactive peptides in vitro, and the use of peptides as reagents to disrupt the localization or inhibition of enzyme activity in cells.

Cytoskeletal Regulation of Membrane Function © 1997 by The Rockefeller University Press

The Targeting Hypothesis

Numerous studies have demonstrated that the regulation of reversible phosphorylation events is tightly controlled. In fact, a focus of this article is to highlight the importance of subcellular location as a means to control the actions of multifunctional protein kinases and phosphatases. Recent advances from a variety of laboratories indicate that the subcellular location of these enzymes is maintained by association with specific targeting proteins. For example, the cAMP dependent protein kinase (PKA), protein kinase C (PKC), and the Ca^{2+}/calmodulin-dependent protein kinase (CaM KII) are localized by specific binding proteins, and likewise, protein phosphatases 1, 2A, and 2B are positioned through association with "phosphatase targeting subunits" (Faux and Scott, 1996*a*). This has led to the formulation of a "targeting hypothesis" by Hubbard and Cohen which proposes that "targeting" subunits or proteins specify the location and catalytic and regulatory properties of protein kinases and phosphatases, and thereby play a key role in ensuring the fidelity of protein phosphorylation events (Hubbard and Cohen, 1993). The targeting subunit is defined as that part of a kinase or phosphatase which directs the catalytic subunit to a subcellular location (Fig. 1 *A*). A variation on this theme is the formation of kinase/phosphatase signaling complexes to modulate the phosphorylation state of specific target substrates (Fig. 1 *B*). Coordination of these complexes is achieved by multivalent targeting proteins which serve as platforms for the assembly of these signaling units (Faux and Scott, 1996*b*). The following sections describe the use of peptides as reagents to probe the function of kinase/phosphatase signaling complexes.

Compartmentalization of the cAMP-dependent Protein Kinase

The work of numerous investigators has shown that many hormone-stimulated signaling cascades emanate from transmembrane receptors at the plasma membrane and proceed through intermediary G-proteins to promote the stimulation of adenylyl cyclase (Wedegaertner et al., 1995). The net effect of this transduction system is the generation of the diffusible second messenger cAMP which binds and activates the cAMP-dependent protein kinase (PKA) (Walsh et al., 1968). The kinase is activated by the release of two catalytic subunits (C) from the regulatory (R) subunit-cAMP complex. An array of PKA isozymes are expressed in mammalian cells, and

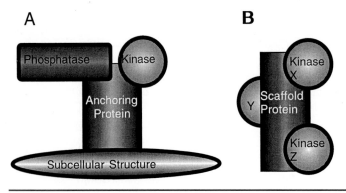

Figure 1. Localization of kinases and phosphatases through scaffold or anchoring proteins. (*A*) A schematic diagram depicting the topology of a prototypic kinase signaling scaffold and (*B*) an anchored kinase-phosphatase complex.

genes encoding three C subunits (Cα, Cβ and Cγ) and four R subunits (RIα, RIβ, RIIα and RIIβ) have been identified (reviewed in Scott, 1991). Two holoenzyme subtypes called type I and type II are formed by the combination of RI or RII with the C subunits (Brostrom et al., 1971; Corbin et al., 1973).

Although many hormones use parallel pathways to activate PKA, some measure of specificity must be cryptically built into each signaling cascade to ensure that the correct pool of kinase becomes active in the right place and at the right time (Harper et al., 1985). Compartmentalization of the kinase has been proposed as a regulatory mechanism that may increase the selectivity and intensity of a cAMP-mediated hormonal response. To facilitate this process, up to 75% of type II PKA is targeted through association of the regulatory subunit (RII) with A-kinase anchoring proteins, called AKAPs (reviewed in Rubin, 1994 and Scott and McCartney, 1994). Additionally, there is now some indication that the type I PKA is also compartmentalized (Rubin et al., 1972; Skalhegg et al., 1994). In recent years, numerous AKAPs have been identified that target PKA to a variety of subcellular locations (Coghlan et al., 1993).

A-Kinase Anchoring Proteins (AKAPs)

Initially, anchoring proteins were identified as contaminating proteins that copurified with the RII after affinity chromatography on cAMP-sepharose (Sarkar et al., 1984). However, detailed study of AKAPs was made possible by the original observation of Lohmann et al. (1984) that many, if not all, of these associated proteins retain their ability to bind RII after they have been immobilized to nitrocellulose or similar solid-phase supports. As a result, the standard technique for detecting AKAPs is an overlay method which is essentially a modification of the Western blot (reviewed by Carr and Scott, 1992). Using the overlay technique we have surveyed various mouse, bovine, and human tissues and have detected AKAPs ranging in size from 21 to 300 kD (Carr et al., 1992a). From this type of study it would appear that a typical cell expresses 5–10 distinct AKAPs that presumably adapt the type II PKA for specific functions. Also, the expression of some AKAPs may be hormonally regulated, as treatment of granulosa cells with FSH induces the expression of an 80-kD anchoring protein (Carr et al., 1993). The RII overlay method has also been developed into an efficient interaction cloning strategy wherein cDNA expression libraries have been screened using recombinant RII as a probe. Collectively, these techniques have been used to probe RII/AKAP interactions and have allowed us to develop a model for the topology of the anchored PKA holoenzyme. This model, which is presented in Fig. 2, illustrates the essential features of AKAPs. Each anchoring protein contains two classes of binding site: a conserved "anchoring motif" which binds the R subunit of PKA, and a unique "targeting domain" which compartmentalizes the PKA-AKAP complex through association with structural proteins, membranes, or cellular organelles.

The Conserved RII Anchoring Motif

Since several anchoring proteins apparently bind to the same or overlapping sites on RIIα, it seemed likely that these molecules share a common RII-binding do-

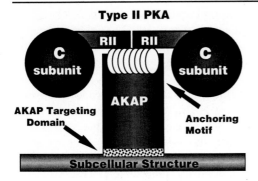

Figure 2. Topology of the anchored PKA holoenzyme complex; topology of anchored type II PKA. This model illustrates the essential features of AKAPs: (*A*) A conserved RII-binding site, and (*B*) a unique targeting domain for localization to intracellular sites.

main. Accordingly, one of the objectives of our expression cloning studies was to define a consensus RII-binding sequence that was common in all AKAPs. However, comparison of these sequences revealed no striking homology (Fig. 3 *A*) leading us to examine the RII-binding site in each anchoring protein for a conserved secondary structure binding motif. Computer aided secondary structure predictions of each putative RII-binding site showed a high probability for amphipathic helix formation (Fig. 3 *B*). The distinction between the hydrophobic and hydrophilic faces can be clearly seen when the sequences are drawn in a helical-wheel configuration (Carr et al., 1991). In each RII-anchoring protein there was a similar alignment of acidic residues throughout the hydrophilic face of each putative helix.

Analysis of Ht 31, a novel human thyroid RII anchoring protein of 1,035 amino acids that we cloned, identified a potential amphipathic helix between residues 494 to 509 (Carr et al., 1991). This sequence (Leu-Ile-Glu-Glu-Ala-Ala-Ser-Arg-Ile-

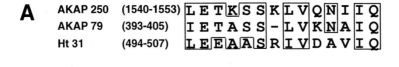

Figure 3. The conserved RII-binding motif on AKAPs. Association with the type II regulatory subunit of PKA occurs through a region of conserved secondary structure on the AKAPs. (*A*) Sequence alignment of the RII-binding regions on three AKAPs. Boxed areas represent regions of homology. (*B*) Each sequence is portrayed in a helical wheel projection (3.6 side-chains per turn) which is used to orient amino-acid side chains in an alpha helical conformation. Shaded areas represent the clustering on hydrophobic residues.

Val-Asp-Ala-Val-Ile-Glu-Gln) was 43% identical to a region within the RII-binding site of MAP 2. To determine whether residues 494-509 of Ht 31 were involved in RII binding, a 318 amino acid fragment representing residues 418 to 736 of Ht 31 was expressed in *E. coli*. Ht 31Δ 418-736 bound RIIα as assessed by solid-phase binding and gel-shift assays (Carr et al., 1991). To determine if an intact amphipathic helix was required for RII binding, a family of Ht 31 point mutants were produced in the Ht 31Δ 418-736 fragment. The introduction of proline into an α-helix conformation disrupts the secondary structure of the region and causes a 20° bend in the peptide backbone. The introduction of proline into the amphipathic helix region of Ht 31 diminished or abolished RIIα binding (Carr et al., 1991). Weak RIIα binding was observed with mutant Ht 31 Pro/Ala 498 which contained a mutation at position 5 in the putative helix region. No RIIα binding was detected with either Ht 31 Pro/Ile 502 or Ht 31 Pro/Ile 507. In contrast, proline substitution of Ala 522, which lies 12 residues downstream of the amphipathic helix region, had no apparent effect on RII binding. These results suggest that disruption of protein secondary structure between residues 498-507 of Ht 31 diminishes or abolishes RII binding. A peptide which spans the putative amphipathic helix region of Ht 31 binds RIIα with an affinity of approximately four, and circular dichromism analysis suggests it can adopt an α-helical conformation (Carr et al., 1992a).

PKA Anchoring In Vivo

So far, the majority of RII/AKAP interactions have been studied in vitro and under nonphysiological conditions. However, the high affinity of the amphipathic helix peptides for RII makes them ideal antagonists of PKA anchoring in vivo. The test system for these studies was the compartmentalization of PKA to the postsynaptic densities in hippocampal neurons where the kinase has easy access to the ionotrophic glutamate receptors. These receptors are central to the process of signal transduction across the synaptic membranes, and PKA-dependent phosphorylation is required to maintain the activity of AMPA/kainate responsive glutamate receptor channels (Greengard et al., 1991; Wang et al., 1991). The role of PKA in maintaining channel activity was confirmed by a gradual decline in whole-cell currents evoked by kainate (20 μM) recorded in the presence of ATP (20 μM) and 1 μM PKI (5-24) peptide, a potent and specific inhibitor of the catalytic subunit of PKA (61.8 ± 3.2% $n = 11$) (Fig. 4 A). To test the role of AKAPs in localizing the kinase near the channel, the anchoring inhibitor peptides (1 μM) were added to the whole cell pipette. The anchoring inhibitor peptide derived from two AKAPs, Ht 31 or AKAP79, inhibited AMPA/kainate currents to the same extent as the PKI peptide (64.9 ± 3.2% $n = 12$ and 68.8 ± 3.3% $n = 12$, respectively) (Fig. 4 B). The effects of PKI and the anchoring inhibitor peptides were not additive. However, the action of the Ht 31 peptide could be overcome by the C subunit of PKA (0.3 μM) suggesting that the anchoring inhibitor peptide interfered with PKA-dependent phosphorylation but did not directly inhibit the kinase (Fig. 4 C). In addition, the control peptide unable to block RII/AKAP interaction had no effect on kainate currents (85 ± 4.1% $n = 7$). Finally, currents evoked by AMPA (1 μM $n = 6$) behaved in the same manner as those evoked by application of kainate. Therefore, these results indicate that PKA localization is required for modulation of AMPA/kainate currents. These studies represented the first physiological evidence of the importance of PKA anchoring in the modulation of a specific cAMP responsive event (Rosenmund et al., 1994). Recently, Catterall and colleagues (Johnson et al., 1994) have used these peptides to demonstrate that disruption of PKA anchoring close to the L-type Ca^{2+} channel is required to maintain the channel in the active state. The functional effect of phosphorylation of the receptors appears to be desensitization for their agonist.

The Delivery of Cell-soluble Bioactive Peptides

To perform the biochemical experiments it is often necessary to introduce bioactive peptides into cells. Two problems often mar this approach: the stability of the peptide and the solubility of the cell-soluble peptide form. A myristylated peptide based upon the Ht31 (493-515) anchoring inhibitor peptide was synthesized according to the methods of Eichholtz et al. (1993). These authors have previously shown that a myristylated pseudosubstrate peptide is cell soluble and inhibits PKC in vivo. The anchoring inhibitor peptide (10 mM) was dissolved in DMSO and diluted to a final concentration of 500 μM in 0.1% DMSO/PBS and incubated with living cultures of hippocampal neurons for 1 h at 37°C. Control experiments were performed with solutions of a nonmyristylated anchoring inhibitor peptide. Cultures were washed extensively in PBS and fixed with 3.7% formaldehyde for 5 min at room temp, washed in 100% acetone at −20°C for 1 min and incubated in a blocking solution of PBS containing 0.1% BSA for 30 min at room temp. Individual coverslips were stained for uptake of Ht31 (493-515) peptide with a peptide-specific antibody to the same sequence (1:200 dilution). Intracellular uptake of anchoring inhibitor peptide was detected by indirect immunofluorescence using a confocal microscope (Fig. 5). Confocal analysis of individual neurons was performed on nine focal planes (0.5 μm) to confirm that detection of immunofluorescence was predomi-

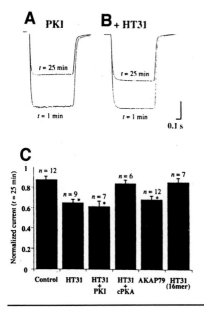

Figure 4. Displacement of PKA by anchoring inhibitor peptides blocks regulation of AMPA/kainate channels by PKA. (*A*) Inward currents evoked by kainate (20 μM) at 1 and 25 min after the start of whole cell recordings are superimposed. (*left*) The current decreased in the presence of the PKI(5-24) peptide (1 μM) which blocks kinase activity. (*B*) The current decreased to a similar extent in the presence of the Ht 31 peptide which is a competitive inhibitor of PKA anchoring. (*C*) Controls showing the amplitude of AMPA/kainate currents in the presence of ATP (5 mM) 25 min after application of various bioactive peptides and C subunit of PKA (0.3 μM). Peptide concentrations were Ht 3.1, (1 μM), PKI 5-24 (1 μM), and AKAP79 388-409 (1 μM). (Adapted from Rosenmund et al., 1994).

nantly intracellular. Increased amounts of the myristylated Ht31 (493-515) peptide were detected inside neurons (Fig. 5 *A*) when compared to cells incubated with the nonmyristylated form (Fig. 5 *B*). Control experiments incubated in the absence of peptide demonstrate the background level of staining (Fig. 5 *C*). This experiment suggests that myristylation is a viable means to introduce these peptides into living cells. Moreover, uptake of the myristylated peptide was primarily in the cell body and neurites which suggests that peptides will be available to compete with attachment to the PSD. It is possible that the myristal group targets these peptides to membranes which will be optimum for disruption of the AKAP79 signaling complex we believe is positioned close to the channels.

Other Enzyme Binding Sites on AKAPs: AKAP79/Calcineurin Interaction

In 1992, we cloned an anchoring protein called AKAP79 which is a component of the postsynaptic densities of neurons (Carr et al., 1992*b*). In accordance with the anchoring hypothesis, AKAP79 associates with the PSD to adapt the PKA for specific roles in postsynaptic events. Therefore, we proposed that AKAP79 must contain at least two functional domains (Scott and McCartney, 1994): a PKA binding region, and a targeting site responsible for association with proteins in the PSD. In an effort to identify AKAP-binding proteins, we used the yeast two-hybrid system (Fields and Song, 1989) to isolate cDNAs that encode proteins that associate with AKAP79 (Coghlan et al., 1995). While we expected to identify structural proteins, the two-hybrid method yielded a cDNA encoding another signaling enzyme, a cDNA for a murine β isoform of the calcineurin (CaN) A subunit. As anticipated, the two-hybrid system positively identified interactions between RII and itself (dimerization) and between RII and AKAP79 or Ht 31. These findings provide evidence for association of AKAP79 with CaN in yeast and suggest, because the AKAP also binds RII, the occurrence of a ternary complex between type II PKA, AKAP79, and CaN.

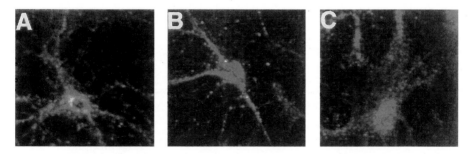

Figure 5. Uptake of myristylated peptide into cultured hippocampal neurons. Neurons were incubated with myristylated Ht31 (493-515) peptide. (*A*) Staining pattern for myr-Ht31 (493-515) peptide-treated neurons using affinity-purified peptide-specific antibodies (10 μg/ml). (*B*) Staining pattern for Ht31 (493-515) peptide-treated neurons using affinity-purified antibodies (10 μg/ml). (*C*) Staining pattern for neurons incubated with buffer alone, using affinity-purified antibodies (10 μg/ml).

Biochemical methods were employed to examine whether PKA and CaN were associated in mammalian brain. In initial experiments, calmodulin-binding proteins were isolated from bovine brain extracts by affinity chromatography and calcineurin was immunoprecipitated using affinity purified antisera against CaN A subunit. Immunoprecipitates were incubated with cAMP and the eluate was assayed for PKA activity by addition of ATP and Kemptide substrate. The specific activity of protein kinase was substantially increased by purification of a 30–60% ammonium sulfate fraction of the extract over calmodulin-agarose (28-fold; ±6, $n = 3$) followed by immunoprecipitation with CaN specific antibodies (123-fold; ±3.6). All of the protein kinase activity in the immunoprecipitate was cAMP-dependent and specifically inhibited by PKI peptide indicating that the catalytic (C) subunit of PKA was a component of the isolated complex. Coimmunoprecipitation of AKAP79 was confirmed by analysis of protein blots using a ^{32}P-RII overlay method. In complementary experiments, R subunits of PKA were isolated by affinity chromatography on cAMP-agarose. A proportion of the CaN present in the brain extract co-purified with R subunit and was eluted from the affinity matrix with 10 μM cAMP. Because recombinant CaN did not display any intrinsic cAMP binding activity, it was likely that the phosphatase was purified as part of a ternary complex with the AKAP and RII. Analysis of purified fraction by RII-overlay confirmed this hypothesis as AKAP79 was also present in the cAMP-agarose elution. Separate experiments were performed using a PKA anchoring inhibitor peptide, specifically displaced AKAPs from RII bound on the affinity column (Fig. 6). The peptide eluted both CaN (Fig. 6 *A*) and AKAP79 (Fig. 6 *B*), while R subunit remained on the affinity column and was subsequently eluted with cAMP (Fig. 6 *C*). Combined, these results confirm the simultaneous association of CaN and type II PKA with the AKAP.

Our model for colocalization of PKA and CaN through AKAP79 implies that the AKAP contains distinct sites for kinase and phosphatase binding. Residues 88 to 102 of AKAP79 were considered likely to comprise the CaN binding site due to homology with a region of the immunophilin FKBP-12 that contains determinants for CaN association. Binding of FKBP-12 to CaN is dependent on the immunosuppressant drug FK-506 and results in inhibition of the phosphatase. Considering our finding that CaN was inactive when isolated as a complex with the AKAP, we ex-

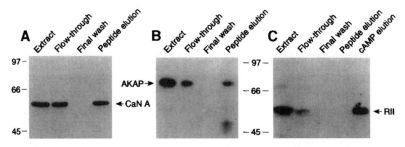

Figure 6.   Copurification of calcineurin and PKA from brain. R subunits were purified from crude extracts of bovine brain by affinity chromatography with cAMP-agarose. The AKAP/CaN complex was eluted with Ht31 (493-515) peptide. (*A*) Protein immunoblots were probed with affinity purified antibodies against calcineurin. (*B*) ^{32}P RII; (*C*) affinity-purified RII antibodies.

amined the effects of recombinant AKAP79 on CaN phosphatase activity. The AKAP inhibited both full-length CaN (Ca^{2+}/calmodulin dependent) and a truncated, constitutively active form of CaN, called CaN_{420} (Ca^{2+}/calmodulin independent), in a noncompetitive manner ($K_i = 4.2 \pm 1.8 \mu M$; $n = 3$) with respect to phosphorylated RII-substrate peptide. Furthermore, a synthetic peptide, corresponding to AKAP79 residues 81-102, inhibited both CaN forms, whereas the Ht31 (493-515) peptide was not an inhibitor of CaN. The observed inhibition was specific for calcineurin; AKAP79 (81-102) peptide did not significantly affect the activity of protein phosphatases 1 or 2A at peptide concentrations as high as 0.4 mM. Although CaN-binding sites on AKAP79 and FKBP-12 are similar, their differences may have functional significance: FK-506 (2 μM) did not affect the potency of inhibition, and recombinant AKAP79 did not display peptidyl prolyl isomerase activity toward a fluorescent peptide substrate. Collectively, these findings suggest that CaN is localized by AKAP79 in its inactive state in a manner analogous to AKAP-bound PKA. While the mechanism for activation of PKA by cAMP is apparent, regulation of AKAP-CaN association is likely to involve multiple factors.

AKAP79/Protein Kinase C Interaction

In neurons, PKA and CaN are both localized to postsynaptic densities (PSD) by association with AKAP79 which positions both enzymes close to key neuronal substrates. Because other neuronal signaling enzymes are present at the PSD (Wolf et al., 1986), we investigated their potential to associate with the anchoring protein AKAP79. Protein kinase C (PKC), a family of ser-thr kinases, is tethered to the PSD through association with binding proteins (Wolf et al., 1986). We used a solid phase binding assay (overlays) with PKC as a probe on bovine brain extracts to detect several PKC-binding proteins, including a protein that migrated with the same mobility as a prominent RII-binding protein of 75 kD. This band corresponds to AKAP75, the bovine homolog of AKAP79. This result indicated that the anchoring protein could bind both RII and PKC. Indeed, recombinant AKAP79 bound to PKC in the presence of Ca^{2+} and phosphatidylserine. This suggests that the AKAP binds in a different manner to other PKC-substrate/binding proteins such as MARCKS (myristylated alanine-rich C kinase substrate) and g-adducin, for which phosphorylation regulates association with PKC (Hyatt et al., 1994). We mapped the PKC-binding region by screening a family of AKAP79 fragments. Fragments encompassing the first 75 residues of AKAP79 bound PKC, whereas COOH-terminal fragments containing the RII and CaN-binding regions did not, thus implying that PKC binds to AKAP79 at a site that is distinct from those bound by RII and CaN.

Because basic and hydrophobic regions are determinants for binding of certain proteins to PKC (Liao et al., 1994) we focused our attention on a region located between residues 31 to 52 of AKAP79 (Fig. 7 *A*). A peptide encompassing this region specifically blocked the interaction of AKAP79 with PKC in the overlay assay but did not affect RII-binding to the AKAP (Fig. 7 *B*). Conversely, the RII anchoring inhibitor peptide (AKAP79 388-409) did not affect PKC-binding but did block interaction with RII (Fig. 7 *C*). This indicates that residues 31 to 52 represent the principal determinants for PKC-binding. Many kinases or phosphatases bound to anchoring proteins are maintained in an inactive state (Faux and Scott, 1996*b*). Ac-

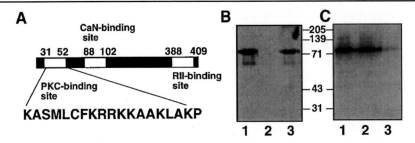

Figure 7. Inhibition of PKC activity by AKAP79. (*A*) Schematic representation of AKAP79 showing putative binding sites for PKC, CaN, and RII. The amino acid sequence for residues 31 to 52 is indicated. (*B*) Recombinant AKAP79 was blotted and PKC overlays were done in the absence (lane *1*) and presence of either 1.5 μM AKAP79 (31-52) (lane *2*) or 1.5 μM RII-anchoring inhibitor peptide AKAP79 (390-412) (lane *3*) with ∼12.5 nM PKC. (*C*) ^{32}P-RII (100 cpm/μl) overlays were done under the same conditions as in *B*.

cordingly, recombinant AKAP79 protein inhibited PKC activity with a half-maximal inhibition (IC_{50}) value of 0.35 ± 0.06 μM ($n = 3$). In addition, the AKAP79 peptide (residues 31 to 52) and a recombinant AKAP79 fragment (residues 1 to 75) inhibited PKC activity with IC_{50} values of 2.0 ± 0.6 μM ($n = 4$) and 1.6 ± 0.3 μM ($n = 4$), respectively. In contrast, the 31-52 peptide did not inhibit the activity of the catalytic subunit of PKA. Inhibition of PKC activity by the 31-52 peptide was mixed with an apparent inhibition constant (K_i) of 1.41 ± 0.28 μM ($n = 3$). The secondary plot of

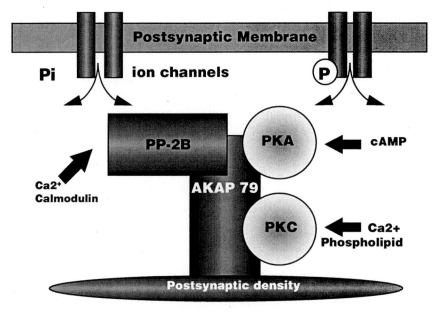

Figure 8. The AKAP79 signaling complex. The anchoring protein AKAP79 is attached to the postsynaptic densities and maintains PKA, PKC, and calcineurin close to substrates in the postsynaptic membrane.

the Michaelis constant divided by the maximal velocity (K_m/V_{max}) as a function of inhibitor concentration was nonlinear, suggesting binding at more than one site. Therefore, we propose that PKC is localized and inhibited by AKAP79 in a Ca^{2+}- and phosphatidylserine-dependent manner, and peptide studies suggest that residues 31-52 represent a principal site of contact.

Conclusions

Anchoring proteins have emerged as a new class of regulatory molecule which contribute to the organization and specificity of signal transduction pathways. The broad specificity kinases and phosphatases seem to use these proteins not only to direct their catalytic subunits to particular parts of the cell but also to enhance specificity towards preferred substrates. This may represent a mechanism to prevent indiscriminate phosphorylation or dephosphorylation of phosphoproteins by these broad specificity enzymes. A variation on this theme seems to be multifunctional anchoring proteins such as AKAP79, which package several signaling enzymes into multiprotein complexes. This may prove to be an efficient means for controlling the phosphorylation state of a given protein in response to multiple intracellular signals such as Ca^{2+}, phospholipid, and cAMP. Moreover, the structure of AKAP79 is modular, in that deletion analysis, peptide studies and coprecipitation techniques have demonstrated that each enzyme binds to a distinct region of the anchoring protein. Targeting of the AKAP79 signaling complex to the postsynaptic densities suggests a model for reversible phosphorylation in which the opposing effects of kinase and phosphatase action are colocalized in a signal transduction complex by association with a common anchor protein (Fig. 8). Potential substrates for the AKAP79 transduction complex are likely to be synaptic receptor/channels and may include AMPA/kainate receptors and Ca^{2+} channels which have recently been shown to be modulated by AKAP-targeted PKA, and NMDA receptors which are activated by PKC and attenuated by calcineurin (Klauck and Scott, 1995).

Acknowledgments

This work was funded by NIH grant No. GM 48231.

References

Brostrom, C.O., J.D. Corbin, C.A. King, and E.G. Krebs. 1971. Interaction of the subunits of adenosine 3':5'-cyclic monophosphate-dependent protein kinase of muscle. *Proc. Natl. Acad. Sci. USA.* 68:2444–2447.

Carr, D.W., D.A. DeManno, A. Atwood, M. Hunzicker-Dunn, and J.D. Scott. 1993. FSH-induced regulation of A-kinase anchoring protein in granulosa cells. *J. Biol. Chem.* 268: 20729–20732.

Carr, D.W., Z.E. Hausken, I.D.C. Fraser, R.E. Stofko-Hahn, and J.D. Scott. 1992a. Association of the type II cAMP-dependent protein kinase with a human thyroid RII-anchoring protein cloning and characterization of the RII-binding domain. *J. Biol. Chem.* 267:13376–13382.

Carr, D.W., and J.D. Scott. 1992. Blotting and band-shifting: techniques for studying protein-protein interactions. *TIBS.* 17:246–249.

Carr, D.W., R.E. Stofko-Hahn, I.D.C. Fraser, S.M. Bishop, T.S. Acott, R.G. Brennan, and J.D. Scott. 1991. Interaction of the regulatory subunit (RII) of cAMP-dependent protein kinase with RII-anchoring proteins occurs through an amphipathic helix binding motif. *J. Biol. Chem.* 266:14188–14192.

Carr, D.W., R.E. Stofko-Hahn, I.D.C. Fraser, R.D. Cone, and J.D. Scott. 1992*b*. Localization of the cAMP-dependent protein kinase to the postsynaptic densities by A-kinase anchoring proteins. *J. Biol. Chem.* 24:16816–16823.

Coghlan, V., B.A. Perrino, M. Howard, L.K. Langeberg, J.B. Hicks, W.M. Gallatin, and J.D. Scott. 1995. Association of protein kinase A and protein phosphatase 2B with a common anchoring protein. *Science.* 267:108–111.

Coghlan, V.M., S.E. Bergeson, L. Langeberg, G. Nilaver, and J.D. Scott. 1993. A-Kinase Anchoring Proteins: a key to selective activation of cAMP-responsive events? *Mol. Cell. Biochem.* 127:309–319.

Corbin, J.D., T.R. Soderling, and C.R. Park. 1973. Regulation of adenosine 3′,5′-monophosphate-dependent protein kinase. *J. Biol. Chem.* 248:1813–1821.

Eichholtz, T., D.B.A. de Bont, J. de Widt, R.M.J. Liskamp, and H.L. Ploegh. 1993. A myristolated pseudosubstrate peptide, a novel protein kinase C inhibitor. *J. Biol. Chem.* 268:1982–1986.

Faux, M.C., and J.D. Scott. 1996*a*. More on target with protein phosphorylation. *TIBS.* 21:312–315.

Faux, M.C., and J.D. Scott. 1996*b*. Molecular glue: kinase anchoring and scaffold proteins. *Cell.* 70:8–12.

Fields, S., and O. Song. 1989. A novel genetic system to detect protein-protein interactions. *Nature.* 340:245–246.

Fischer, E.H., and E.G. Krebs. 1955. Conversion of phosphorylase *b* to phosphorylase *a* in muscle extracts. *J. Biol. Chem.* 216:121–132.

Greengard, P., J. Jen, A.C. Nairn, and C.F. Stevens. 1991. Enhancement of glutamate response by cAMP-dependent protein kinase in hippocampal neurons. *Science.* 253:1135–1138.

Harper, J.F., M.K. Haddox, R. Johanson, R.M. Hanley, and A.L. Steiner. 1985. Compartmentation of second messenger action: immunocytochemical and biochemical evidence. *Vitamins and Hormones.* 42:197–252.

Hubbard, M., and P. Cohen. 1993. On target with a mechanism for the regulation of protein phosphorylation. *TIBS.* 18:172–177.

Hyatt, S.L., L. Liao, A. Aderem, A.C. Nairn, and S. Jaken. 1994. Correlation between protein kinase C binding proteins and substrates in REF52 cells. *Cell Growth Diff.* 5:495–502.

Johnson, B.D., T. Scheuer, and W.A. Catterall. 1994. Voltage-dependent potentiation of L-type Ca^{2+} channels in skeletal muscle cells requires anchored cAMP-dependent protein kinase. *Proc. Natl. Acad. Sci. USA.* 91:11492–11496.

Kemp, B.E., and R.B. Pearson. 1990. Protein kinase recognition sequence motifs. *Trends Biochem. Sci.* 15:342–346.

Klauck, T., and J.D. Scott. 1995. The postsynaptic density: a subcellular anchor for signal transduction enzymes. *Cell Signaling.* 7:747–757.

Krebs, E.G., and J.A. Beavo. 1979. Phosphorylation–dephosphorylation of enzymes. *Annu. Rev. Biochem.* 43:923–959.

Krebs, E.G., D.J. Graves, and E.H. Fischer. 1959. Factors affecting the activity of muscle phosphorylase *b* kinase. *J. Biol. Chem.* 234:2867–2873.

Liao, L., S.L. Hyatt, C. Chapline, and S. Jaken. 1994. Protein kinase C domains involved in interactions with other proteins. *Biochemistry.* 33:1229–1233.

Lohmann, S.M., P. DeCamili, I. Enig, and U. Walter. 1984. High-affinity binding of the regulatory subunit (RII) of cAMP-dependent protein kinase to microtubule-associated and other cellular proteins. *Proc. Natl. Acad. Sci. USA.* 81:6723–6727.

Rosenmund, C., D.W. Carr, S.E. Bergeson, G. Nilaver, J.D. Scott, and G.L. Westbrook. 1994. Anchoring of protein kinase A is required for modulation of AMPA/kainate receptors on hippocampal neurons. *Nature.* 368:853–856.

Rubin, C.S. 1994. A kinase Anchor proteins and the intracellular targeting of signals carried by cAMP. *Biochim. Biophys. Acta.* 1224:467–479.

Rubin, C.S., J. Erlichman, and O.M. Rosen. 1972. Cyclic adenosine 3′,5′-monophosphate-dependent protein kinase of human erythrocyte membranes. *J. Biol. Chem.* 247:6135–6139.

Sarkar, D., J. Erlichman, and C.S. Rubin. 1984. Identification of a calmodulin-binding protein that co-purifies with the regulatory subunit of brain protein kinase II. *J. Biol. Chem.* 259:9840–9846.

Scott, J.D. 1991. Cyclic nucleotide-dependent protein kinases. *Pharmacol. Ther.* 50:123–145.

Scott, J.D., and S. McCartney. 1994. Localization of A-kinase through anchoring proteins. *Mol. Endocrinol.* 8:5–13.

Skalhegg, B.S., K. Tasken, V. Hansson, H.S. Huitfeldt, T. Jahnsen, and T. Lea. 1994. Location of cAMP-dependent protein kinase type I with the TCR-CD3 complex. *Science.* 263:84–87.

Walsh, D.A., J.P. Perkins, and E.G. Krebs. 1968. An adenosine 3′,5′-monophosphate-dependent protein kinase from rabbit skeletal muscle. *J. Biol. Chem.* 243:3763–3765.

Wang, L.Y., M.W. Salter, and J.F. MacDonald. 1991. Regulation of kainate receptors by cAMP-dependent protein kinase and phosphatases. *Science.* 253:1132–1134.

Wedegaertner, P.B., P.T. Wilson, and H.R. Bourne. 1995. Lipid modifications of trimeric G proteins. *J. Biol. Chem.* 270:503–506.

Wolf, M., S. Burgess, U.K. Misra, and N. Sahyoun. 1986. Postsynaptic densities contain a subtype of protein kinase C. *Biochem. Biophys. Res. Commun.* 140:691–698.

Signaling through the Focal Adhesion Kinase

Michael D. Schaller

Department of Cell Biology and Anatomy, University of North Carolina at Chapel Hill, Chapel Hill, North Carolina 27599

Many cell types adhere to the highly organized extracellular matrix (ECM) that surrounds them both in situ and in tissue culture via receptors called the integrins. These are heterodimeric complexes comprised of two membrane spanning subunits, α and β, each of which contains a large extracellular domain and short cytoplasmic tail (2, 39). Obviously one important structural function of the integrins is to provide attachment to the ECM substrate upon which the cells grow. The short cytoplasmic domain of the β_1 subunit binds to a number of cytoskeletal proteins, including talin and α-actinin, and through these interactions fulfills another structural function, anchorage of the cytoskeleton to the membrane (10, 42, 96). As a key molecule bridging the actin cytoskeleton, the integrins play an important role in maintaining cell shape and through their interaction with ECM proteins are fundamental for cell migration.

Integrin Signaling

The ECM provides a substrate for cell attachment and also provides instructions to the cell controlling biological events, e.g., cell growth, differentiation, and apoptosis (9, 39, 75, 88). The observation that the integrins can alter the physiological state of cells, including regulating gene expression, suggested that they transmit signals into the cell. Over the last five years a number of cytoplasmic signaling pathways involved in integrin-regulated signal transduction have been identified (recently reviewed in 9, 23, 88). Activation of integrins, by crosslinking with antibodies or allowing attachment to its cognate ligand, triggers changes in intracellular pH, intracellular Ca^{2+}, and tyrosine phosphorylation of cellular proteins (9, 23, 88). Ligation of integrins causes coclustering of a very large array of signaling molecules suggesting that the integrins may influence the activity of a number of signaling pathways (57). There is biochemical evidence documenting the activation of the GTP binding protein p21[ras] (24, 45), a number of serine/threonine kinases including Raf-1 (19), MEK (19), and MAPK (18, 58, 86, 105), the three of which comprise a protein kinase signaling cascade and phosphatidylinositol 5′ kinase which phosphorylates phosphatidylinositol to generate phosphatidylinositol bisphosphate (PIP$_2$) (21). Cell adhesion also modulates the actions of the cyclin-dependent kinases which control progression of cells through the cell cycle (30, 92, 106). It is apparent from this expanding list of regulatory molecules whose actions are modified by integrin-dependent adhesion that transduction of extracellular cues into cytoplasmic signals is an important function of the integrins.

pp125[FAK], The Focal Adhesion Kinase

Avian pp125[FAK] was first isolated as a candidate substrate for the oncogenic protein tyrosine kinase (PTK) pp60[v-src] (81). Independent efforts led to the isolation of cDNAs encoding the murine and human homologues of pp125[FAK] (4, 20, 33, 102). pp125[FAK] is a highly conserved protein exhibiting more than 90% amino acid identity among the *Xenopus*, avian, and human homologues (35). It has a central catalytic domain flanked by large (~400 amino acid) NH$_2$- and COOH-terminal domains (Fig. 1). Sequences within the COOH-terminal 150 residues of pp125[FAK] are responsible for targeting the protein to discrete regions within the cell called focal adhesions (36). These are sites of close contact between the cell and underlying ECM substrate and regions where the cytoskeleton is anchored to the cytoplasmic domain of the integrins (10, 42, 96). pp125[FAK] is a major substrate for integrin induced tyrosine phosphorylation and becomes enzymatically activated in an integrin dependent manner (11, 31, 33, 47, 53). A number of other stimuli induce tyrosine phosphorylation of pp125[FAK] including platelet-derived growth factor, hepatocyte growth factor, macrophage colony stimulating factor, neuropeptides like bombesin, lysophosphatidic acid, crosslinking cell surface FcεRI, and hyaluronic acid (Fig. 2) (80).

CAKβ/Pyk2/RAFTK

A pp125[FAK]-related PTK, called CAKβ/Pyk2/RAFTK, was recently isolated (5, 51, 79). Like pp125[FAK], CAKβ contains a central catalytic domain flanked by large NH2- and COOH-terminal noncatalytic domains. While the two PTKs exhibit ~45% overall sequence identity, two regions are more highly related. The catalytic domain and the COOH-terminal 150 residues, i.e., the region containing the focal adhesion targeting (FAT) sequence of pp125[FAK], both show 60% identity between the two PTKs. Despite the sequence similarity with the COOH-terminal targeting sequence of pp125[FAK], CAKβ was originally reported to localize to sites of cell–cell contact rather than focal adhesions and its tyrosine phosphorylation was cell adhesion independent suggesting that it was regulated differently than pp125[FAK] (79). A recent study challenges these original findings. Both the CAKβ endogenous to a

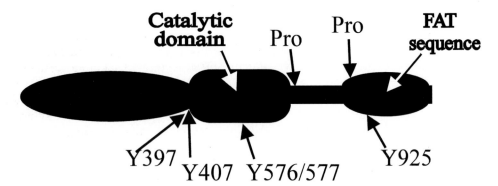

Figure 1.   Schematic depiction of pp125[FAK]. The central catalytic domain, the COOH-terminal focal adhesion targeting (*FAT*) sequence, sites of tyrosine phosphorylation (*Y*), and proline-rich SH3 binding sites (*Pro*) are indicated.

megakaryocytic cell line and CAKβ exogenously expressed in COS cells colocalized with vinculin by immunofluorescence and became tyrosine phosphorylated and enzymatically activated in response to integrin-dependent cell adhesion (52). We have examined the localization of CAKβ exogenously expressed in chicken embryo cells. The COOH-terminal noncatalytic domain, engineered to be expressed autonomously, was found in focal adhesions by immunofluorescence (M.D. Schaller and T. Sasaki, manuscript submitted for publication). These findings suggest that COOH-terminal sequences shared between CAKβ and pp125[FAK] mediate localization and that the functions of these two PTKs may partially overlap.

Rho, the Cytoskeleton, and pp125[FAK]

The integrity of the actin cytoskeleton is required for tyrosine phosphorylation of pp125[FAK] since cytochalasin D treatment ablates pp125[FAK] phosphorylation in response to integrin-dependent adhesion, LPA, bombesin, and PDGF (53, 69, 90, 91). Rho family proteins (Rac, Cdc42, and Rho), members of a branch of the ras superfamily of small GTP binding proteins, regulate the architecture of the actin cytoskeleton (61, 72, 73). Perturbation of the function of endogenous Rho using the C3 toxin of *C. botulinum*, which specifically ADP-ribosylates and inactivates Rho, blocks cytoskeletal changes induced by extracellular stimuli such as LPA (72) and impairs tyrosine phosphorylation of pp125[FAK] in response to LPA (48, 68). Introduction of GTPγS into permeabilized Swiss 3T3 cells triggers tyrosine phosphorylation of pp125[FAK], an event that is blocked by C3 toxin or synthetic peptides de-

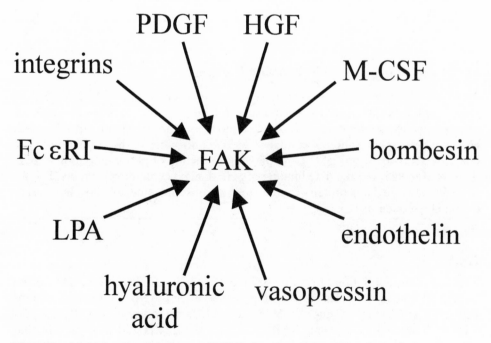

Figure 2. Multiple stimuli induce pp125[FAK] phosphorylation. Stimuli that induce tyrosine phosphorylation of pp125[FAK] are depicted.

signed to prevent the interaction of Rho with downstream effectors (89). These data support the hypothesis that the Rho functions as an intermediary signaling protein linking pp125^{FAK} to upstream stimuli.

Rho is regulated by guanine nucleotide binding (8, 60). Rho-GTP can bind directly to downstream effector molecules to alter their function resulting in the propagation of a signal. Recently candidate effectors for Rho have been identified and include serine/threonine protein kinases and a lipid kinase (9, 59). One of these, Rho-kinase, may regulate the formation of focal adhesions and stress fibers by controlling contractility (3, 22, 46). Phosphatidylinositol 4-phosphate 5-kinase (PI5K) is activated by GTP-bound Rho in cell lysates (21) and phosphorylates phosphatidylinositol 4-phosphate to generate phosphatidylinositol bisphosphate (PIP$_2$). PIP$_2$ is the substrate for phospholipase C, an enzyme activated upon cellular stimulation with growth factors, and thus Rho may function in controlling availability of a substrate required for receptor PTK signaling (50). PIP$_2$ has also been implicated in regulating the actin cytoskeleton since it modulates the activities of a number of proteins that control actin polymerization (41). Most recently PIP$_2$ was shown to bind vinculin, inducing a conformation change exposing talin and actin binding sites on vinculin (27). Thus Rho may control assembly of the actin cytoskeleton via several mechanisms.

The mechanism by which Rho may affect tyrosine phosphorylation of pp125^{FAK} has not been elucidated. pp125^{FAK} could lie downstream of one of the Rho regulated protein kinases and receive signals via conventional signaling mechanisms, e.g., a protein kinase cascade. Alternatively pp125^{FAK} could be regulated through alterations in the architecture of the cytoskeleton. Given the rapidity of recent advances in elucidating Rho signaling, the resolution of these issues should be forthcoming.

SH2 Containing pp125^{FAK} Binding Partners

Tyrosine 397 is the major site of autophosphorylation of pp125^{FAK} (14, 26, 83). Other sites of phosphorylation, tyrosines 407, 576, 577, and 925, are substrates for phosphorylation for another PTK, likely a member of the Src family of PTKs (12, 86). While phosphorylation of residues 576/577 contributes to the regulation of the enzymatic activity of pp125^{FAK} (12), the major function of phosphorylation at residues 397 and 925 is to regulate binding of SH2 domain containing proteins (Fig. 3). SH2 domains mediate protein–protein interactions and bind to phosphotyrosine containing sequences (25, 63).

Grb2

Tyrosine phosphorylation of residue 925 creates a high affinity binding site for the Grb2 SH2 domain (86). This sequence is conserved in CAKβ and Grb2 associates with CAKβ in vivo (51). Thus pp125^{FAK} and CAKβ may utilize similar effector proteins to transmit downstream signals. Grb2 is an SH2/SH3 adaptor protein that binds SOS, a guanine nucleotide exchange factor for Ras (25, 63). The recruitment of Grb2/SOS suggests that tyrosine phosphorylation of pp125^{FAK} may be a mechanism to regulate activation of the ras pathway culminating in the activation and translocation of MAP kinase to the nucleus.

pp60[src]

Phosphorylation of tyrosine 397 creates a high affinity binding site for the SH2 domain of pp60[src] (80, 88). The association of pp60[src] with pp125[FAK] in *src*-transformed cells requires the SH2 domain of pp60[src] and tyrosine phosphorylation of residue 397 within pp125[FAK]. A consensus SH3 binding site lies ∼30 residues to the NH$_2$-terminal side of tyrosine 397 (104). BIAcore analysis using synthetic peptides indicates that this site can bind the Src SH3 domain (B. Ellis and M.D. Schaller, unpublished observations). Furthermore a peptide containing both the SH2 and SH3 binding sites of pp125[FAK] binds with higher affinity to recombinant Src than a peptide containing the SH2 site alone. Therefore the pp125[FAK]/pp60[src] interaction may be mediated by both SH3 and SH2 domain interactions.

pp125[FAK] and pp60[src]

We have established two model systems to study pp125[FAK] signaling. Vanadate treatment of pp125[FAK] overexpressing chicken embryo cells, or coexpression of pp125[FAK] and an Src-like PTK, induces pp125[FAK]-dependent tyrosine phosphoryla-

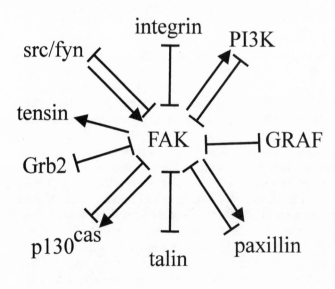

Figure 3. pp125[FAK]-binding proteins and substrates. Schematic illustration of pp125[FAK] and its relationship with other proteins. Arrows radiate from the PTK and point to substrates. Protein–protein interactions are indicated by the lines.

tion of proteins including paxillin (85) (Schaller et al., manuscript submitted). In each case mutation of tyrosine 397, i.e., the pp60[src] binding site, abrogates this response. Thus assembly of a complex between pp125[FAK] and an Src-like PTK may be critical for signaling. By coexpressing pp125[FAK] and pp60[src] in CE cells, we have identified two potential consequences of formation of a pp125[FAK]/pp60[src] complex. In these cells the subcellular localization of pp60[src] is redirected to focal adhesions following cell adhesion to fibronectin, and the pp60[src] molecules that associate with pp125[FAK] are hypophosphorylated at their COOH-terminal negative regulatory site, implying that they are enzymatically active (Schaller et al., manuscript submitted). These observations suggest that the activity and location of Src-like PTKs within the cell might be regulated through interactions with pp125[FAK]. An additional, intriguing observation is that a catalytically inactive variant of pp125[FAK] can induce tyrosine phosphorylation of paxillin in both of these scenarios (M.D. Schaller et al., manuscript submitted and unpublished data). This suggests that a major function of pp125[FAK] might be recruitment of the Src-like PTKs into complex, rather than directly phosphorylating downstream substrates.

SH3 Containing pp125[FAK] Binding Partners

p130cas

p130[cas] was first identified as a phosphotyrosine containing protein associated with pp60[src] and p47[gag-crk] in cells transformed by the *src* and *crk* oncogenes (43, 70, 76). It contains an NH2-terminal SH3 domain, a large central domain containing multiple consensus binding sites for the Crk SH2 domain, and a COOH-terminal domain (76). SH3 domains function to mediate protein–protein interactions by binding to proline rich sequences (25, 63). p130[cas] complexes with pp125[FAK] in vivo, and the interaction is mediated by binding of SH3 domain of p130[cas] to the sequence at proline residues 712 and 715 within pp125[FAK] (Fig. 3) (34, 66). A p130[cas]-related protein called Hef1 is also able to bind pp125[FAK] in an SH3-dependent interaction (49). p130[cas] localizes to cellular focal adhesions and becomes phosphorylated upon cellular adhesion to fibronectin (34, 62, 64, 100). Tyrosine phosphorylation of p130[cas] and pp125[FAK] is also coordinately induced in response to LPA further suggesting an enzyme/substrate relationship (90). However, genetic evidence using *fak*[−/−] and *src*[−/−] cell lines suggests that the Src-like PTKs, rather than pp125[FAK], is the critical kinase for phosphorylation of p130[cas] (7, 95, 99). The caveat to this finding is that CAKβ could potentially fulfill the functions of pp125[FAK] in the *fak*[−/−] cells. Following integrin-induced tyrosine phosphorylation, p130[cas] binds to two SH2/SH3 adaptor proteins, Crk and CrkL (65, 99). The SH3 domains of Crk bind to two guanine nucleotide exchange factors, SOS and C3G, that regulate the activity of ras and rap1 (29, 54, 94). These results imply that the p130[cas] branch of the integrin signaling pathway(s) may control the activation of the ras family of GTP binding proteins.

Phosphatidylinositol 3-kinase

Phosphatidylinositol 3-kinase (PI3K) is a heterodimeric enzyme comprised of an SH2/SH3 containing regulatory subunit and a catalytic subunit (44). PI3K associates with pp125[FAK] following cell adhesion or stimulation with platelet-derived growth factor (16, 17). The SH3 domain of the regulatory subunit of PI3K may mediate this interaction by binding to the same proline containing site as p130^{cas} (32). This interaction may serve to recruit PI3K to the cytoskeleton or to facilitate tyrosine phosphorylation of PI3K by pp125^{FAK} (16, 32). In these scenarios PI3K functions downstream of pp125^{FAK}. Paradoxically it has been suggested that PI3K may lie upstream of pp125^{FAK} (67). Further analysis is obviously required to irrefutably determine if PI3K functions upstream or downstream of pp125^{FAK} or both.

GRAF

GRAF was recently identified as a pp125^{FAK}-binding, SH3 domain containing regulatory protein for the rho family of GTP binding proteins and can specifically stimulate the intrinsic GTPase activity of rhoA and cdc42 (38). The sequence around proline 878 within pp125^{FAK} functions as a binding site for the SH3 domain of GRAF. The function of GRAF in pp125^{FAK} signaling is speculative but could serve as an effector for Rho and/or Cdc42 or alternatively may function as a negative regulator to terminate Rho/Cdc42 signaling.

Focal Adhesion-associated pp125^{FAK}-binding Partners

The COOH-terminal focal adhesion targeting (FAT) sequence of pp125^{FAK} contains binding sites for talin (15) and paxillin (37). Binding to talin is not responsible for targeting pp125^{FAK} to focal adhesions since a mutant of pp125^{FAK} that fails to localize to focal adhesions retains the ability to bind talin (15, 36). CAKβ also binds paxillin suggesting sequences shared with pp125^{FAK} make contact with paxillin (M.D. Schaller and T. Sasaki, manuscript submitted for publication). A similarity in sequence between pp125^{FAK} and vinculin, another paxillin binding protein (97, 103), has been identified (93). Mutation of these sequences within pp125^{FAK} abolishes binding to paxillin in vitro and focal adhesion targeting (93). Although these results suggest that pp125^{FAK} targets to focal adhesions through interactions with paxillin, another study has apparently dissociated these functions (37, 82).

Paxillin

Paxillin is a tyrosine phosphorylated, focal adhesion-associated protein. Its phosphotyrosine content is elevated in *src-* and *crk*-transformed cells and paxillin is a major binding partner for the *crk* oncogene product, p47$^{gag-crk}$ (80, 88). In cells overexpressing the Csk PTK, tyrosine-phosphorylated paxillin binds Csk (80, 88). These interactions are mediated by the SH2 domains of Crk and Csk. In untransformed cells tyrosine phosphorylation of paxillin is induced by integrin-dependent cell adhesion or stimulation with a variety of agents including PDGF, bombesin, vasopressin, endothelin, LPA, angiotensin II, and crosslinking cell surface FcεRI (80, 88). Thus in many instances tyrosine phosphorylation of pp125^{FAK} and paxillin occurs coordinately.

The amino acid sequence of paxillin, deduced from the nucleotide sequence of the cDNA, exhibits a number of interesting features including four COOH-terminal LIM domains (77, 98). LIM domains mediate protein–protein interactions (87), but the proteins that bind to the LIM domains of paxillin are unknown. Near the NH$_2$ terminus of paxillin there is a proline rich sequence that can mediate binding to the SH3 domain of pp60^{src} in vitro (101). The NH$_2$ half of paxillin also contains the binding site(s) for pp125[FAK] and vinculin (98). Four sites of tyrosine phosphorylation reside in the NH$_2$-terminal quarter of paxillin (6, 85, and unpublished observations).

Tyrosine phosphorylation of paxillin is not required for localization to focal adhesions (6 and unpublished observations) but likely regulates binding to SH2 domain containing proteins. Phosphorylation of tyrosine residues 31 and 118 create high affinity binding sites for the SH2 domains of the adaptor proteins Crk and CrkL (78, 85). Paxillin might therefore function, like p130[cas], in the recruitment of these signaling molecules into focal adhesions.

Integrins

The NH$_2$-terminal noncatalytic domain of pp125[FAK] can directly bind to synthetic peptides mimicking the cytoplasmic domains of β subunits of integrins in vitro (84). Furthermore, pp125[FAK] and tensin cocap with integrins in the absence of cocapping of other focal adhesion-associated proteins (56). These observations are consistent with the hypothesis that integrins bind pp125[FAK]. The function of this proposed interaction remains speculative.

Biological Function of pp125[FAK]

pp125[FAK] was first proposed to regulate the assembly of focal adhesions when inhibitors of PTKs were found to impair focal adhesion assembly and cell spreading (11). Expression of a dominant negative variant of pp125[FAK] in chicken embryo cells inhibits tyrosine phosphorylation of candidate pp125[FAK] substrates and retards focal adhesion formation and cell spreading when cells adhered to fibronectin (71). However, microinjection of a similar variant into human umbilical vein endothelial cells (HUVEC) abolishes tyrosine phosphorylation in focal adhesions, yet does not alter focal adhesion structure (28). *fak*$^{-/-}$ murine fibroblasts spread normally and exhibit well formed focal adhesions, suggesting that pp125[FAK] is dispensable for focal adhesion formation (40). The discovery that CAKβ may be functionally redundant with pp125[FAK] raises the possibility that CAKβ may replace pp125[FAK] in these cells.

Several lines of evidence originally suggested that pp125[FAK] might regulate cell migration. Stimulation of cells with nonmitogenic doses of PDGF induces tyrosine phosphorylation of pp125[FAK] suggesting that pp125[FAK] may participate in an alternate response to PDGF, such as migration (1, 69). Hepatocyte growth factor, a motogenic stimulus, also induces pp125[FAK] phosphorylation (55). Induction of HUVEC migration by wounding tissue culture monolayers triggered tyrosine phosphorylation of pp125[FAK] and PTK inhibitors blocked migration (74). Fibroblasts derived from *fak*$^{-/-}$ mouse embryos migrate slower than wild-type fibroblasts over short periods of time (3 hours) although over longer periods of time (9 hours) *fak*$^{-/-}$ and *FAK*$^{+/+}$ fibroblasts exhibit similar mobilities (40). Microinjection of a dominant negative pp125[FAK] construct dramatically impairs the migration of HUVEC cells in a wound healing assay (28). Finally, overexpression of pp125[FAK] in Chinese hamster ovary cells enhances their rate of migration, and the interaction between pp125[FAK] and Src-like kinases is required for this function (13). Intriguingly, a catalytically defective pp125[FAK] variant also enhances cell migration, again suggesting that recruitment of Src-like PTKs may be the crucial event for pp125[FAK] signaling.

Concluding Remarks

The last few years have witnessed great strides in elucidating how GTP binding proteins regulate the actin cytoskeleton and how integrins transmit intracellular signals. pp125[FAK] is a recipient of signals generated by the integrins and by other stimuli that induce cytoskeletal alterations, e.g., LPA. The cast of characters that likely function in pp125[FAK] signaling are rapidly being identified, and emerging evidence has implicated pp125[FAK] in regulating cell migration. The next step in solving the pp125[FAK] puzzle is defining which effectors of Rho might regulate pp125[FAK], the role of individual pp125[FAK] binding proteins/substrates in controlling motility, and the mechanisms by which these proteins operate.

Acknowledgment

Research in the author's laboratory is supported by grants from the NIH (#5R29GM53666-02) and the American Cancer Society (#CB-173).

References

1. Abedi, H., K.E. Dawes, and I. Zachary. 1995. Differential effects of platelet-derived growth factor BB on p125 focal adhesion kinase and paxillin tyrosine phosphorylation and on cell migration in rabbit aortic vascular smooth muscle cells and Swiss 3T3 fibroblasts. *J. Biol. Chem.* 270:11367–11376.

2. Albelda, S.M., and C.A. Buck. 1990. Integrins and other cell adhesion molecules. *FASEB J.* 4:2868–2880.

3. Amano, M., M. Ito, K. Kimura, Y. Fukata, K. Chihara, T. Nakano, Y. Matsuura, and K. Kaibuchi. 1996. Phosphorylation and activation of myosin by rho-associated kinase (rho-kinase). *J. Biol. Chem.* 271:20246–20249.

4. Andre, E., and M. Becker-Andre. 1993. Expression of an N-terminally truncated form of human focal adhesion kinase in the brain. *Biochem. Biophys. Res. Commun.* 190:140–146.

5. Avraham, S., R. London, Y. Fu, S. Ota, D. Hiregowdara, J. Li, S. Jiang, L.M. Pasztor, R.A. White, J.E. Groopman, and H. Avraham. 1995. Identification and characterization of a novel related adhesion focal tyrosine kinase (RAFTK) from megakaryocytes and brain. *J. Biol. Chem.* 270:27742–27751.

6. Bellis, S.L., J.T. Miller, and C.E. Turner. 1995. Characterization of tyrosine phosphorylation of paxillin in vitro by focal adhesion kinase. *J. Biol. Chem.* 270:17437–17441.

7. Bockholt, S.M., and K. Burridge. 1995. An examination of focal adhesion formation and tyrosine phosphorylation in fibroblasts isolated from src, fyn and yes mice. *Cell Adhesion and Communication.* 3:91–100.

8. Boguski, M.S., and F. McCormick. 1993. Proteins regulating ras and its relatives. *Nature.* 366:643–654.

9. Burridge, K., and M. Chrzanowska-Wodnicka. 1996. Focal adhesions, contractility, and signaling. *Annu. Rev. Cell Dev. Biol.* 12:463–518.

10. Burridge, K., K. Fath, T. Kelly, G. Nuckolls, and C. Turner. 1988. Focal adhesions: transmembrane junctions between the extracellular matrix and the cytoskeleton. *Annu. Rev. Cell Biol.* 4:487–525.

11. Burridge, K., C.E. Turner, and L.H. Romer. 1992. Tyrosine phosphorylation of paxillin and pp125[FAK] accompanies cell adhesion to extracellular matrix: a role in cytoskeletal assembly. *J. Cell Biol.* 119:893–903.

12. Calalb, M.B., T.R. Polte, and S.K. Hanks. 1995. Tyrosine phosphorylation of focal adhesion kinase at sites in the catalytic domain regulates kinase activity: a role for src family kinases. *Mol. Cell. Biol.* 15:954–963.

13. Cary, L.A., J.F. Chang, and J.L. Guan. 1996. Stimulation of cell migration by overexpression of focal adhesion kinase and its association with src and fyn. *J. Cell Sci.* 109:1787–1794.

14. Chan, P.-Y., S.B. Kanner, G. Whitney, and A. Aruffo. 1994. A transmembrane-anchored chimeric focal adhesion kinase is constitutively activated and phosphorylated at tyrosine residues identical to pp125[FAK]. *J. Biol. Chem.* 269:20567–20574.

15. Chen, H.-C., P.A. Appeddu, J.T. Parsons, J.D. Hildebrand, M.D. Schaller, and J.-L. Guan. 1995. Interaction of FAK with cytoskeletal protein talin. *J. Biol. Chem.* 270:16995–16999.

16. Chen, H.-C., and J.-L. Guan. 1994. Association of focal adhesion kinase with its potential substrate phosphatidylinositol 3-kinase. *Proc. Natl. Acad. Sci. USA.* 91:10148–10152.

17. Chen, H.-C., and J.-L. Guan. 1994. Stimulation of phosphatidylinositol 3'-kinase association with focal adhesion kinase by platelet-derived growth factor. *J. Biol. Chem.* 269:31229–31333.

18. Chen, Q., M.S. Kinch, T.H. Lin, K. Burridge, and R.L. Juliano. 1994. Integrin-mediated cell adhesion activates mitogen-activated protein kinases. *J. Biol. Chem.* 269:26602–26605.

19. Chen, Q., T.H. Lin, C.J. Der, and R.L. Juliano. 1996. Integrin-mediated activation of MEK and mitogen-activated protein kinase is independent of ras. *J. Biol. Chem.* 271:18122–18127.

20. Choi, K., M. Kennedy, and G. Keller. 1993. Expression of a gene encoding a unique protein-tyrosine kinase within specific fetal- and adult-derived hematopoietic lineages. *Proc. Natl. Acad. Sci. USA.* 90:5747–5751.

21. Chong, L.D., A. Traynor-Kaplan, G.M. Bokoch, and M.A. Schwartz. 1994. The small GTP-binding protein rho regulates a phosphatidylinositol 4-phosphate 5-kinase in mammalian cells. *Cell.* 79:507–513.

22. Chrzanowska-Wodnicka, M., and K. Burridge. 1996. Rho-stimulated contractility drives the formation of stress fibers and focal adhesions. *J. Cell Biol.* 133:1403–1415.

23. Clark, E.A., and J.S. Brugge. 1995. Integrins and signal transduction pathways: the road taken. *Science.* 268:233–239.

24. Clark, E.A., and R.O. Hynes. 1996. Ras activation is necessary for integrin-mediated activation of extracellular signal-regulated kinase 2 and cytosolic phospholipase A_2 but not for cytoskeletal organization. *J. Biol. Chem.* 271:14814–14818.

25. Cohen, G.B., R. Ren, and D. Baltimore. 1995. Modular binding domains in signal transduction proteins. *Cell.* 80:237–248.

26. Eide, B.L., C.W. Turck, and J.A. Escobedo. 1995. Identification of Tyr-397 as the primary site of tyrosine phosphorylation and pp60[src] association in the focal adhesion kinase, pp125[FAK]. *Mol. Cell. Biol.* 15:2819–2827.

27. Gilmore, A.P., and K. Burridge. 1996. Regulation of vinculin binding to talin and actin by phosphatidylinositol-4-5-bisphosphate. *Nature.* 381:531–535.

28. Gilmore, A., and L.H. Romer. 1996. Inhibition of FAK signaling in focal adhesions decreases cell motility and proliferation. *Mol. Biol. Cell.* 7:1209–1224.

29. Gotoh, T., S. Hattori, S. Nakamura, H. Kitayama, M. Noda, Y. Takai, K. Kaibuchi, H. Matsui, O. Hatase, H. Takahashi, T. Kurata, and M. Matsuda. 1995. Identification of Rap1 as a target for the crk SH3 domain-binding guanine nucleotide-releasing factor C3G. *Mol. Cell. Biol.* 15:6746–6753.

30. Guadagno, T.M., M. Ohtsubo, J.M. Roberts, and R.K. Assoian. 1993. A link between cyclin A expression and adhesion-dependent cell cycle progression. *Science.* 262:1572–1575.

31. Guan, J.-L., and D. Shalloway. 1992. Regulation of pp125^{FAK} both by cellular adhesion and by oncogenic transformation. *Nature.* 358:690–692.

32. Guinebault, C., B. Payrastre, C. Racaud-Sultan, H. Mazarguil, M. Breton, G. Mauco, M. Plantavid, and H. Chap. 1995. Integrin-dependent translocation of phosphoinositide 3-kinase to the cytoskeleton of thrombin-activated platelets involves specific interactions of p85 alpha with actin filaments and focal adhesion kinase. *J. Cell Biol.* 129:831–842.

33. Hanks, S.K., M.B. Calalb, M.C. Harper, and S.K. Patel. 1992. Focal adhesion protein tyrosine kinase phosphorylated in response to cell spreading on fibronectin. *Proc. Natl. Acad. Sci. USA.* 89:8487–8489.

34. Harte, M.T., J.D. Hildebrand, M.R. Burnham, A.H. Bouton, and J.T. Parsons. 1996. p130^{cas}, a substrate associated with v-src and v-crk, localizes to focal adhesions and binds to focal adhesion kinase. *J. Biol. Chem.* 271:13649–13655.

35. Hens, M.D., and D.W. DeSimone. 1995. Molecular analysis and developmental expression of the focal adhesion kinase pp125^{FAK} in Xenopus laevis. *Dev. Biol.* 170:274–288.

36. Hildebrand, J.D., M.D. Schaller, and J.T. Parsons. 1993. Identification of sequences required for the efficient localization of the focal adhesion kinase, pp125^{FAK}, to cellular focal adhesions. *J. Cell Biol.* 123:993–1005.

37. Hildebrand, J.D., M.D. Schaller, and J.T. Parsons. 1995. Paxillin, a tyrosine phosphorylated focal adhesion-associated protein binds to the carboxyl terminal domain of focal adhesion kinase. *Mol. Biol. Cell.* 6:637–647.

38. Hildebrand, J.D., J.M. Taylor, and J.T. Parsons. 1996. An SH3 domain-containing GTPase-activating protein for rho and cdc42 associates with focal adhesion kinase. *Mol. Cell. Biol.* 16:3169–3178.

39. Hynes, R.O. 1992. Integrins: versatility, modulation, and signaling in cell adhesion. *Cell.* 69:11–25.

40. Ilic, D., Y. Furuta, S. Kanazawa, N. Takeda, K. Sobue, N. Nakatsuji, S. Nomura, J. Fujimoto, M. Okada, T. Yamamoto, and S. Aizawa. 1995. Reduced cell motility and enhanced focal adhesion contact formation in cells from FAK-deficient mice. *Nature.* 377:539–544.

41. Janmey, P.A. 1994. Phosphoinositides and calcium as regulators of cellular actin assembly and disassembly. *Annu. Rev. Physiol.* 56:169–191.

42. Jockusch, B.M., P. Bubeck, K. Giehl, M. Kroemker, J. Moschner, M. Rothkegel, M. Rudiger, K. Schluter, G. Stanke, and J. Winkler. 1995. The molecular architecture of focal adhesions. *Annu. Rev. Cell Dev.* Biol. 11:379–416.

43. Kanner, S.B., A.B. Reynolds, H.C. Wang, R.R. Vines, and J.T. Parsons. 1991. The SH2 and SH3 domains of pp60^{src} direct stable association with tyrosine phosphorylated proteins p130 and p110. *EMBO J.* 10:1689–1698.

44.   Kapeller, R., and L.C. Cantley. 1994. Phosphatidylinositol 3-kinase. *Bioessays.* 16:565–576.

45.   Kapron-Bras, C., L. Fitz-Gibbon, P. Jeevaratnam, J. Wilkins, and S. Dedhar. 1993. Stimulation of tyrosine phosphorylation and accumulation of GTP-bound p21^{ras} upon antibody-mediated alpha$_2$beta$_1$ integrin activation in T-lymphoblastic cells. *J. Biol. Chem.* 268:20701–20704.

46.   Kimura, K., M. Ito, M. Amano, K. Chihara, Y. Fukata, M. Nakafuku, B. Yamamori, J. Feng, T. Nakano, K. Okawa, A. Iwamatsu, and K. Kaibuchi. 1996. Regulation of myosin phosphatase by rho and rho-associated kinase (rho-kinase). *Science.* 273:245–248.

47.   Kornberg, L., H.S. Earp, J.T. Parsons, M.D. Schaller, and R.L. Juliano. 1992. Cell adhesion or integrin clustering increases phosphorylation of a focal adhesion-associated tyrosine kinase. *J. Biol. Chem.* 267:23439–23442.

48.   Kumagai, N., N. Morii, K. Fujisawa, Y. Nemoto, and S. Narumiya. 1993. ADP-ribosylation of rho p21 inhibits lysophosphatidic acid-induced protein tyrosine phosphorylation and phosphatidylinositol 3-kinase activation in cultured Swiss 3T3 cells. *J. Biol. Chem.* 268:24535–24538.

49.   Law, S.F., J. Estojak, B. Wang, T. Mysliwiec, G. Kruh, and E.A. Golemis. 1996. Human enhancer of filamentation 1, a novel p130^{cas}-like docking protein, associates with focal adhesion kinase and induces pseudohyphal growth in saccharomyces cerevisiae. *Mol. Cell. Biol.* 16:3327–3337.

50.   Lee, S.B., and S.G. Rhee. 1995. Significance of PIP$_2$ hydrolysis and regulation of phospholipase C isozymes. *Curr. Opin. Cell Biol.* 7:183–189.

51.   Lev, S., H. Moreno, R. Martinez, P. Canoll, E. Peles, J.M. Musacchio, G.D. Plowman, B. Rudy, and J. Schlessinger. 1995. Protein tyrosine kinase PYK2 involved in Ca^{2+}-induced regulation of ion channel and MAP kinase functions. *Nature.* 376:737–745.

52.   Li, J., H. Avraham, R.A. Rogers, S. Raja, and S. Avraham. 1996. Characterization of RAFTK, a novel focal adhesion kinase, and its integrin-dependent phosphorylation and activation in megakaryocytes. *Blood.* 88:417–428.

53.   Lipfert, L., B. Haimovich, M.D. Schaller, B.S. Cobb, J.T. Parsons, and J.S. Brugge. 1992. Integrin-dependent phosphorylation and activation of the protein tyrosine kinase pp125^{FAK} in platelets. *J. Cell Biol.* 119:905–912.

54.   Matsuda, M., Y. Hashimoto, K. Muroya, H. Hasegawa, T. Kruata, S. Tanaka, S. Nakamura, and S. Hattori. 1994. Crk protein binds to two guanine nucleotide-releasing proteins for the ras family and modulates nerve growth factor-induced activation of ras in PC12 cells. *Mol. Cell. Biol.* 14:5495–5500.

55.   Matsumoto, K., K. Matsumoto, T. Nakamura, and R.H. Kramer. 1994. Hepatocyte growth factor/scatter factor induces tyrosine phosphorylation of focal adhesion kinase (p125^{FAK}) and promotes migration and invasion by oral squamous cell carcinoma cells. *J. Biol. Chem.* 269:31807-31813.

56.   Miyamoto, S., S.K. Akiyama, and K.M. Yamada. 1995. Synergistic roles for receptor occupancy and aggregation in integrin transmembrane function. *Science.* 267:883–885.

57.   Miyamoto, S., H. Teramoto, O.A. Coso, J.S. Gutkind, P.D. Burbelo, S.K. Akiyama, and K.M. Yamada. 1995. Integrin function: molecular hierarchies of cytoskeletal and signaling molecules. *J. Cell Biol.* 131:791–805.

58.   Morino, N., T. Mimura, K. Hamasaki, K. Tobe, K. Ueki, K. Kikuchi, K. Takehara, T. Kadowaki, Y. Yazaki, and Y. Nojima. 1995. Matrix/integrin interaction activates the mitogen-activated protein kinase, p44^{erk-1} and p42^{erk-2}. *J. Biol. Chem.* 270:269–273.

59. Nagata, K., and A. Hall. 1996. The rho GTPase regulates protein kinase activity. *Bioessays.* 18:529–531.

60. Nobes, C., and A. Hall. 1994. Regulation and function of the Rho subfamily of small GTPases. *Curr. Opin. Genet. Dev.* 4:77–81.

61. Nobes, C.D., and A. Hall. 1995. Rho, rac and cdc42 GTPases regulate the assembly of multimolecular focal complexes associated with actin stress fibers, lamellipodia, and filopodia. *Cell.* 81:53–62.

62. Nojima, Y., N. Morino, T. Mimura, K. Hamasaki, H. Furuya, R. Sakai, T. Sato, K. Tachibana, C. Morimoto, Y. Yazaki, and H. Hirai. 1995. Integrin-mediated cell adhesion promotes tyrosine phosphorylation of p130Cas, a src homology 3-containing molecule having multiple src homology 2-binding motifs. *J. Biol. Chem.* 270:15398–15402.

63. Pawson, T. 1995. Protein modules and signalling networks. *Nature.* 373:573–580.

64. Petch, L.A., S.M. Bockholt, A. Bouton, J.T. Parsons, and K. Burridge. 1995. Adhesion-induced tyrosine phosphorylation of the p130 src substrate. *J. Cell Sci.* 108:1371–1379.

65. Petruzzelli, L., M. Takami, and R. Herrera. 1996. Adhesion through the interaction of lymphocyte function-associated antigen-1 with intracellular adhesion molecule-1 induces tyrosine phosphorylation of p130[cas] and its association with c-crkII. *J. Biol. Chem.* 271:7796–7801.

66. Polte, T.R., and S.K. Hanks. 1995. Interaction between focal adhesion kinase and crk-associated kinase substrate p130[cas]. *Proc. Natl. Acad. Sci. USA.* 92:10678–10685.

67. Rankin, S., R. Hooshmand-Rad, L. Claesson-Welsh, and E. Rozengurt. 1996. Requirement for phosphatidylinositol 3′-kinase activity in platelet-derived growth factor-stimulated tyrosine phosphorylation of p125 focal adhesion kinase and paxillin. *J. Biol. Chem.* 271:7829–7834.

68. Rankin, S., N. Morii, S. Narumiya, and E. Rozengurt. 1994. Botulinum C3 exoenzyme blocks the tyrosine phosphorylation of p125[FAK] and paxillin induced by bombesin and endothelin. *FEBS Lett.* 354:315–319.

69. Rankin, S., and E. Rozengurt. 1994. Platelet-derived growth factor modulation of focal adhesion kinase (p125[FAK]) and paxillin tyrosine phosphorylation in Swiss 3T3 cells. *J. Biol. Chem.* 269:704–710.

70. Reynolds, A.B., S.B. Kanner, H.-C. Wang, and J.T. Parsons. 1989. Stable association of activated pp60[src] with two tyrosine phosphorylated cellular proteins. *Mol. Cell. Biol.* 9:3951–3958.

71. Richardson, A., and J.T. Parsons. 1996. A mechanism for regulation of the adhesion-associated protein tyrosine kinase pp125[FAK]. *Nature.* 380:538–540.

72. Ridley, A.J., and A. Hall. 1992. The small GTP-binding protein rho regulates the assembly of focal adhesions and actin stress fibers in response to growth factors. *Cell.* 70:389–399.

73. Ridley, A.J., H.F. Paterson, C.L. Johnston, D. Diekmann, and A. Hall. 1992. The small GTP-binding protein rac regulates growth factor-induced membrane ruffling. *Cell.* 70:401–410.

74. Romer, L.H., N. McLean, C.E. Turner, and K. Burridge. 1994. Tyrosine kinase activity, cytoskeletal organization, and motility in human vascular endothelial cells. *Mol. Biol. Cell.* 5:349–361.

75. Ruoslahti, E., and J.C. Reed. 1994. Anchorage dependence, integrins and apoptosis. *Cell.* 77:477–478.

76.   Sakai, R., A. Iwamatsu, N. Hirano, S. Ogawa, T. Tanaka, H. Mano, Y. Yazaki, and H. Hirai. 1994. A novel signaling molecule, p130, forms stable complexes in vivo with v-Crk and v-Src in a tyrosine phosphorylation-dependent manner. *EMBO J.* 13:3748–3756.

77.   Salgia, R., J.-L. Li, S.H. Lo, B. Brunkhorst, G.S. Kansas, Y. Sun, E. Pisick, T. Ernst, L.B. Chen, and J.D. Griffin. 1995. Molecular cloning of paxillin, a focal adhesion protein phosphorylated by p210 BCR/ABL. *J. Biol. Chem.* 270:5039–5047.

78.   Salgia, R., N. Uemura, K. Okuda, J.L. Li, E. Pisick, M. Sattler, R. de Jong, B. Druker, N. Heisterkamp, L.B. Chen, J. Groffen, and J.D. Griffen. 1995. CrkL links p210$^{BCR/ABL}$ with paxillin in chronic myelogenous leukemia cells. *J. Biol. Chem.* 270:29145–29150.

79.   Sasaki, H., K. Nagura, M. Ishino, H. Tobioka, K. Kotani, and T. Sasaki. 1995. Cloning and characterization of cell adhesion kinase beta, a novel protein-tyrosine kinase of the focal adhesion kinase subfamily. *J. Biol. Chem.* 270:21206–21219.

80.   Schaller, M.D. 1996. The focal adhesion kinase. *J. Endocrinol.* 150:1–7.

81.   Schaller, M.D., C.A. Borgman, B.S. Cobb, R.R. Vines, A.B. Reynolds, and J.T. Parsons. 1992. pp125^{FAK}, a structurally distinctive protein tyrosine kinase associated with focal adhesions. *Proc. Natl. Acad. Sci.* USA. 89:5192–5196.

82.   Schaller, M.D., C.A. Borgman, and J.T. Parsons. 1993. Autonomous expression of a non-catalytic domain of the focal adhesion associated protein tyrosine kinase pp125^{FAK}. *Mol. Cell. Biol.* 13:785–791.

83.   Schaller, M.D., J.D. Hildebrand, J.D. Shannon, J.W. Fox, R.R. Vines, and J.T. Parsons. 1994. The autophosphorylation site of the focal adhesion kinase, pp125^{FAK}: a high affinity binding site for pp60^{src}. *Mol. Cell. Biol.* 14:1680–1688.

84.   Schaller, M.D., C.A. Otey, J.D. Hildebrand, and J.T. Parsons. 1995. pp125^{FAK} and paxillin physically complex with beta integrins in vitro. *J. Cell Biol.* 130:1181–1187.

85.   Schaller, M.D., and J.T. Parsons. 1995. pp125^{FAK}-dependent tyrosine phosphorylation of paxillin creates a high-affinity binding site for Crk. *Mol. Cell. Biol.* 15:2635–2645.

86.   Schlaepfer, D.D., S.K. Hanks, T. Hunter, and P. van der Geer. 1994. Integrin-mediated signal transduction linked to ras pathway by GRB2 binding to focal adhesion kinase. *Nature.* 372:786–791.

87.   Schmeichel, K.L., and M.C. Beckerle. 1994. The LIM domain is a modular protein-binding interface. *Cell.* 79:211–219.

88.   Schwartz, M.A., M.D. Schaller, and M.H. Ginsberg. 1995. Integrins: emerging paradigms of signal transduction. *Annu. Rev. Cell Dev. Biol.* 11:549–599.

89.   Seckl, M.J., N. Morii, S. Narumiya, and E. Rozengurt. 1995. Guanosine 5′-3-O-(thio)triphosphate stimulates tyrosine phosphorylation of p125^{FAK} and paxillin in permeabilized Swiss 3T3 cells. Role of p21rho. *J. Biol. Chem.* 270:6984–6990.

90.   Seufferlein, T., and E. Rozengurt. 1994. Lysophosphatidic acid stimulates tyrosine phosphorylation of focal adhesion kinase, paxillin and p130. *J. Biol. Chem.* 269:9345–9351.

91.   Sinnett-Smith, J., I. Zachary, A.M. Valverde, and E. Rozengurt. 1993. Bombesin stimulation of pp125 focal adhesion kinase tyrosine phosphorylation. *J. Biol. Chem.* 268:14261–14268.

92.   Symington, B.E. 1992. Fibronectin receptor modulates cyclin-dependent kinase activity. *J. Biol. Chem.* 267:25744–25747.

93.   Tachibana, K., T. Sato, N. D'Avirro, and C. Morimoto. 1995. Direct association of pp125[FAK] with paxillin, the focal adhesion-targeting mechanism of pp125[FAK]. *J. Exp. Med.* 182:1089–1100.

94.   Tanaka, S., T. Morishita, Y. Hashimoto, S. Hattori, S. Nakamura, M. Shibuya, K. Matuoka, T. Takenawa, T. Kurata, K. Nagashima, and M. Matsuda. 1994. C3G, a guanine nucleotide-releasing protein expressed ubiquitously, binds to the Src homology 3 domains of CRK and GRB2/ASH proteins. *Proc. Natl. Acad. Sci. USA.* 91:3443–3447.

95.   Thomas, S.M., P. Soriano, and A. Imamoto. 1995. Specific and redundant roles of src and fyn in organizing the cytoskeleton. *Nature.* 376:267–271.

96.   Turner, C.E., and K. Burridge. 1991. Transmembrane molecular assemblies in cell-extracellular matrix interactions. *Curr. Opin. Cell Biol.* 3:849–853.

97.   Turner, C.E., J.R. Glenney, and K. Burridge. 1990. Paxillin: a new vinculin-binding protein present in focal adhesions. *J. Cell Biol.* 111:1059–1068.

98.   Turner, C.E., and J.T. Miller. 1994. Primary sequence of paxillin contains putative SH2 and SH3 binding motifs and multiple LIM domains: identification of a vinculin and pp125[FAK]-binding region. *J. Cell Sci.* 107:1583–1591.

99.   Vuori, K., H. Hirai, S. Aizawa, and E. Ruoslahti. 1996. Induction of p130[cas] signaling complex formation upon integrin-mediated cell adhesion: a role for src family kinases. *Mol. Cell. Biol.* 16:2606–2613.

100.   Vuori, K., and E. Ruoslahti. 1995. Tyrosine phosphorylation of p130[cas] and cortactin accompanies integrin-mediated cell adhesion to extracellular matrix. *J. Biol. Chem.* 270:22259–22262.

101.   Weng, Z., J.A. Taylor, C.E. Turner, J.S. Brugge, and C. Seidel-Dugan. 1993. Detection of Src homology 3-binding proteins, including paxillin, in normal and v-src transformed Balb/ c 3T3 cells. *J. Biol. Chem.* 268:14956–14963.

102.   Whitney, G.S., P.-Y. Chan, J. Blake, W.L. Cosand, M.G. Neubauer, A. Aruffo, and S.B. Kanner. 1993. Human T and B lymphocytes express a structurally conserved focal adhesion kinase, pp125[FAK]. *DNA Cell Biol.* 12:823–830.

103.   Wood, C.K., C.E. Turner, P. Jackson, and D.R. Critchley. 1994. Characterisation of the paxillin-binding site and the C-terminal focal adhesion targeting sequence in vinculin. *J. Cell Sci.* 107:709–717.

104.   Yu, H., J.K. Chen, S. Feng, D.C. Dalgarno, A.W. Brauer, and S.L. Schreiber. 1994. Structural basis for the binding of proline-rich peptides to SH3 domains. *Cell.* 76:933–945.

105.   Zhu, X., and R.K. Assoian. 1995. Integrin-dependent activation of MAP kinase: a link to shape-dependent cell proliferation. *Mol. Biol. Cell.* 6:273–282.

106.   Zhu, X., M. Ohtsubo, R.M. Bohmer, J.M. Roberts, and R.K. Assoian. 1996. Adhesion-dependent cell cycle progression linked to the expression of cyclin D1, activation of cyclin E-cdk2, and phosphorylation of the retinoblastoma protein. *J. Cell Biol.* 133:391–403.

Involvement of the Actin Network in Insulin Signalling

Theodoros Tsakiridis,*‡ Qinghua Wang,* Celia Taha,* Sergio Grinstein,*
Gregory Downey,‡ and Amira Klip*

*Division of Cell Biology, The Hospital for Sick Children, Toronto, Ontario,
Canada M5G 1X8; and ‡Clinical Science, Department of Medicine, University of
Toronto, Toronto, Ontario, Canada M5S 1A8

The purpose of the studies included in this chapter was to examine the role of the
actin network in the propagation of insulin action leading to stimulation of glucose
transport and activation of the mitogen-activated protein kinase cascade. The active insulin receptor phosphorylates tyrosine residues of intracellular proteins such
as the insulin receptor substrate-1 (IRS-1) which acts as docking sites for molecules
containing Src homology 2 (SH2) domains. One such molecule is phosphatidylinositol
3-kinase (PI 3-kinase) which becomes activated by binding to IRS-1. PI 3-kinase activity is required for the insulin-stimulation of glucose transport and glycogen synthesis.
Grb2, a small adaptor molecule, can bind IRS-1 and, through the guanine nucleotide
exchange factor Sos, leads to the activation of the small GTP binding protein Ras.
Through a cascade of protein kinases, activation of Ras results in activation of the
Erk 1 and 2 mitogen-activated protein kinases (MAPKs) which appear to control
important nuclear and metabolic events. To investigate the role of the actin network in the propagation of insulin action leading to stimulation of glucose transport
and the activation of the Erk MAPKs, we used the fungal metabolite cytochalasin D
which disassembles the actin network. Actin disassembly abolished almost completely the ability of insulin to increase the rate of glucose transport into L6 muscle
cells (myotubes) through prevention of the insulin-induced recruitment of glucose
transporters to the plasma membrane which is the event that mediates the increase in
the rate of transport. Actin disassembly did not affect either the insulin-mediated
phosphorylation of IRS-1, the association of PI 3-kinase with this molecule, or the activation of IRS-1-associated PI 3-kinase. These results were also verified in another
insulin responsive cell line, the 3T3-L1 adipocytes. In these cells, actin disassembly inhibited the insulin-induced recruitment of PI 3-kinase to intracellular membranes
containing glucose transporters. Moreover, actin disassembly abolished the insulin-
mediated phosphorylation of the Erk MAPKs. We conclude that the cellular actin
network of insulin responsive cells is not required for the activation of PI 3-kinase but
prevents its cellular redistribution. In contrast, intact actin filaments are essential for
the propagation of insulin signals leading to the the activation of the MAPKs.

Introduction

The insulin receptor is a tyrosine kinase the activity of which is stimulated upon
binding of its ligand, insulin (White and Khan, 1994). The activated insulin receptor
mediates a wide spectrum of biological responses including (i) membrane transport

events such as stimulation of glucose and amino acid transport, (ii) metabolic events such as stimulation of glycogen and protein sysnthesis, and (iii) nuclear events such as gene expression. Chief among the immediate targets of the insulin receptor tyrosine kinase are cytoplasmic proteins termed the insulin receptor substrate-1 (Keller and Lienhard, 1994) (IRS-1) and the Src and collagen homologous protein (Shc) (Kovacina and Roth, 1993), which, when phosphorylated on tyrosine residues, serve as docking sites for other signalling molecules containing phosphotyrosine binding Src homology 2 (SH2) domains (Kock et al., 1991). An important component of the insulin-signalling network that binds to the tyrosine phosphorylated IRS-1 is phosphatidylinositol 3-kinase (PI 3-kinase), a kinase which phosphorylates phosphoinositides at the D-3 position and also posseses serine kinase activity (Fry, 1994; Lam et al., 1994). PI 3-kinase consists of two subunits, a regulatory 85-kD (p85) subunit which contains two SH2 domains and mediates the interaction of the enzyme with IRS-1, and a catalytic 110-kD (p110) subunit which contains the lipid and protein kinase activity (Fry, 1994). Binding of p85 SH2 domains to IRS-1 leads to the stimulation of PI 3-kinase activity and initiation of a multitude of downstream effects. We and others have shown that PI 3-kinase is a mediator of the insulin stimulation of glucose transport in insulin-responsive cells (Cheatham et al., 1994; Clarke et al., 1994; Okada et al., 1994; Tsakiridis et al., 1995). Stimulation of glucose transport is facilitated by the translocation of glucose transporters from an intracellular storage pool to the plasma membrane (Cushman and Wardzala, 1980; Suzuki and Kono, 1980; Mitsumoto and Klip, 1992; Guma et al., 1995) (see Fig. 1). GLUT4 is the main insulin responsive glucose transporter isoform of skeletal mus-

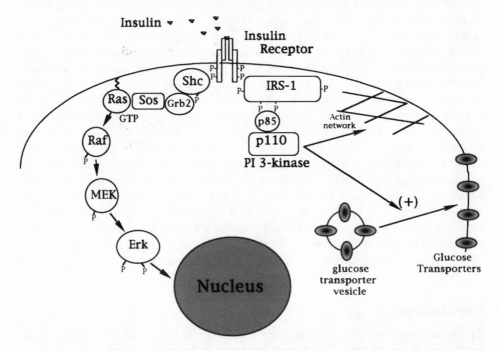

Figure 1.   The insulin signalling network. The activated insulin receptor initiates signalling cascades that result in a wide spectrum of biological responses including cytoskeletal changes, stimulation of glucose uptake, and induction of gene expression (see text).

cle and adipose cells but the GLUT1 (Gould et al., 1989; Calderhead et al., 1990; Mitsumoto and Klip, 1992; Tsakiridis et al., 1994) and the GLUT3 (Bilan et al., 1992; Tsakiridis et al., 1994, 1995) glucose transporter isoforms expressed in muscle and adipose cell lines also respond to insulin by translocation.

PI 3-kinase appears also to be involved in the insulin-induced changes in cell morphology. Fibroblasts as well as muscle and adipose cells in culture respond to insulin with a rapid reorganization of their actin network which induces ruffling of their plasma membrane (Ridley and Hall, 1992; Ridley et al., 1992; Kotani et al., 1994; Tsakiridis et al., 1994). Actin reorganization and membrane ruffling has been shown to be controlled by the small GTP binding protein Rac (Ridley et al., 1992; Hall, 1994). Recent studies have shown that Rac in its GTP-bound form may interact directly with PI 3-kinase (Hawkins, 1995) and function downstream of PI 3-kinase in the mediation of the insulin-induced membrane ruffling (Kotani et al., 1994, 1995) (see Fig. 1). In addition to cortical actin and membrane ruffles, cells maintain stress fibers which allow them to pull against the adhesion substrate and organize their morphology. The number of stress fibers as well as the formation of focal adhesion complexes are regulated by another small GTP binding protein, Rho (Hall, 1994). The role of stress fibers and of actin reorganization and membrane ruffling in the mediation of insulin action has not been elucidated.

Another signalling cascade distinct from the IRS-1/PI 3-kinase route has also been shown to respond to insulin and to mediate events that lead mainly to gene expression and mitogenesis (see Fig. 1). The tyrosine phosphorylation of Shc and IRS-1 by the insulin receptor results in their association with the small adaptor protein Grb2, which is bound to the guanine nucleotide exchange factor Sos (White and Khan, 1994). These events lead to the activation of the small GTP binding protein Ras (Polakis and McCormick, 1993) and initiation of a cascade of protein kinases which ultimately result in the activation of the mitogen-activated protein kinases (MAPKs) Erk 1 and 2 (Cobb and Goldsmith, 1995; Malarkey et al., 1995; Seedorf, 1995). Many of the putative targets of Erk are proteins that are localized in the nucleus and are involved in the regulation of transcription. For example, Erk phosphorylates c-Fos, c-Myc, and p62[TCF] (ternary complex factor) (Gille et al., 1992; Gupta et al., 1993; Marais, 1993).

Recent findings have suggested that the actin network is involved in the propagation of signals activated by integrins (Chen et al., 1994; Ezumi et al., 1995; Morino et al., 1995) and other cell surface receptors such as platelet-derived growth factor (PDGF) and epidermal growth factor (EGF) receptors (Weernink and Rijksen, 1995). However, the involvement of the actin cytoskeleton in the propagation of insulin signalling cascade and the mediation of its effects has not received equal attention. Our work has focused on the involvement of the actin network in the mediation of the insulin-induced stimulation of glucose transport and the insulin-induced activation of the MAPK cascade that may lead to regulation of gene expression. Here we present data suggesting that the actin network of insulin-responsive cells is required for stimulation of glucose transport and for the activation of the Ras/MAPK cascade.

Materials and Methods
Materials

α-Minimum essential medium (α-MEM), fetal bovine serum, and antibiotic/antimycotic solution were obtained from GIBCO/BRL (Burlington, Canada). Cytochalasins D and B, 2-deoxy-D-glucose (2DG), paraformaldehyde, and polyacrylamide were obtained from Sigma Chemical Co. (St. Louis, MO). 2-Deoxy-^{3}H-D-glucose (^{3}H-2DG) was purchased from DuPont NEN (Boston, MA). Human insulin was obtained from Eli Lilly Co., Scarborough, Canada. Rhodamine-labeled phalloidin was purchased from Molecular Probes (Eugene, OR). All electrophoresis and immunoblotting reagents were obtained from BioRad Laboratories (Richmond, CA). All other chemicals were obtained from Sigma Chemical Co. Polyclonal and monoclonal (1F8) anti-GLUT4 glucose transporter antibodies for immunoblotting and immunoprecipitation, respectively, were purchased from East Acres (Southbridge, MA). Polyclonal antibodies for immunoblotting and immunoprecipitations against IRS-1, the 85-kD regulatory subunit of PI 3-kinase (p85), and the monoclonal antibody against phosphotyrosine were purchased from UBI (Lake Placid, NY). Protein A Sepharose was purchased from Pharmacia (Uppsala, Sweden), and M280 magnetic beads were from Dynal (Oslo, Norway).

Cell Culture, Incubations, Glucose Uptake, and Fluorescence Microscopy

Monolayers of L6 muscle cells or 3T3-L1 adipocytes were grown as described earlier (Mitsumoto et al., 1993; Sargeant and Paquet, 1993). Incubations were performed as described in the figure legends. Glucose uptake was performed as described previously (Walker, 1990). Fluorescence microscopy experiments were performed as described earlier (Tsakiridis, 1994).

Membrane Isolation and Immunoblotting

Plasma membranes were isolated as described earlier (Bilan et al., 1992). Membrane samples (20 μg) were subjected to SDS-polyacrylamide gel electrophoresis essentially according to Laemmli (Laemmli, 1970).

Immunoprecipitation of IRS-1, and PI 3-kinase Assay

IRS-1 was immunoprecipitated as described previously (Tsakisidis et al., 1994). The PI 3-kinase activity assay was performed as described earlier (Tsakiridis et al., 1995).

Immunoisolation of GLUT4 Glucose Transporter-containing Vesicles

GLUT4-containing vesicles from 3T3-L1 adipocytes were immunoisolated from low density microsomes (LDM, prepared as described earlier: Bilan et al., 1992) essentially by a modification of the protocol of Laurie et al. (1993), using a monoclonal anti-GLUT4 antibody (1F8) coupled with magnetic beads (Dynal M500) (Volchuk et al., 1995).

Detection of Erk Phosphorylation in Whole Cell Lysates

Whole cell lysates were prepared according to Lamphere et al. (Lamphere and Lienhard, 1992).

Statistical Analysis

Statistical analysis was performed using the ANOVA test (Fisher, multiple comparisons).

Results and Discussion

Effect of Insulin and Cytochalasin D on the Organization of the Myotube Actin Network

In fibroblasts, many growth factors, including EGF, PDGF, as well as insulin, induce a rapid actin filament polymerization and fiber aggregation under the cell surface which cause ruffling of the plasma membrane (Goshima et al., 1984; Ridley and Hall, 1992; Ridley et al., 1992). These events have not been studied in typical insulin-responsive cells such as muscle and adipose tissues. We hypothesized that the cytoskeletal events activated by insulin may be an essential step in insulin action in muscle and fat cells. We first examined the effect of insulin on the organization of the actin network in insulin-responsive differentiated L6 muscle cells (myotubes). As shown in Fig. 2, labeling of myotubes with rhodamine-labeled phalloidin, which specifically binds to filamentous actin, revealed the existence of an organized stress fiber network which spans the entire length of the cells. Insulin caused a marked reorganization of the actin network under the plasma membrane of myotubes which could be detected as large aggregates of phalloidin-labeled filaments. By confocal microscopy, actin aggregates were localized to the surface of the cell. Using differential interference contrast (DIC) microscopy, membrane ruffles could be detected as early as 20 s after addition of insulin (K. Foskett, T.Tsakiridis and A. Klip, unpublished observations).

The above findings led us to hypothesize that actin reorganization may be a generalized action of insulin essential for the transmission of its intracellular signals. Therefore, we investigated the role of the actin network on several actions of insulin using cytochalasin D (CD) as an experimental tool. Fig. 3 shows that CD treatment of myotubes (2 h with 1 μM CD) caused complete disassembly of the actin network. The effect of CD on actin filaments was dose dependent (results not shown). Maximum actin disassembly was observed at a concentration of 1 μM after 2 h treatment. No intact stress fibers could be detected in CD-treated cells. Moreover, no aggregation of filamentous actin could be detected in response to insulin in CD-pretreated myotubes (results not shown). Despite its marked effect on the actin network, CD did not appear to affect cell viability since its effects were fully revers-

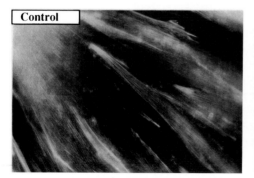 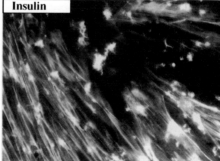

Figure 2. Insulin causes reorganization of the actin network in L6 myotubes. Serum-deprived L6 myotubes were stimulated without (control) or with insulin (10^{-7} M for 30 min). The actin network was stained with rhodamine-labeled phalloidin. Reprinted with permission from *The Journal of Biological Chemistry*.

ible within 4 h of removal of the drug (Fig. 3). Thus CD is an effective tool to disassemble the actin network, and therefore we used this compound to examine the role of actin filaments in insulin action.

Effect of Actin Disassembly on the Insulin Stimulation of Glucose Transport and the Distribution of Glucose Transporters

We first examined the effect of CD on the stimulation of glucose transport by insulin as well as on the basal rate of transport (Tsakiridis et al., 1994). As shown in Fig. 4, pretreatment of myotubes with CD, under conditions that caused complete actin disassembly (1 μM for 2 h), almost completely prevented insulin stimulation of glucose transport. CD pretreatment did not have any significant effect on the basal rate of glucose transport. These results suggested that an intact actin network is required for insulin to mediate stimulation of glucose uptake.

As mentioned above, insulin stimulates the rate of glucose transport into muscle and fat cells by a vesicle-mediated translocation of glucose transporters from intracellular storage sites to the plasma membrane. Involvement of the actin network in the stimulation of membrane transport events such as secretion, endocytosis, and intracellular organelle traffic has been reported for different cell types (Adams and Pollard, 1986; Hirokawa et al., 1989; Castellino et al., 1992; Kuznetsov et al., 1992; Gottlieb et al., 1993; Reizman, 1993). Hence, we hypothesized that actin disassembly might inhibit the insulin stimulation of glucose transport through inhibition of

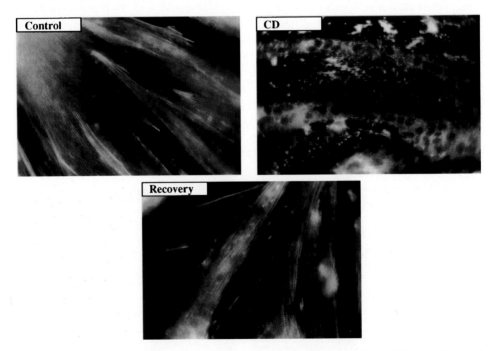

Figure 3. Disassembly of the actin network of L6 myotubes by cytochalasin D. Serum-deprived L6 myotubes were treated without (*Control*), with 1 μM cytochalasin D (*CD*) for 2 h or were recovering for 4 h in serum-containing medium after a 2-h treatment with 1 μM cytochalasin D (*Recovery*). The actin network was stained with rhodamine-labeled phalloidin. Reprinted with permission from *The Journal of Biological Chemistry*.

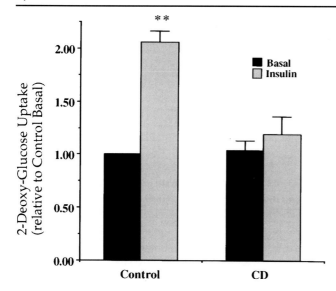

Figure 4. Inhibition of the insulin stimulation of glucose transport by cytochalasin D pretreatment. Serum-deprived L6 myotubes were incubated without (*Control*) or with 1 μM cytochalasin D (*CD*) for 2 h, without (*basal*) or with 10^{-7} M insulin during the last 30 min of this period. 2-Deoxy-D-glucose transport was then measured. All glucose transport values are expressed in relative units relative to the basal control condition. **$P < 0.01$ relative to the basal control.

glucose transporter recruitment to the plasma membrane. To investigate this possibility we performed subcellular fractionation studies which allowed isolation of plasma membranes from control, insulin-, and CD and insulin-treated myotubes. As shown in Fig. 5, insulin induced the recruitment of the GLUT4 glucose transporters to the plasma membrane. Actin disassembly by CD did not affect significantly the basal amount of GLUT4 at the plasma membrane; however, it abolished the insulin-induced recruitment of transporters. Similar effects of CD and insulin were observed for the GLUT1 and GLUT3 glucose transporter isoforms expressed in these cells (not shown, see Tsakiridis et al., 1994).

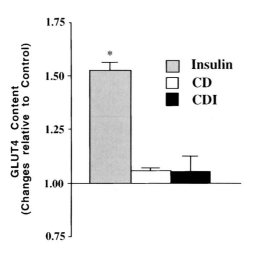

Figure 5. Inhibition of the insulin stimulation of GLUT4 glucose transporter recruitment to the plasma membrane by cytochalasin D pretreatment. Serum-deprived L6 myotubes were either treated with insulin for 30 min (*insulin*), 1 μM cytochalasin D (*CD*) for 2 h, or pretreated with cytochalasin D for 2 h followed by insulin treatment in the last 30 min of this incubation (*CDI*). Subcellular fractionation and isolation of plasma membranes were performed. 20 μg of protein from the isolated plasma membrane were subjected to SDS-PAGE and immunoblotted with anti-GLUT4 antibody. The immunoblots were quantitated, normalized to the value of the untreated control, and are expressed as changes above this value. *$P < 0.05$ relative to the untreated control.

It is conceivable that the effects of CD treatment on the insulin stimulation of glucose transport were due to generalized effects of the drug on the organization of all subcellular membrane structures and not due to a specialized effect on glucose transporters. Two lines of evidence indicate that this is probably not the case: (i) CD does not affect the subcellular distribution of a typical plasma membrane protein, the $\alpha 1$ subunit of the Na^+/K^+ ATPase, and (ii) the drug did not alter the general protein profile in either the plasma membrane or the internal membrane fraction (not shown, see Tsakiridis et al., 1994).

Effect of Actin Disassembly on the Propagation of Insulin Signalling Cascade

Studies using inhibitors of PI 3-kinase such as wortmannin (Clarke et al., 1994; Shimizu and Shimizu, 1994; Tsakiridis et al., 1995; Yang et al., 1996) and LY294002 (Cheatham et al., 1994) have shown that the stimulation of glucose transport by insulin and the recruitment of glucose transporters to the plasma membrane requires the activity of PI 3-kinase. Furthermore, microinjection of a dominant negative mutant of the p85 subunit of PI 3-kinase, which lacks a binding site for the catalytic p110 subunit, inhibited glucose transporter translocation induced by insulin (Kotani et al., 1995). These observations indicate that PI 3-kinase is necessary for insulin-stimulated glucose transporter translocation and glucose transport.

To investigate whether the effect of CD on glucose transport and glucose transporter translocation could be due to inhibition of the insulin-signalling cascade, we examined the effect of CD on the insulin-induced tyrosine phosphorylation of IRS-1 and its association with PI 3-kinase. Fig. 6 shows that insulin resulted in tyrosine phosphorylation of IRS-1, and CD did not alter this event. Furthermore, CD did not affect the insulin-mediated coimmunoprecipitation of p85 PI 3-kinase with IRS-1, suggesting that actin disassembly does not affect the initiation of insulin signalling leading to PI 3-kinase activation. However, these results did not ascertain whether the stimulation of PI 3-kinase activity, which indeed takes place through its association with IRS-1, is still intact after actin disassembly. Therefore, we immunoprecipitated IRS-1 and examined the IRS-1-associated PI 3-kinase activity in vitro.

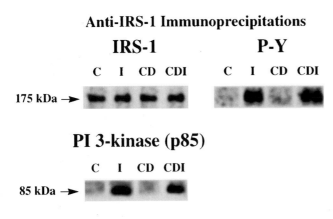

Figure 6. The effect of insulin and cytochalasin D treatment on the tyrosine phosphorylation of IRS-1 and binding of phosphatidylinositol 3-kinase (p85) to tyrosine phosphorylated IRS-1. IRS-1 was immunoprecipitated from control (*C*), insulin-stimulated (*I*), cytochalasin D–treated (*CD*), and cytochalasin D–treated cells stimulated with insulin (*CDI*). The cells were treated with cytochalasin D for 2 h and stimulated with insulin for only 3 min. The pellets of the immunoprecipitation were split into two equal parts, separated on two gels by SDS-PAGE, and probed by immunoblotting using antibodies against IRS-1, phosphotyrosine (*P-Y*), and p85. Reprinted with permission from *The Journal of Biological Chemistry*.

Table I shows that insulin caused a marked stimulation of the IRS-1-associated PI 3-kinase activity and that CD did not inhibit this event.

Effect of Insulin and CD on the Association of PI 3-kinase Activity with the GLUT4 Vesicle in 3T3-L1 Adipocytes

The above results showed that the most distal signalling event known to be involved in the insulin stimulation of glucose transport, namely, the activation of PI 3-kinase and its association with IRS-1, was unaffected by actin disassembly. However, recent studies (Heller-Harrison et al., 1996; Yang et al., 1996) have proposed a model for the stimulation of glucose transport which includes not only activation of PI 3-kinase by binding to IRS-1 but also translocation of the IRS-1/PI 3-kinase complex to the intracellular glucose transporter vesicle (see Fig. 7 *A*). Additionally, a recent study (Ricort et al., 1996) showed that this translocation of PI 3-kinase is not induced by PDGF which also activates PI 3-kinase very potently. PDGF does not stimulate glucose transport, supporting the notion that the PI 3-kinase recruitment to the glucose transporter vesicle may be required for glucose transporter translocation to the plasma membrane. Thus, we investigated whether the occurence of PI 3-kinase recruitment to intracellular GLUT4-containing vesicles is affected by actin disassembly. For these studies we used 3T3-L1 adipocytes, which is the system used in the studies quoted above. GLUT4 glucose transporter-containing vesicles were immunoisolated from control, insulin-, CD- alone, or CD and insulin-treated adipocytes. The GLUT4-associated PI 3-kinase activity was measured in vitro and the results of these experiments are presented in Table II. Insulin treatment of adipocytes for 3.5 min significantly increased the levels of PI 3-kinase activity associated with immunoisolated GLUT4 vesicles. Interestingly, this event was blocked by actin disassembly with CD. Fig. 7 *B* illustrates a model for the effects of actin disassembly on the mechanism of stimulation of glucose transport. It is tempting to speculate that the mechanism by which CD inhibits the stimulation of glucose transport may be the prevention of recruitment of PI 3-kinase to the glucose transporter vesicles which in turn prevents intracellular glucose transporters from departing from the storage compartment. Although no direct evidence is currently available to support such a hypothesis, the production of D-3 phosphorylated inositides may be essential for vesicle budding and release of glucose transports from their storage compartment.

TABLE I
IRS-1–associated PI 3-kinase Activity

Control	Insulin	CD	CDI
1.0 ± 0.0	10.1 ± 0.1	1.2 ± 0.1	17.6 ± 3.6

Effect of insulin and cytochalasia D treatments on the level of phosphatidylinositol 3-kinase activity associated with IRS-1. IRS-1 was immunoprecipitated from control, insulin-, cytochalasin D- (CD) alone, or CD and insulin (CDI)-treated myotubes as described in Materials and Methods. In Vitro PI 3-kinase activity assays were then performed and the products were separated by thin layer chromatography. Quantitation of ^{32}P-phosphatidylinositol 3-phosphate produced during the assay was performed with a PhosphorImager system. Average values of three independent experiments are expressed in relative units as fold change relative to the control untreated condition.

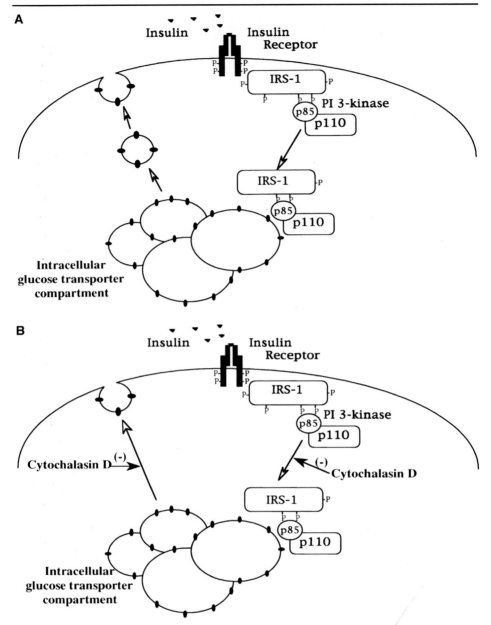

Figure 7. (*A*) A proposed model for the role of PI 3-kinase in glucose transporter translocation and stimulation of glucose uptake. IRS-1 becomes tyrosine-phosphorylated upon insulin stimulation and associates with PI 3-kinase. A population of IRS-1/PI 3-kinase complexes are delivered to an intracellular glucose transporter compartment. The activity of PI 3-kinase in glucose transporter vesicles may cause or allow budding, fusion, or movement of these membranes which regulates glucose transporter translocation to the plasma membrane. (*B*) A proposed model for the effects of actin disassembly on glucose transporter translocation and stimulation of glucose uptake. Cytochalasin D inhibits the delivery of IRS-1/PI 3-kinase complexes to an intracellular glucose transporter compartment, thus inhibiting the appearance of glucose transporters at the plasma membrane.

TABLE II
GLUT4-associated P I3-kinase Activity

Control	Insulin	CD	CDI
1.0 ± 0.0	$2.2 \pm 0.2^*$	0.85 ± 0.1	$1.3 \pm 0.1^{\ddagger}$

Effect of insulin and cytochalasin D treatment on the level of phosphatidylinositol 3-kinase activity associated with intracellular GLUT4 glucose transporter containing vesicles. GLUT4-containing vesicles were immunoisolated from low density microsomal fractions of control, insulin-, cytochalasin D- (CD) alone, or CD and insulin- (CDI)-treated 3T3-L1 adipocytes, by monoclonal anti-GLUT4 antibody-coupled magnetic beads as described in Materials and Methods. In vitro PI 3-kinase activity assays were then performed and the products were separated by thin layer chromatography. Quantitation of ^{32}P-phosphatidylinositol 3-phosphate produced during the assay was performed with a PhosphorImager system. Average values of three independent experiments are expressed in relative units as fold change relative to the control untreated condition. *Significantly different ($P < 0.02$) compared to control. ‡Significantly different ($P < 0.02$) compared to CD.

Effect of Cytochalasin D on Erk Phosphorylation

As shown in Fig. 1, in addition to PI 3-kinase and its downstream targets, insulin also activates the Ras/Erk MAPK pathway resulting in stimulation of gene expression and mitogenesis. Therefore, we investigated the effect of actin disassembly on

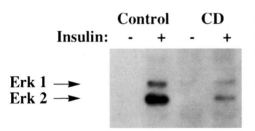

Anti-PY immunoblot

Figure 8. Inhibition of the insulin-stimulated Erk1 and Erk2 tyrosine phosphorylation by cytochalasin D treatment. Serum-deprived L6 myotubes were incubated without (*Control*) or with 1 μM cytochalasin D (*CD*) for 2 h; where indicated, cells were stimulated with 10^{-7} M insulin in the last 5 min of this incubation. Cells were then lysed, and Erk phosphorylation was detected in whole cell lysates using an anti-phosphotyrosine (*PY*) antibody.

the insulin-stimulated tyrosine phosphorylation of Erks. Tyrosine phosphorylation correlates with enzymatic activation (Skolnik et al., 1993). Fig. 8 shows that in L6 myotubes, insulin induces a rapid (in 5 min) tyrosine phosphorylation of the Erk 1 and Erk 2 MAPKs. Erk1 and Erk2, shown in Fig. 8, are detected with an antiphosphotyrosine antibody. The bands shown have been tentatively identified as Erk1 and Erk2 using immunoprecipitation experiments combined with immunoblotting, and activity assays using myelin basic protein as a substrate (Tsakiridis, T., and A. Klip, unpublished observations). Disassembly of the actin network with CD almost completely abolished the ability of insulin to induce phosphorylation of these enzymes. Hence, an intact actin network is required for the phosphorylation of the Erk MAPKs by insulin and may therefore be needed for the insulin-mediated stimulation of mitogenesis and gene expression.

Conclusion

The results of these studies showed for the first time that an organized actin network is required to mediate the propagation of the insulin signalling cascade in muscle cells. Actin filaments appear to facilitate (i) the insulin-induced redistribution of PI 3-kinase to the intracellular glucose transporter vesicles and (ii) the activation of MAPK by insulin. Future studies should investigate which specific intracellular actin-containing structures are involved in the propagation of insulin action as well as the exact molecular interactions between insulin-signalling molecules and actin filaments.

Acknowledgments

This work was supported by grant MT-7307 from the Medical Research Council of Canada to A. Klip .

References

Adams, R.J., and T.D. Pollard. 1986. Propulsion of organelles isolated from Acanthamoeba along actin filaments by myosin-I. *Nature.* 322:754–756.

Bilan, P.J., Y. Mitsumoto, F. Maher, I.A. Simpson, and A. Klip. 1992. Detection of the GLUT3 glucose transporter in rat L6 muscle cells: regulation by cellular differentiation, insulin and insulin-like growth factor I. *Biochem. Biophys. Res. Commun.* 186:1129–1137.

Bilan, P.J., Y. Mitsumoto, T. Ramlal, and A. Klip. 1992. Acute and long-term effects of insulin-like growth factor I on glucose transporters in muscle cells. Translocation and biosynthesis. *FEBS Lett.* 298:285–290.

Calderhead, D.M., K. Kitagawa, L.I. Tanner, G.D. Holman, and G.E. Lienhard. 1990. Insulin regulation of the two glucose transporters in 3T3-L1 adipocytes. *J. Biol. Chem.* 265:13801–13808.

Castellino, F., J. Heuser, S. Marchetti, B. Bruno, and A. Luini. 1992. Glucocorticoids stabilization of actin filaments: a possible mechanism for inhibition of corticotropin release. *Proc. Natl. Acad. Sci. USA.* 89:3775–3779.

Cheatham, B., C.J. Vlahos, L. Cheatham, L. Wang, J. Blenis, and C.R. Khan. 1994. Phosphatidylinositol 3-kinase activation is required for insulin stimulation of pp70S6K, DNA synthesis, and glucose transporter translocation. *Mol. Cell Biol.* 14:4902–4911.

Chen, Q., M.S. Kinch, T.H. Lin, K. Burridges, and R.L. Juliano. 1994. Integrin-mediated cell adhesion activates mitogen-activated protein kinases. *J. Biol. Chem.* 269:26602–26605.

Clarke, J.F., P.W. Young, K. Yonezawa, M. Kasuga, and G.D. Holman. 1994. Inhibition of the translocation of GLUT1 and GLUT4 in 3T3-L1 cells by the phosphatidylinositol 3-kinase inhibitor, wortmannin. *Biochem. J.* 300:631–635.

Cobb, M.H., and E.J. Goldsmith. 1995. How MAP kinases are regulated. *J. Biol. Chem.* 270:14843–14846.

Cushman, S.W., and L.J. Wardzala. 1980. Potential mechanism of insulin action on glucose transport in the isolated rat adipose cell. *J. Biol. Chem.* 255:4758–4762.

Ezumi, Y., H. Takayama, and M. Okuma. 1995. Differential regulation of protein-tyrosine phosphatases by integrin $\alpha IIb\beta 3$ through cytoskeletal reorganization and tyrosine phosphorylation in human platelets. *J. Biol. Chem.* 270:11927–11934.

Fry, M.J. 1994. Stucture, regulation and function of phosphoinositide 3-kinases. *Biochim. Biophys. Acta.* 1226:237–268.

Gille, H., A.D. Sharrocks, and P.E. Shaw. 1992. Phosphorylation of transcription factor p62 TCF by MAP kinase stimulates ternary complex formation by c-fos promoter. *Nature.* 358:414–417.

Goshima, K., A. Masuda, and K. Qwaribe. 1984. Insulin-induced formation of ruffling membranes of KB cells and its correlation with enhancement of amino acid transport. *J. Cell Biol.* 98:801–809.

Gottlieb, T.A., I.E. Ivanov, M. Adensik, and D. Sabatini. 1993. Actin microfilaments play a critical role in endocytosis at the apical but not the basolateral surface of polarized epithelial cells. *J. Cell Biol.* 120:695–710.

Gould, G.W., V. Derechin, D.E. James, K. Tordjman, S. Ahern, E.M. Gibbs, G.E. Lienhard, and M. Mueckler. 1989. Insulin-stimulated translocation of the HepG2/erythrocyte-type glucose transporter expressed in 3T3-L1 adipocytes. *J. Biol. Chem.* 264:2180–2184.

Guma, A., J.R. Zierath, H. Wallberg-Henriksson, and A. Klip. 1995. Insulin induces translocation of GLUT-4 glucose transporters in human skeletal muscle. *Am. J. Physiol.* 268:E613–E622.

Gupta, S., A. Seth, and R.J. Davis. 1993. Transactivation of gene expression by Myc is inhibited by mutation at the phosphorylation sites Thr-58 and Ser-62. *Proc. Natl. Acad. Sci. USA.* 90:3216–3220.

Hall, A. 1994. Small GTP-binding proteins and the regulation of the actin cytoskeleton. *Annu. Rev. Cell. Biol.* 10:31–54.

Hawkins, P.T., A. Equinoa, R.-Q. Qiu, D. Stokoe, F. Cooke, R. Walters, S. Wennstrom, L. Claesson-Welch, T. Evans, M. Symons, and L. Stephens. 1995. PDGF stimulates an increase in GTP-Rac via activation of phosphoinositide 3-kinase. *Curr. Biol. Opin.* 5:393–403.

Heller-Harrison, R.A., M. Morin, A. Guilherme, and M.P. Czech. 1996. Insulin-mediated targeting of phosphatidylinositol 3-kinase to GLUT4-containing vesicles. *J. Biol. Chem.* 271:10200–10204.

Hirokawa, N., K. Sobue, K. Kanda, A. Harada, and H. Yorifuji. 1989. The cytoskeletal architecture of the presynaptic terminal and molecular structure of synapsin 1. *J. Cell. Biol.* 108:111–126.

Keller, S.R., and G.E. Lienhard. 1994. Insulin signalling: the role of insulin receptor substrate 1. *Trends Cell Biol.* 4:115–118.

Koch, C.A., D. Anderson, M.F. Moran, C. Ellis, and T. Pawson. 1991. SH2 and SH3 domains: elements that control interactions of cytoplasmic signalling proteins. *Science.* 252:668–674.

Kotani, K., A.J. Carozzi, H. Sakaue, K. Hara, L.J. Robinson, S.F. Clark, K. Yonezawa, D.E. James, and M. Kasuga. 1995. Requirement for phosphoinositide 3-kinase in insulin-stimulated GLUT4 translocation in 3T3-L1 adipocytes. *Biochem. Biophys. Res. Commun.* 209:343–348.

Kotani, K., K. Hara, K. Kotani, K. Yonezawa, and M. Kasuga. 1995. Phosphoinositide 3-kinase as an upstream regulator of the small GTP-binding protein Rac in the insulin signaling of membrane ruffling. *Biochem. Biophys. Res. Commun.* 208:985–990.

Kotani, K., K. Yonezawa, K. Hara, H. Ueda, Y. Kitamura, H. Sakaue, A. Ando, A. Chavanieu, B. Calas, F. Grigorescu et al. 1994. Involvement of phosphoinositide 3-kinase in insulin- or IGF-1-induced membrane ruffling. *EMBO J.* 13:2313–2321.

Kovacina, K.S., and R.A. Roth. 1993. Identification of SHC as a substrate of the insulin receptor kinase distinct from the GAP-associated 62 KDa tyrosine phosphoprotein. *Biochem. Biophys. Res. Commun.* 192:1303–1311.

Kuznetsov, S., G.M. Langford, and D.G. Weiss. 1992. Actin-dependent organelle movement in squid axoplasm. *Nature.* 356:722–725.

Laemmli, U.K. 1970. Cleavage of structural proteins during the assembly of the head of bactereophage T4. *Nature.* 227:680–685.

Lam, K., C.L. Carpenter, N.B. Ruderman, J.C. Friel, and K.L. Kelly. 1994. The phosphatidylinositol 3-kinase serine kinase phosphorylates IRS-1. *J. Biol. Chem.* 269:20648–20652.

Lamphere, L., and G.E. Lienhard. 1992. Components of signaling pathways for insulin and insulin-like growth factor-1 in muscle myoblasts and myotubes. *Endocrinology.* 131:2196–2202.

Laurie, S.M., C.C. Cain, G.E. Lienhard, and J.D. Castle. 1993. The glucose transporter GLUT A and secreting carrier membrane proteins (SCAMPS) colocalize in rat adipocytes and partially segregate during insulin stimulation. *J. Biol. Chem.* 268:19110–19117.

Malarkey, K., C.M. Belham, A. Paul, A. Graham, A. McLees, P.H. Scott, and R. Plevin. 1995. The regulation of tyrosine kinase signalling pathways by growth factor and G-protein-coupled receptors. *Biochem. J.* 309:361–375.

Marais, R., J. Wynne, and R. Treisman. 1993. The SRF accessory pyotein Elk-1 contains a growth factor-regulated transcriptional activation domain. *Cell.* 73:381–393.

Mitsumoto, Y., and A. Klip. 1992. Developmental regulation of the subcellular distribution and glycosylation of GLUT1 and GLUT4 glucose transporters during myogenesis of L6 muscle cells. *J. Biol. Chem.* 267:4957–4962.

Mitsumoto, Y., Z. Liu, and A. Klip. 1993. Regulation of glucose and ion transporter expression during controlled differentiation and fusion of rat skeletal muscle cells. *Endocrine J.* 1:307–315.

Morino, N., T. Mimura, K. Hamasaki, K. Tobe, K. Ueki, K. Kikuchi, K. Takehara, T. Kadowaki, Y. Yazaki, and Y. Nojima. 1995. Matrix/integrin interaction activates the mitogen-activated protein kinase, p44erk-1 and p42erk-2. *J. Biol. Chem.* 270:269–273.

Okada, T., Y. Kawano, T. Sakakibara, O. Hazeki, and M. Ui. 1994. Essential role of phosphatidylinositol 3-kinase in insulin-induced glucose transport and antilipolysis in rat adipocytes. *J. Biol. Chem.* 269:3568–3573.

Polakis, P., and F. McCormick. 1993. Structural requirements for the interaction of p21ras with GAP, exchange factors, and its biological effector target. *J. Biol. Chem.* 268:9157–9160.

Reizman, H. 1993. Yeast endocytosis (Review). *Trends Cell Biol.* 3:273–277.

Ricort, J.-M., J.-F. Tanti, E. Van Obberghen, and Y. Le Marchand-Brustel. 1996. Different effects of insulin and platelet-derived growth factor on phosphatidylinositol 3-kinase at the subcellular level in 3T3-L1 adipocytes: a possible explanation for their specific effects on glucose transport. *Eur. J. Biochem.* 239:17–22.

Ridley, A.J., and A. Hall. 1992. The small GTP-binding protein rho regulates assembly of focal adhesions and actin stress fibers in response to growth factors. *Cell.* 70:389–399.

Ridley, A.J., H.F. Paterson, C.L. Johnston, D. Diekmann, and A. Hall. 1992. The small GTP-binding protein rac regulates growth factor-induced membrane ruffling. *Cell.* 70:401–410.

Sargeant, R.J., and M.R. Paquet. 1993. Effect of insulin on the rates of synthesis and degradation of GLUT1 and GLUT4 glucose transporters in 3T3-L1 adipocytes. *Biochem. J.* 290:913–919.

Seedorf, K. 1995. Intracellular signalling by growth factors. *Metabolism.* 44:24–32.

Shimizu, Y., and T. Shimizu. 1994. Effects of wortmannin on increased glucose transport by insulin and norepinephrine in primary culture of brown adipocytes. *Biochem. Biophys. Res. Commun.* 202:660–665.

Skolnik, E.Y., A. Batzer, N. Li, C.H. Lee, E. Lowenstein, M. Mohammadi, B. Margolis, and J. Schlessinger. 1993. The function of GRB2 in linking the insulin receptor to Ras signalling pathways. *Science.* 260:1953–1955.

Suzuki, K., and T. Kono. 1980. Evidence that insulin causes translocation of glucose transport activity to the plasma membrane from an intracellular storage site. *Proc. Natl. Acad. Sci. USA.* 77:2542–2545.

Tsakiridis, T., H.E. McDowell, T. Walker, C.P. Downes, H.S. Hundal, M. Vranic, and A. Klip. 1995. Multiple roles of phosphatidylinositol 3-kinase in regulation of glucose transport, amino acid transport, and glucose transporters in L6 skeletal muscle cells. *Endocrinology.* 136:4315–4322.

Tsakiridis, T., M. Vranic, and A. Klip. 1994. Disassembly of the actin network inhibits insulin-dependent stimulation of glucose transport and prevents recruitment of glucose transporters to the plasma membrane. *J. Biol. Chem.* 269:29934–29942.

Volchuk, A., R. Sargeant, S. Sumitani, Z. Liu, L. He, and A. Klip. 1995. Cellubrevin is a resident protein of insulin-sensitive GLUT4 glucose transporter vesicles in 3T3-L1 adipocytes. *J. Biol. Chem.* 270:8233–8240.

Walker, P.S., T. Ramlal, V. Sarabia, U.M. Koivisto, P.J. Bilan, J.E. Pessin, and A. Klip. 1990. Glucose transport activity in L6 muscle cells is regulated by the coordinate control of subcellular glucose transporter distribution, biosynthesis, and mRNA transcription. *J. Biol. Chem.* 265:1516–1523.

Weernink, P.A.O., and G. Rijksen. 1995. Activation and translocation of c-Src to the cytoskeleton by both platelet-derived growth factor and epidermal growth factor. *J. Biol. Chem.* 270:2264–2267.

White, M.F., and C.R. Khan. 1994. The insulin signaling system. *J. Biol. Chem.* 269:1–4.

Yang, J., J.F. Clarke, C.J. Ester, P.W. Young, M. Kasuga, and G.D. Holman. 1996. Phosphatidylinositol 3-kinase acts at an intracellular membrane site to enhance GLUT4 exocytosis in 3T3-L1 cells. *Biochem. J.* 313:125–131.

List of Contributors

Marvin E. Adams, Department of Physiology, University of North Carolina, Chapel Hill, North Carolina 27599-7545

Kurt J. Amann, Graduate Program in Cellular and Molecular Biology, University of Wisconsin Medical School, Madison, Wisconsin 53706

Kenneth A. Beck, Department of Molecular and Cellular Physiology, Beckman Center for Molecular and Genetic Medicine, Stanford University School of Medicine, Stanford, California 94305-5426

Vann Bennett, Howard Hughes Medical Institute and Departments of Cell Biology and Biochemistry, Duke University Medical Center, Durham, North Carolina 27710

Gabriela Bezakova, Department of Physiology, School of Medicine, University of Maryland at Baltimore, Baltimore, Maryland 21201

Robert J. Bloch, Department of Physiology, School of Medicine, University of Maryland at Baltimore, Baltimore, Maryland 21201

Elayne A. Bornslaeger, R.H. Lurie Cancer Center, Northwestern University Medical School, Chicago, Illinois 60611

Jeffrey S. Chamberlain, Department of Human Genetics, The University of Michigan Medical School, Ann Arbor, Michigan 48109-0618

Richard E. Cheney, Department of Biology, Yale University, New Haven, Connecticut 06520

Catherine P. Chia, Worcester Foundation for Biomedical Research, Shrewsbury, Massachusetts 01545

Kathleen Corrado, Department of Human Genetics, The University of Michigan Medical School, Ann Arbor, Michigan 48109-0618

Gregory A. Cox, Department of Human Genetics, The University of Michigan Medical School, Ann Arbor, Michigan 48109-0618

Jonathan Q. Davis, Howard Hughes Medical Institute and Departments of Cell Biology and Biochemistry, Duke University Medical Center, Durham, North Carolina 27710

Patrick Doherty, Department of Experimental Pathology, UMDS, Guy's Hospital, London SE1 9RT, United Kingdom

James Dowling, Howard Hughes Medical Institute, Department of Molecular Genetics and Cell Biology, The University of Chicago, Chicago, Illinois 60637

Gregory Downey, Clinical Science, Department of Medicine, University of Toronto, Toronto, Canada M5S 1A8

Ronald R. Dubreil, Department of Pharmacological & Physiological Sciences and the Committees on Cell Physiology and Developmental Biology, University of Chicago, Chicago, Illinois 60637

James M. Ervasti, Department of Physiology, University of Wisconsin Medical School, Madison, Wisconsin 53706

Marcus Fechheimer, Department of Cellular Biology, University of Georgia, Athens, Georgia 30602

Richard G. Fehon, Developmental, Cell, and Molecular Biology Group, Department of Zoology, Duke University, Durham, North Carolina 27708-1000

Velia M. Fowler, Department of Cell Biology, MB24, The Scripps Research Institute, La Jolla, California 92037

Stanley C. Froehner, Department of Physiology, University of North Carolina, Chapel Hill, North Carolina 27599-7545

Elaine Fuchs, Howard Hughes Medical Institute, Department of Molecular Genetics and Cell Biology, The University of Chicago, Chicago, Illinois 60637

Heinz Furthmayr, Department of Pathology, Stanford University, Stanford, California 94305

Mikio Furuse, Department of Cell Biology, Faculty of Medicine, Kyoto University, Yoshida-Konoe, Sakyo-ku, Kyoto 606, Japan

Stephen H. Gee, Department of Physiology, University of North Carolina, Chapel Hill, North Carolina 27599-7545

Kathleen J. Green, R.H. Lurie Cancer Center, Northwestern University Medical School, Chicago, Illinois 60611

Sergio Grinstein, Division of Cell Biology, The Hospital for Sick Children, Toronto, Ontario, Canada M5G 1X8

Lifei Guo, Howard Hughes Medical Institute, Department of Molecular Genetics and Cell Biology, The University of Chicago, Chicago, Illinois 60637

Michael Hauser, Department of Human Genetics, The University of Michigan Medical School, Ann Arbor, Michigan 48109-0618

Anne L. Hitt, Worcester Foundation for Biomedical Research, Shrewsbury, Massachusetts 01545

Masahiko Itoh, Department of Cell Biology, Faculty of Medicine, Kyoto University, Yoshida-Konoe, Sakyo-ku, Kyoto 606, Japan

Amira Klip, Division of Cell Biology, The Hospital for Sick Children, Toronto, Ontario, Canada M5G 1X8

Panos Kouklis, Howard Hughes Medical Institute, Department of Molecular Genetics and Cell Biology, The University of Chicago, Chicago, Illinois 60637

Andrew P. Kowalczyk, R.H. Lurie Cancer Center, Northwestern University Medical School, Chicago, Illinois 60611

Johannes J. Krupp, Vollum Institute, Oregon Health Sciences University, Portland, Oregon 97201

Dennis LaJeunesse, Developmental, Cell, and Molecular Biology Group, Department of Zoology, Duke University, Durham, North Carolina 27708-1000

Rebecca Lamb, Developmental, Cell, and Molecular Biology Group, Department of Zoology, Duke University, Durham, North Carolina 27708-1000

Stephen Lambert, Howard Hughes Medical Institute and Departments of Cell Biology and Biochemistry, Duke University Medical Center, Durham, North Carolina 27710

Tze Hong Lu, Worcester Foundation for Biomedical Research, Shrewsbury, Massachusetts 01545

Carey Lumeng, Department of Human Genetics, The University of Michigan Medical School, Ann Arbor, Michigan 48109-0618

Elizabeth J. Luna, Worcester Foundation for Biomedical Research, Shrewsbury, Massachusetts 01545

Gary R. MacVicar, Department of Pharmacological & Physiological Sciences and the Committees on Cell Physiology and Developmental Biology, University of Chicago, Chicago, Illinois 60637

Pratumtip Boontrakulpoontawee Maddux, Department of Pharmacological & Physiological Sciences and the Committees on Cell Physiology and Developmental Biology, University of Chicago, Chicago, Illinois 60637

Brooke M. McCartney, Developmental, Cell, and Molecular Biology Group, Department of Zoology, Duke University, Durham, North Carolina 27708-1000

Karina Meiri, Department of Experimental Pathology, UMDS, Guy's Hospital, London SE1 9RT, United Kingdom

Mark S. Mooseker, Department of Biology, Yale University, New Haven, Connecticut 06520

Anne Müsch, Margaret Dyson Vision Research Institute, Cornell University Medical College, New York, New York 10021

W. James Nelson, Department of Molecular and Cellular Physiology, Beckman Center for Molecular and Genetic Medicine, Stanford University School of Medicine, Stanford, California 94305-5426

Suzanne M. Norvell, R.H. Lurie Cancer Center, Northwestern University Medical School, Chicago, Illinois 60611

Helena L. Palka, R.H. Lurie Cancer Center, Northwestern University Medical School, Chicago, Illinois 60611

Kersi N. Pestonjamasp, Worcester Foundation for Biomedical Research, Shrewsbury, Massachusetts 01545

Matthew F. Peters, Department of Physiology, University of North Carolina, Chapel Hill, North Carolina 27599-7545

Peter A. Piepenhagen, Department of Molecular and Cellular Physiology, Beckman Center for Molecular and Genetic Medicine, Stanford University School of Medicine, Stanford, California 94305-5426

David W. Pumplin, Department of Anatomy and Neurobiology, School of Medicine, University of Maryland at Baltimore, Baltimore, Maryland 21201

Jill A. Rafael, Department of Human Genetics, The University of Michigan Medical School, Ann Arbor, Michigan 48109-0618

Enrique Rodriguez-Boulan, Margaret Dyson Vision Research Institute, Cornell University Medical College, New York, New York 10021

Inna N. Rybakova, Department of Physiology, University of Wisconsin Medical School, Madison, Wisconsin 53706

Frederick Sachs, Biophysical Sciences, SUNY Buffalo, Buffalo, New York 14214

Michael D. Schaller, Department of Cell Biology and Anatomy, University of North Carolina at Chapel Hill, Chapel Hill, North Carolina 27599

Liang Schweizer, Developmental, Cell, and Molecular Biology Group, Department of Zoology, Duke University, Durham, North Carolina 27708-1000

John D. Scott, Vollum Institute, Portland, Oregon 97201-3098

Elizabeth Smith, Howard Hughes Medical Institute, Department of Molecular Genetics and Cell Biology, The University of Chicago, Chicago, Illinois 60637

Christopher P. Strassel, Worcester Foundation for Biomedical Research, Shrewsbury, Massachusetts 01545

Celia Taha, Division of Cell Biology, The Hospital for Sick Children, Toronto, Ontario, Canada M5G 1X8

Theodoros Tsakiridis, Division of Cell Biology, The Hospital for Sick Children, Toronto, Ontario, Canada M5G 1X8

Shoichiro Tsukita, Department of Cell Biology, Faculty of Medicine, Kyoto University, Yoshida-Konoe, Sakyo-ku, Kyoto 606, Japan

Jeanine A. Ursitti, Department of Physiology, School of Medicine, University of Maryland at Baltimore, Baltimore, Maryland 21201

Bryce Vissel, Salk Institute, La Jolla, California 92186

Frank S. Walsh, Department of Experimental Pathology, UMDS, Guy's Hospital, London SE1 9RT, United Kingdom

Qinghua Wang, Division of Cell Biology, The Hospital for Sick Children, Toronto, Ontario, Canada M5G 1X8

Robert E. Ward, Developmental, Cell, and Molecular Biology Group, Department of Zoology, Duke University, Durham, North Carolina 27708-1000

Gary L. Westbrook, Vollum Institute, Oregon Health Sciences University, Portland, Oregon 97201

Yanmin Yang, Howard Hughes Medical Institute, Department of Molecular Genetics and Cell Biology, The University of Chicago, Chicago, Illinois 60637

Qian-Chun Yu, Howard Hughes Medical Institute, Department of Molecular Genetics and Cell Biology, The University of Chicago, Chicago, Illinois 60637

Xu Zhang, Howard Hughes Medical Institute and Departments of Cell Biology and Biochemistry, Duke University Medical Center, Durham, North Carolina 27710

Daixing Zhou, Department of Physiology, School of Medicine, University of Maryland at Baltimore, Baltimore, Maryland 21201

Subject Index